THE CYTOSKELETON
A Target for Toxic Agents

ROCHESTER SERIES ON ENVIRONMENTAL TOXICITY
Series Editors: Thomas W. Clarkson and Morton W. Miller

REPRODUCTIVE AND DEVELOPMENTAL TOXICOLOGY OF METALS
Edited by Thomas W. Clarkson, Gunnar F. Nordberg, and Polly R. Sager

THE CYTOSKELETON: A Target for Toxic Agents
Edited by Thomas W. Clarkson, Polly R. Sager, and Tore L. M. Syversen

A Continuation Order Plan is available for this series. A continuation order will bring delivery of each new volume immediately upon publication. Volumes are billed only upon actual shipment. For further information please contact the publisher.

THE CYTOSKELETON
A Target for Toxic Agents

Edited by
Thomas W. Clarkson
University of Rochester School of Medicine
Rochester, New York

Polly R. Sager
University of Connecticut Health Center
Farmington, Connecticut

and

Tore L. M. Syversen
University of Trondheim Faculty of Medicine
Trondheim, Norway

PLENUM PRESS • NEW YORK AND LONDON

Library of Congress Cataloging in Publication Data

Rochester International Conference on Environmental Toxicity (16th: 1984)
The cytoskeleton.

(Rochester series on environmental toxicity)
"Proceedings of the 16th Rochester International Conference on Environmental Toxicity, held June 4-6, 1984, in Rochester, New York"–T.p. verso.
Includes bibliographies and indexes.
1. Toxicology, Experimental–Congresses. 2. Cell organelles–Research–Congresses. 3. Poisons–Physiological effect–Congresses. I. Clarkson, Thomas W. II. Sager, Polly R. III. Syversen, Tore L. M. IV. Title. V. Series. [DNLM: 1. Cytoplasmic Filaments–drug effects–congresses. 2. Environmental Pollutants–adverse effects–congresses. W3 RO6683 16th 1984c/QH 591 R 676 1984c]
RA1199.R62 1984   615.9   85-24470
ISBN 0-306-42205-0

Proceedings of the 16th Rochester International Conference on
Environmental Toxicity, held June 4-6, 1984, in Rochester, New York

© 1986 Plenum Press, New York
A Division of Plenum Publishing Corporation
233 Spring Street, New York, N.Y. 10013

All rights reserved

No part of this book may be reproduced, stored in a retrieval system, or transmitted, in any form or by any means, electronic, mechanical, photocopying, microfilming, recording, or otherwise, without written permission from the Publisher

Printed in the United States of America

Editorial Board:

J. B. Cavanagh	M. A. Lichtman
T. W. Clarkson	N. K. Mottet
A. Elgsaeter	J. B. Olmsted
H. C. Guldberg	P. R. Sager
S. D. Lee	T. L. M. Syversen

Publications Editor:

M. W. Miller
University of Rochester School of Medicine
Rochester, New York

Conference and Publications Coordinator:

M. A. Terry
University of Rochester School of Medicine
Rochester, New York

PREVIOUS ANNUAL PUBLICATIONS FROM
ROCHESTER INTERNATIONAL CONFERENCES ON ENVIRONMENTAL TOXICITY

1968 Conference: **Chemical Fallout: Current Research on Persistent Pesticides,**
(eds. Miller and Berg),
Charles C. Thomas Publishers, Inc., 1969.

1969 Conference: **Effects of Metals on Cells, Subcellular Elements, and Macromolecules,**
(eds. Maniloff, Coleman and Miller),
Charles C. Thomas Publishers, Inc., 1970.

1970 Conference: **Assessment of Airborne Particles,**
(eds. Mercer, Morrow and Stoeber),
Charles C. Thomas Publishers, Inc., 1971.

1971 Conference: **Mercury, Mercurials and Mercaptans,**
(eds. Miller and Clarkson),
Charles C. Thomas Publishers, Inc., 1973.

1972 Conference: **Behavioral Toxicology,**
(eds. Weiss, Laties and Miller),
Plenum Press, 1975.

1973 Conference: **Molecular and Environmental Aspects of Mutagenesis,**
(eds. Prakash, Sherman, Miller, Lawrence and Taber),
Charles C. Thomas Publishers, Inc., 1973.

1974 Conference: **Fundamental and Applied Aspects of Non-Ionizing Radiation,**
(eds. Michaelson, Miller, Magin and Carstensen),
Plenum Press, 1975.

1975 Conference: **Environmental Toxicity of Aquatic Radionuclides: Models and Mechanisms,**
(eds. Miller and Stannard),
Ann Arbor Science Publications, 1976.

1976 Conference: **Membrane Toxicity,**
(eds. Miller, Shamoo and Brand),
Plenum Press, 1977.

1977 Conference: **Environmental Pollutants: Detection and Measurement,**
(eds. Toribara, Coleman, Dahneke and Feldman),
Plenum Press, 1978.

1978 Conference: **Neurotoxicity of the Visual System,**
(eds. Merigan and Weiss),
Raven Press, 1980.

1979 Conference: **Polluted Rain,**
(eds. Toribara, Miller and Morrow),
Plenum Press, 1980.

1980 Conference: **Measurement of Risks,**
(eds. Berg and Maillie),
Plenum Press, 1981.

1981 Conference: **Induced Mutagenesis: Molecular Mechanisms and Their Implications for Environmental Protection,**
(eds. Lawrence, Prakash, Sherman)
Plenum Press, 1983.

1982 Conference: **Reproductive and Developmental Toxicity of Metals,**
(eds. Clarkson, Nordberg and Sager),
Plenum Press, 1983.

1984 Conference: **The Cytoskeleton: A Target for Toxic Agents,**
(eds. Clarkson, Sager and Syversen),
Plenum Press, 1985.

PREFACE

This book is based on reviews and research presentations given at the 16th Rochester International Conference on Environmental Toxicity, entitled "The Cytoskeleton: A Target for Toxic Agents," held on June 4, 5 and 6 in 1984.

The conference provided an in-depth discussion of the effects and mechanism of action of some toxic agents on the cytoskeleton. Mammalian and other eukaryotic cells contain protein networks within the cytoplasm comprised of microfilaments, intermediate filaments and microtubules. These components of the cytoskeleton play a key role in cell shape, motility, intracellular organization and transport, and cell division. Furthermore, the cytoskeleton, via associations with the cell membrane, appears to function in intracellular communication and cellular responses to membrane events. Because of the complex functional roles of the cytoskeleton which vary with cell type, degree of differentiation, and cell cycle, its disruption may result in a variety of cellular changes.

This expanding field in cell biology has already attracted the interest of toxicologists and environmental health scientists as a potentially fruitful area of research. Indeed, there is mounting evidence that certain toxic and chemotherapeutic compounds, as well as physical agents such as radiation and hydrostatic pressure, disrupt the normal structure and function of the cytoskeleton. This may be an important step in the overall expression of their action. It was, therefore, an opportune time to hold a conference to encourage the development of this area of toxicology and to suggest directions for future research.

The approach taken in the organization of the conference was to bring together scientists in the basic biology of the cytoskeleton and toxicologists to promote an exchange of ideas for future possibilities in toxicology. The conference involved presentations of plenary review lectures, invited papers and workshop discussion groups. An Editorial Board was established to oversee the conference publication and to prepare overview chapters on the topic of the conference.

The conference was sponsored by the Universities of Rochester (USA) and Trodeim (Norway). The plenary lectures and invited papers were presented in three main sessions. The conference was introduced by Deans Hans Cato Guldberg (University of Trondheim) and Marshall A. Lichtman (University of Rochester). The contents of each session including the title and the author who presented each paper are indicatd below:

PLENARY LECTURE: THE CYTOSKELETON
    J.B. Olmsted, University of Rochester

SESSION 1: MICROTUBULES AND MICROTUBULE INHIBITORS

SESSION 1A

Session Chairperson: J.B. Olmsted, University of Rochester
Session Rapporteur: P.R. Sager, University of Connecticut Health Center

Microtubules as Targets for Drug and Toxic Chemical Action: The Mechanisms of Action of Colchicine and Vinblastine
    L. Wilson, University of California at Santa Barbara

Taxol: A Probe for Studying the Structure and Function of Microtubules
    S.B. Horwitz, Albert Einstein College of Medicine

SESSION 1B

Session Chairperson: M.W. Anders, University of Rochester School of Medicine
Session Rapporteur: T.L.M. Syversen, University of Trondheim School of Medicine

Reorganization of Axonal Cytoskeleton Following β,β'-Iminodipropionitrile (IDPN) Intoxication
    Sozos C. Papasozomenos, University of Texas Health Science Center at Houston

Molecular and Cellular Aspects of the Interaction of Benzimidazole Carbamate Pesticides with Microtubules
    K. Gull, University of Kent

Disruption of Microtubules by Methylmercury
    P.R. Sager, University of Connecticut Health Center

SESSION 2: CYTOSKELETON OF THE NERVOUS SYSTEM

Session Chairperson: T.L.M. Syversen, University of Trondheim School of Medicine
Session Rapporteur: J.B. Cavanagh, Institute of Neurology, Queen Square

Primary and Secondary Changes in Axonal Transport in Neurofibrillary Disorders
    J.W. Griffin, The Johns Hopkins School of Medicine

Chemical Neurotoxins Accelerating Axonal Transport of Neurofilaments
    P. Gambetti, Case Western Reserve University, School of Medicine

The Effect of Some Dithiocarbamates, Disulfiram and 2,5-Hexanedione on the Cytoskeleton of Neuronal Cells In Vivo and In Vitro
    V.P. Lehto, University of Helsinki

Neurofibrillary Degeneration: The Role of Aluminum
    M. Shelanski, New York University School of Medicine

Pathogenetic Studies of the Neurofilamentous Neuropathies
    D. Graham, Duke University Medical Center

SESSION 3: CYTOSKELETON AND MEMBRANE RELATED EVENTS

SESSION 3A

    Session Chairperson: P.L. La Celle, University of Rochester School of Medicine
    Session Rapporteur: M.A. Lichtman, University of Rochester School of Medicine

Effects of Calcium on Structure and Function of the Human Red Blood Cell Membrane
    H. Passow, Max Planck Institute, Frankfurt/Main

Spectrin and the Mechanochemical Properties of the Erythrocyte Membrane
    A. Elgsaeter, University of Trondheim, School of Medicine

Effects of Local Anesthetics and pH Change on the Platelet "Cytoskeleton" and on Platelet Activation
    V. Nachmias, University of Pennsylvania

SESSION 3B

    Session Chairperson: A.C. Notides, University of Rochester School of Medicine
    Session Rapporteur: T.L.M., Syversen, University of Trondheim, School of Medicine

Inhibition of Motility by Inactivated Myosin Heads
    M. Sheetz, University of Connecticut Health Center

Intracellular Translocation of Inorganic Particles
    A. Brody, National Institute of Environmental Health Sciences

Receptor Cycling in Pituitary Cells: Biological Consequences of Inhibition of Receptor-Mediated Endocytosis
    P. Hinkle, University of Rochester School of Medicine

Conference Summary
    N.K. Mottet, University of Washington School of Medicine

    The conference co-chairpersons were Thomas W. Clarkson, University of Rochester, and Tore L.M. Syversen, University of Trondheim. The conference organizing committee was, in addition to the co-chairpersons, Hans Cato Guldberg, Paul L. La Celle, Morton W. Miller, Polly R. Sager and Michael A. Terry.

Each session was associated with a workshop, chaired by the chairperson of the session and consisting of the invited speakers, the rapporteur and other members as appropriate. The rapporteur prepared a written summary of toxic agents. The rapporteurs subsequently joined the Editorial Board to participate in the preparation of the overview chapters. Karle Mottet also prepared an overall summary of the conference which is published as a separate chapter in the book. The Editorial Board reviewed the chapters which comprise the book; however, only the authors themselves are responsible for their views, statements and opinions in the invited papers.

The Editorial Board gratefully acknowledges financial support from the following:

- The University of Trondheim
- The Royal Norwegian Research Council of Science and Technology
- The National Institute of Environmental Health Sciences
- The United States Environmental Protection Agency
- The United States Department of Energy
- The Mobil Oil Company
- The United States Air Force
- The United States Navy
- The University of Rochester

The Board is deeply indebted to Muriel Bank Klein for her untiring efforts in processing the papers for publication, to Joyce Morgan for her inspired handling of travel and accommodations and to both for their assistance during the meeting itself. A special thank you is also in order to Nancy Scott and the staff of the Word Processing Center of the Department of Radiation Biology and Biophysics of the University of Rochester for their contribution to the preparation of this book.

The Editorial Board wishes to express its appreciation to Thomas A. Eskin, Robert M. Herndon, William H. Merigan, Paul L. La Celle, and John L. O'Donoghue for reviewing the overview chapters.

The Editors and the Editorial Board wish to thank each contributor for her/his work and cooperation in completing the document.

On behalf of the Editorial Board,

Tom Clarkson
Polly Sager
Tore Syversen

CONTENTS

OVERVIEW

Structure and Function of the Cytoskeleton 3
    The Editorial Board

The Cytoskeleton as a Target for Toxic Agents 25
    The Editorial Board

INVITED PAPERS

SESSION 1. MICROTUBULES AND MICROTUBULE INHIBITORS 35
    Chairpersons: J.B. Olmsted, M.W. Anders
    Rapporteurs: P.R. Sager, T.L.M. Syversen

Microtubules as Targets for Drug and Toxic Chemical Action:
The Mechanisms of Action of Colchicine and Vinblastine 37
    Leslie Wilson

Taxol: A Probe for Studying the Structure and Function of
Microtubules 53
    Susan B. Horwitz, Peter B. Schiff, Jerome Parness,

Reorganization of Axonal Cytoskeleton Following
β,β'-Iminodipropionitrile (IDPN) Intoxication 67
    Sozos Ch. Papasozomenos

Molecular and Cellular Aspects of the Interaction of
Benzimidazole Carbamate Pesticides with Microtubules 83
    Edward H. Byard and Keith Gull

Disruption of Microtubules by Methylmercury 97
    Polly R. Sager and Tore L.M. Syversen

SESSION 2. CYTOSKELETON OF THE NERVOUS SYSTEM 117
    Chairperson: Tore L.M. Syversen
    Rapporteur: John B. Cavanagh

Primary and Secondary Changes in Axonal Transport in
Neurofibrillary Disorders 119
    Bruce G. Gold, John W. Griffin, Donald L. Price
    and Paul N. Hoffman

Chemical Neurotoxins Accelerating Axonal Transport of
Neurofilaments 129
    Pierluigi Gambetti, Salvatore Monaco,
    Lucila Autilio-Gambetti, and Lawrence M. Sayre

The Effect of Some Dithiocarbamates, Disulfiram and
2,5-Hexanedione on the Cytoskeleton of Neuronal Cells
In Vivo and In Vitro 143
    Veli-Pekka Lehto, Ismo Virtanen, and Kai Savolainen

Neurofibrillary Degeneration: The Role of Aluminum 159
    Michael L. Shelanski

Pathogenetic Studies of the Neurofilamentous Neuropathies 167
    Doyle G. Graham, Gyongyi Szakal-Quin, Leslie Milam,
    Marcia R. Gottfried and D. Carter Anthony

SESSION 3. CYTOSKELETON AND MEMBRANE RELATED EVENTS 175
    Chairpersons: Paul L. La Celle, Angelo C. Notides
    Rapporteurs: Marshall A. Lichtman, Arnjolt Elgsaeter

Effects of Calcium on Structure and Function of the Human
Red Blood Cell Membrane 177
    Hermann Passow, Melanie Shields, Paul La Celle,
    Ryszard Grygorczyk, Wolfgang Schwarz and Reiner Peters

Spectrin and the Mechanochemical Properties of the
Erythrocyte Membrane 187
    Arnljot Elgsaeter, Arne Mikkelsen and Bjorn T. Stokke

Effects of Local Anesthetics and pH Change on the Platelet
"Cytoskeleton" and on Platelet Activation 199
    Vivianne T. Nachmias and Robert M. Leven

Inhibition of Motility by Inactivated Myosin Heads 213
    Robin Jones and Michael Sheetz

Intracellular Translocation of Inorganic Particles 221
    Arnold R. Brody, Lila H. Hill, Thomas W. Hesterberg,
    J. Carl Barrett and Kenneth B. Adler

Receptor Cycling in Pituitary Cells: Biological Consequences
of Inhibition of Receptor-Mediated Endocytosis 229
    Patricia M. Hinkle, Jane Halpern, Patricia A. Kinsella
    and Risa A. Freedman

CONFERENCE SUMMARY 243
    N. Karle Mottet

PERSPECTIVES 253

AUTHOR INDEX 255

SUBJECT INDEX 257

THE STRUCTURE AND FUNCTION OF THE CYTOSKELETON

Authors: The Editorial Board

Contents:

- INTRODUCTION 3
- COMPONENTS OF THE CYTOSKELETON 3
  - Microfilaments 4
  - Microtubules 6
  - Intermediate Filaments 6
  - Associations Between Components of the Cytoskeleton . . . 7
- ORGANIZATION OF THE CYTOSKELETON 8
  - Cultured cells 8
  - Neurons 10
  - Erythrocytes 11
  - Platelets 12
  - Leukocytes 13
- FUNCTIONS OF THE CYTOSKELETON 14
  - Structural Support and Cell Shape 14
  - Intracellular Transport 15
  - Membrane Properties and Functions 17
  - Phagocytosis 17
  - Motility 18
- SUMMARY 18
- REFERENCES 18

STRUCTURE AND FUNCTION OF THE CYTOSKELETON

Polly R. Sager[a], Tore L.M. Syversen[b], Thomas W. Clarkson[c], John B. Cavanagh[d], Arnljot Elgsaeter[b], Hans Cato Guldberg[b], Si Duk Lee[e], Marshall A. Lichtman[c], N. Karle Mottet[f], and Joanna B. Olmsted[c]

[a]University of Connecticut, Farmington, Connecticut, USA
[b]University of Trondheim, Norway
[c]University of Rochester, Rochester, New York, USA
[d]Institute of Neurology, Queen Square, London, UK
[e]EPA, Research Triangle Park, North Carolina, USA
[f]University of Washington, Seattle, Washington, USA

INTRODUCTION

Organization within the cytoplasm has been apparent from the earliest observations of living cells. As techniques improved, so has our appreciation of the complexity involved in the ordered arrangement of cytoplasmic components. The view of cytoskeleton has evolved from one of a static or rigid structure to a more dynamic scaffolding capable of rapid and dramatic restructuring. Several components and functions of the cytoskeleton have been defined in cells. Indeed, the number and diversity of cellular processes subserved by the cytoskeleton are impressive, and it is not surprising that numerous toxic agents have been found to act upon it. For some of these agents, their selective interaction with specific components of the cytoskeleton has led to their use as probes for normal structure and function. For other compounds, the interaction may be less well defined or less specific, but of relevance from an environmental or human health perspective.

In these first two chapters, we have attempted to set a background for the invited chapters which follow. First, the organization and functions of the cytoskeleton have been described briefly; this is to provide a general background rather than a detailed review. In the second chapter examples of the use of well-defined, selective agents that act on cytoskeleton components have been discussed. Finally, we have focused on some of the toxic compounds that have effects on the cytoskeleton. For many of these, the exact nature of the interaction is not clear. The following invited chapters describe several different research approaches from basic cell biology to toxicology. We hope that this volume will encourage continued research resulting in a better understanding of the cytoskeleton and its role in mechanisms of damage.

COMPONENTS OF THE CYTOSKELETON

The cytoskeleton consists of three types of filamentous structures—microfilaments, intermediate filaments and microtubules - each of which is comprised of polymers of protein subunits, as summarized in Table 1. For

Table 1. Components of the Cytoskeleton

Structure	Diameter	Composition
Microfilaments	5 - 7 nm	G-actin, 42 K daltons actin-binding proteins (see Table 2)
Intermediate Filaments	8 - 10 nm	Variable (see Table 3)
Microtubles	25 nm	α and β tubulin 55 K daltons (α,β dimers) microtubles-associated proteins

microfilaments and microtubules, a dynamic equilibrium exists between subunits and polymers. Although regulation of this equilibrium is little understood, many functions of these cytoskeletal components appear to depend on dynamic rearrangements. For example, motive forces can result from preferential growth and disassembly at opposite ends of actin filaments, as well as by interaction with myosin. In addition, changes in cell shape and locomotion involve the redistribution of both microfilaments and microtubules. In contrast, no monomeric subunits of intermediate filaments have been detected in cells. These filaments are much more stable and proteases may be required for their cleavage.

Actin, from which microfilaments form, and tubulin, from which microtubules form, are highly evolutionary conserved proteins and the multiple genes coding for these proteins show extensive homologues from gene to gene. Tubulins from different species or different tissues of the same species may show quite variable isoelectric points and some tubulin subspecies show different responses to anti-microtubule drugs (see Byard and Gull, this volume). Many forms of muscle and non-muscle actin have also been described; these arise from multiple genes rather than post-translational processing of a single gene product. Intermediate filaments, however, include a large class of polymers of uniform structure and properties that may vary greatly in their subunit composition (see Table 3).

Microfilaments

Actin exists as the monomeric, globular form G-actin and the filamentous polymer F-actin. The polymer is two-stranded and forms in the presence of ATP, Mg^{2+} and Ca^{2+}. Actin polymers start to form from a nucleus of three or more G-actin monomers. Labelling of F-actin with myosin fragments reveals that the filaments have polarity; one end appears pointed and the other, "barbed". Elongation of actin polymers can occur at either the pointed or the barbed end, although addition to the latter site is generally more efficient. In vivo, elongation may be determined in part by proteins that bind to the pointed or barbed end of actin and block addition of further actin monomers (capping or blocking proteins).

Actin filaments participate in a number of cellular functions including motility, cytokinesis, cytoplasmic structure, membrane topology regulation and intracellular transport. Rings of actin filaments with opposite polarities are often associated with contractile functions, such as cytokinesis. Meshworks of short actin filaments appear to be prevalent in the peripheral or cortical regions of cells and specifically localized in the leading lamellipodia of motile cultured cells and the growth cone of extending neurites and axons. These microfilaments are particularly sensitive to the action of cytochalasins, drugs which bind specifically to one end of actin filaments preventing further assembly. Another drug, phalloidin, binds to actin filaments and prevents their disassembly by preventing monomeric dissociation. Fluorescent-labelled analogues of phalloidin can be used to stain F-actin selectively.

The role of actin in diverse functions appears to be mediated by actin associated proteins, which serve to regulate the assembly and distribution of actin. The actions of some of these actin associated proteins are summarized in Table 2. Several, such as fimbrin, α-actinin and filamin, act to crosslink or bind actin into a gel or bundles. For example, parallel bundles of actin are localized in the microvilli of brush borders of epithelial cells, and these are linked with α-actinin. These bundles display uniform polarity when coated with myosin, having their pointed end oriented towards the center of the cell. Less tightly packed are the stress fibers, containing actin and other proteins, observed in fibroblasts and cultured epithelial cells; actin in stress fibers is not organized with the same polarity.

Other proteins favor net disassembly of actin filaments either by binding to G-actin (profilin) or by fragmenting or capping F-actin (villin, gelsolin). Actin sol-gel transitions, coordinated by various actin associated proteins, together with gel contraction, which occurs in gels containing myosin, appear to be a mechanism by which cells generate forces for intracellular transport (see Jones and Sheetz, this volume). The presence of such actin associated proteins in the cell membranes supports the suggestion of actin participation in membrane functions such as

Table 2. Effects of Actin Associated Proteins

ACTIN SUBUNITS (G-actin)	$\rightleftarrows$	ACTIN FILAMENTS (F-actin)
NET DISASSEMBLY		**NET ASSEMBLY/STABILIZATION**
Polymerization-inhibition		Cofilamentous protein
profilin		tropomysin
		Contraction
		myosin
Fragmenting proteins		Membrane association
villin		vinculin
gelsolin		spectrin
		fodrin
		acting-binding protein
"Capping" proteins		Bundling, cross-linking gel formation
α-actinin		α-actinin
acumentin		filamin
		fimbrin
		spectrin
		fodrin
		protein 4.9

endocytosis, exocytosis, membrane topology regulation and particle movement. For example, vinculin appears to anchor actin fibers to membranes.

Microtubules

Microtubles are formed by the polymerization of α and β tubulin and a small number of microtubule-associated proteins (MAPs). Dimers (α,β tubulin) or perhaps oligomers, assemble into hollow tubules, with protofilaments aligned along the axis of the tubule; MAPs, which comprise 5-15 percent of the mass of brain microtubule protein isolated in vitro decorate the surface of the microtubules. Like actin filaments, microtubules express kinetic polarity in that the two ends of the polymer add and lose subunits at different rates such that, at steady state, "treadmilling" is though to occur (see Wilson, this volume). Microtubules are also the major components of specialized structures including the axonemes of cilia and flaggella and the mitotic spindle.

Much of our understanding of microtubules and drugs which affect them has come from studies using the in vitro assembly of microtubules. Since Weisenberg and coworkers first reported that microtubules could be formed from soluble extracts of brain, this system has been refined and exploited to purify microtubule proteins, to study mechanisms of microtubule assembly, and to elucidate the actions of antimicrotubule drugs. The system is based on the fact that microtubules will form when solutions containing tubulin and MAPs are warmed; the formed microtubules disassemble when cooled (see Wilson, this volume). Thus, successive cycles of temperature-dependent assembly and disassembly of microtubules can be used to purify tubulin and MAPs which copolymerize. The assembly of microtubules, which is initiated by warming and requires GTP, Mg^{2+} and the absence of Ca^{2+}, can be monitored by viscometry or by measuring light scatter. Microtubule formation is thought to be a nucleation- elongation reaction, and the presence of MAPs decreases the critical concentration of tubulin for polymerization. Taxol, a drug which stabilizes microtubules, also promotes assembly by reducing the critical concentration of tubulin (see Horwitz et al., this volume). In vitro polymerization has provided a convenient method for assessing agents which inhibit or enhance polymerization of microtubules, and for studying their mechanism of action. The substoichiometric action of colchicine and colcemid and the stoichiometric inhibition of microtubule formation by vinca alkaloids have been investigated extensively (see Chapter 2 and Wilson, this volume).

A number of MAPs have been identified to date. The distribution of MAPs varies both within cells and among cell types or tissues, and the most extensively studied have been those from mammalian brain (MAP1, MAP2, and tau). For example, high molecular weight MAP2 seems to be located preferentially in dendrites of neurons while MAP1 is more widely distributed in axons and cell bodies. Two functions have been suggested for MAPs, namely spacing and stabilization of microtubules. As tubulin assembles into microtubules, MAPs copolymerize and form sidearms that may serve to regulate spacing between microtubules. Indeed, recent studies have shown that the packed volume of microtubules assembled in vitro can be modulated by the amount of MAPs present. Additionally, in vitro, the presence of MAPs will shift the subunit/polymer equilibrium towards assembly. A similar action has been demonstrated in cultured cells. For example, as neuroblastoma cells undergo differentiation, there is an increase in the ratio of polymerized to unpolymerized tubulin which coincides with the appearance of a specific MAP.

Intermediate Filaments

These filaments, named because of their size, may represent the true "skeleton" of the cell. They are generally insoluble in aqueous solutions containing salts or non-ionic detergents and provide rigidity within cells, especially in areas of mechanical stress. Little is known of their function other than providing tensile strength, since no drugs have been described which act specifically on intermediate filaments with the possible exception of diketones (see Chapter 2 and Graham et al., this volume). However, certain perturbations, such as incubation of cultured cells with colchicine or vanadate can cause a collapse of intermediate filaments containing vimentin, while leaving keratin filaments unaltered. The significance of the large number of constituent proteins (Table 3) is not understood, especially since they all form similar α-helical structures with variable terminal sequences and structures. More recently, a number of intermediate filament associated protein have been identified. As for microfilaments and microtubules, these associated proteins probably serve to regulate distribution or assembly of filaments. Some of these proteins are very large and may function in the cross-linking of intermediate filaments and other cytoskeletal elements.

Neurofilaments, which are the intermediate filaments particular to neurons, have been described especially well. They are the only intermediate filaments known to contain sidearms (non-α-helical terminal portions of the protein) that may be involved in lateral associations with other filaments or microtubules via MAPs. It has been suggested that neurofilaments contribute to both fast and slow axonal transport, but this role has been challenged recently in reports showing fast transport to occur in areas of axoplasm devoid of neurofilaments. The neurofilament subunit triplet (approximately 68, 145 and 200 K daltons), however, remains a marker for slow axonal transport. Another role of neurofilaments involves regulation of axonal caliber and shape. This is based on studies of regenerating axons where neurofilament content was determined by neurofilament transport and was correlated with axon caliber. These observations are supported indirectly by evidence that several chemically-induced axonal neuropathies involving axonal swelling seem to be due to impaired transport, and subsequent accumulation, of neurofilaments (see Gold et al., Graham et al., and Papasozomenos, this volume).

Associations Between Components of the Cytoskeleton

There are several lines of evidence suggesting that biochemical and structural associations exist among the various components of the

Table 3. Intermediate Filaments

Filament Protein Type	Molecular weight[a] (K daltons)	Distribution: Cell Type
Keratins	45 - 68	Epithelial cells
Neurofilament	68, 145, 200	Neurons (central and peripheral)
Glial fibrillary acidic protein	51	Glial cells (astrocytes and Bergmann glia and ependymal cells)
Vimentin (Decamin)	57	Mesenchymal cells (fibroblasts, chondrocytes, lymphocytes, macrophages); also acquired by non-mesenchymal cells lines
Desmin	53	Muscle

[a] Approximate, exact molecular weights may vary among species.

cytoskeleton and other organelles. For example, in rapid freeze-etch preparations of axons, there appear to be connections between neurofilaments and microtubules. Certain disease or toxic states also seem to involve a loss of the connections between microtubules and neurofilaments thus altering their distribution (see Papasozomenos, this volume). Microtubule disrupting drugs such as colchicine, colcemid and nocodazole cause changes in the distribution of axonal neurofilaments, and alter the distribution of vimentin intermediate filaments in some cultured cells. For example, MAP2 serves to crosslink microtubules and vimentin-containing intermediate filaments which are present in neuronal precursor cells and disappear as neurons mature. When vinblastine was used to disrupt the microtubules in these immature neuronal cells, MAP2 antibodies colocalized with the vimentin filaments.

Biochemical evidence also exists for interactions among some cytoskeletal elements. Limited evidence exists for actin-microtubule interactions. Purified microtubules and actin filaments have been reported to form a viscous three-dimensional network. This association of actin and microtubules is mediated by MAPs. These observations are supported by evidence that agents which disrupt actin filaments also inhibit fast axonal transport, a process which is dependent on microtubules. Furthermore, short filaments, with a diameter similar to actin, have been described in association with microtubules in neurites. Thus, indirect evidence suggests that actin-microtubule interactions may underline at least some microtubule-dependent movements in cells.

ORGANIZATION OF THE CYTOSKELETON

The cellular organization of the cytoskeleton has been studied mainly by two methods: electron microscopy and immunofluorescence microscopy. Electron microscopy has been used extensively to describe the ultrastructure of the cytoskeleton in specialized cells or parts of cells such as axons. However, this method proves difficult for three-dimensional reconstruction of the cell. Immunofluorescence staining using antibodies directed against various components of the cytoskeleton has been more useful both in defining the whole-cell distribution of the cytoskeleton components and in following alterations related to cell function or changes induced by drugs. Non-ionic detergents such as Triton X-100, which remove many soluble components and some membranes, have been used in preparing "cytoskeletons" consisting of polymerized microfilaments, microtubules, intermediate filaments, associated organelles and nuclei. Cellular functions related to the cytoskeleton have been dissected with the use of drugs interfering with specific elements of the cytoskeleton (see Chapter 2). More recently, microinjection of antibodies against cytoskeletal proteins has provided an additional tool for correlating structure and function. While these techniques have yielded valuable information about the distribution of cytoskeletal proteins among and within tissues and cells, they present a relatively static view of a dynamic system. Several systems are described below that have been used extensively in studies discussed in subsequent chapters.

Cultured Cells

The organization of the cytoskeleton has been well defined for many cultured cell lines (see Fig. 1). In general, for well-spread, adherent fibroblasts or epithelial cells, actin filaments are observed in stress fibers (Fig. 1a). In motile cells, the cortex and leading edge of cytoplasm contain a meshwork of microfilaments. Microtubules (Fig. 1b) appear throughout the cytoplasm emanating from one or more perinuclear microtubule organizing centers (MTOCs). Intermediate filament distribution varies with the filament type. Vimentin filaments, present in most established cell

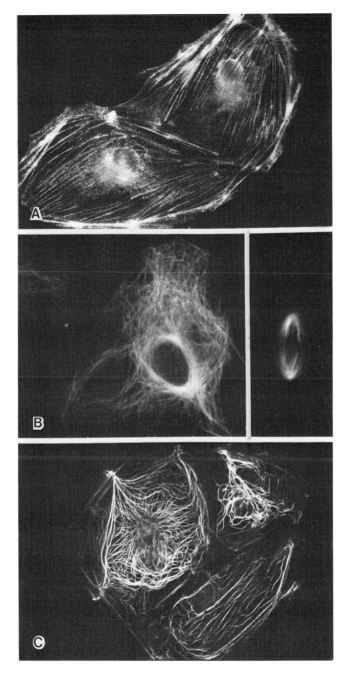

Fig. 1. Organization of cytoskeletal components in cultured cells. Rat kangaroo PtK$_2$ cells, were permeabilized and stained using antibodies against: (a) actin, arranged in stress fibers; (b) tubulin, showing microtubules emanating from a perinuclear organizing center, and (insert) in the mitotic spindle; and (c) cytokeratin intermediate filaments, showing points of attachment at the cell membrane. (Micrographs by L.S. Daniels, P.R. Sager, R.D. Berlin, cytokeratin monoclonal antibody, gift of V.-P. Lehto).

lines, are spread diffusely throughout the cytoplasm. Keratin filament (Fig. 1c) networks appear to attach to the plasma membrane at localized regions which correspond to desmosomes.

The organization and association of the cytoskeleton can change rapidly. One example is the rearrangement of the cytoskeleton that occurs as cells enter mitosis. Stress fibers and cytoplasmic microtubules disappear as the cells round up at the onset of mitosis. Using preexisting tubulin subunits, the mitotic spindle forms with microtubules anchored at the poles of the spindle in the dense pericentriolar material. In some cell types, intermediate filaments condense into a ring or basket around the mitotic spindle; keratin filaments also may collapse. At the completion of chromosome separation, actin filaments form the contractile ring responsible for cytokinesis. With the completion of mitosis, the spindle dissociates, and cytoplasmic microtubules reform. As the daughter cells spread, actin filaments again bundle to form stress fibers, and the intermediate filaments redistribute.

Neurons

The nerve cell perikaryon is responsible for providing the materials needed for the development and maintenance of its dendrites and axon. Transport and supply problems become particularly critical as both the length and surface area of dendrites and axons increase. For example, in the dorsal root ganglion cells of large animals, the ratio of axon volume to cell volume is about 100 to 1. In such cases, axons may attain lengths in excess of one meter. In these highly elongated cells, the cytoskeleton appears to provide both structural support and a framework for intracellular transport in neuronal cells.

As in other cell types, the neuronal perikara contain microtubules and filaments through the cytoplasm. Under normal circumstances there is no evidence for specific centers of cytoskeleton formation, although MTOCs have been shown to appear under certain pathological conditions, e.g., after acrylamide poisoning. The microtubule associated proteins, MAP1 and MAP2, have been found to be abundant throughout the perikaryon of large neurons.

Ultrastructural studies have revealed microtubules to be aligned parallel to the axes of dendrites and these microtubules appear widely spaced in an orderly array. MAP2 is abundant in dendrites and its physical characteristics - as a sidearm bound to the microtubule surface - support its proposed role in regulating microtubule spacing.

Along the length of an individual axon from the perikaryon to the terminal, there is a slight, progressive decrease in axonal diameter. There is, however, considerable variation in diameter among axons and this appears to be due partly to cytoskeletal elements. Smaller, unmyelinated axons have very few neurofilaments and therefore, a higher ratio of microtubules to neurofilaments than the larger diameter axons. Indeed, neurofilament content is related closely to axonal diameter.

Individual microtubules apparently do not run through the entire axon. The length of a single axonal microtubule has been measured in cultured, unmyelinated sensory neurons and was shown to be, on an average, more than 100 μm; whether they are longer in neurons in situ is not known. Neurofilaments may also be discontinuous along the axon, but there are even fewer experimental data on this as compared to microtubules. However, neurofilaments must be able to negotiate the constrictions at the nodes of Ranvier which become more pronounced with increasing size of the axon.

Between the neurofilaments and microtubules, there is a network of tubular membranes which have been demonstrated by immunological methods and by high voltage electron microscopy of thick sections. Crossbridges among microtubules, neurofilaments, microfilaments and membranous organelles have been demonstrated also by transmission electron microscopy of thin sections of axon. The biochemical nature of these links is unknown.

While MAP1 is abundant in axons as well as in neuronal cell bodies and dendrites it has not yet been shown to crosslink to, or colocalize with neurofilaments in vivo. In contrast, MAP2 has not been detected in central axons by immunofluorescence techniques, although small amounts of MAP2 have been detected biochemically in white matter. Furthermore, MAP2 immunoreactivity has been shown to localize with neurofilaments in peripheral axons when the normal cytoskeleton is disrupted by β,β'-iminodipropionitrile (IDPN). Neurofilaments also tend to crosslink with other neurofilaments. This is most likely mediated by the terminal (non-α-helical) portion of the 200 K dalton subunit of the neurofilament triplet.

The organizaton of the axon underscores the mechanical structure provided by the cytoskeleton. The stability of the components of the axonal cytoskeleton is dependent on the maintenance of a low concentration of free calcium in their environment. Under these conditions, neurofilaments are virtually insoluble and some axonal microtubules may remain polymerized after conditions that would normally cause depolymerization (e.g., colchicine treatment). In addition, microtubules and neurofilaments crosslink to themselves and each other. While it is not known exactly which molecules are responsible for the interfilament bridges, MAPs, (MAP1, MAP2 and tau proteins), the terminal portion of the 200 K dalton neurofilament protein, and fodrin (brain spectrin) have all been implicated. This stabilized cytoskeletal matrix provides not only the mechanical integrity and circularity of the axonal profile, but also the framework along which transport of materials can be conducted. This efficient transport system is necessary for generating and maintaining the large volume of cytoplasm and surface area of membrane comprising the axon.

Erythrocytes

Mammalian erythrocytes are highly specialized cells containing a membrane-associated cytoskeleton but no transcellular cytoskeleton in their mature, anucleate form. In contrast, the nucleated red blood cells of other vertebrates and the immature erythrocytes of mammals contain a prominent ring of microtubules, the marginal band. The current view of the organization of the human erythrocyte membrane skeleton is illustrated schematically in Fig. 2. Spectrin, actin, protein 4.1 and probably tropomyosin and protein 4.9 constitute the major components of the membrane skeleton in these cells. Spectrin can exist as α, β-dimers, α_2,β_2-tetramers, as well as higher order oligomers (see Elgsaeter, et al. this volume). Electron microscopy has shown that the spectrin α, β-dimer consists of two highly elongated wormlike subunits connected at both ends, spectrin tetramers consist of two dimers connected head-to-head. Spectrin is bound to the erythrocyte membrane via ankyrin, which in turn binds to the main integral membrane protein, band III. Band III acts in anion transport and constitutes the majority of intramembrane particles (freeze-etch) on human erythrocytes. The spectrin α-subunit has one ankyrin binding site whereas the β-subunit has no such binding site. Thus, only one ankyrin molecule is bound per spectrin tetramer in the human erythrocyte suggesting only one linkage per tetramer to the intramembrane particles.

The details of the membrane skeleton topology (Fig. 3) are still somewhat unclear. Under normal physiological conditions, tetramers are believed to be the dominant species of spectrin in the membrane skeleton, linked into a locally two-dimensional network covering the entire cytoplasmic side of the erythrocyte plasma membrane. The spectrin network junctions

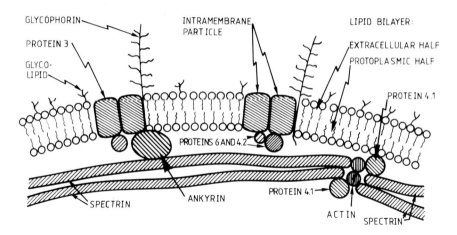

Fig. 2. Schematic cross-section through the plasma membrane of the human erythrocyte.

consist of actin oligomers containing 10-15 actin monomers and possibly the actin bundling protein 4.9 and tropomysin. The spectrin tetramers are bound to the junctions via protein 4.1 which binds both to the actin oligomers and a specific binding site for protein 4.1 at both ends of each spectrin tetramer.

Platelets

Blood platelets are small discoid, anucleate cells that are derived from hemopoietic stem cells, as are leukocytes and erythrocytes. During differentiation in the bone marrow, megakaryocytes proliferate and mature to form polyploid giant cells whose cytoplasm demarcates and fragments into cellules which become the blood platelets. The primary function of platelets is to form a syncytial aggregate at the site of the vascular disruptions and in so doing provide the primary hemostatic mechanism along with vasoconstriction. A major shape alteration from a disc to a spiny sphere precedes platelet aggregation (see Nachmias and Leven, this volume). As platelets adhere to a surface, the platelets spread to form large flat structures. These shape changes are mediated by contractile platelet proteins. Contractile proteins also play a role in the secretion of platelet granules and their contents into the canalicular system which is in communication with the extracellular milieu.

The platelet contractile system is composed of actin, predominantly of the β-type (85 percent) and the remainder of the γ-type, as classified by their isoelectric points. Two forms of actin are also apparent from the conditions required for polymerization in vitro, one similar to that found in muscle and one that is functionally dissimilar to muscle actin since it polymerizes in the absence of ATP.

Platelets also contain proteins that regulate the assembly of actin filaments: profilin, actin-binding protein, α-actinin and gelsolin. Unlike muscle, the motive forces generated in blood cells require the rapid assembly and disassembly of contractile filaments and the development of contractile forces in different loci in the cell. The accessory proteins regulate the assembly and disassembly process. In resting platelets about 45 percent of actin is filamentous whereas after trombin stimulation about 70 percent is filamentous. Thus, activation of the platelet causes a shift

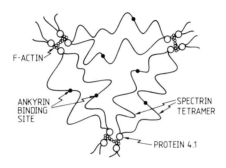

Fig. 3. Schematic of the skeletal topology of the human erythrocyte membrane.

in the G-actin to F-actin equilibrium; profilin, α-actinin and other actin-binding proteins may be involved in regulating this net assembly. Gelsolin binds to actin filaments in the presence of a relatively high free calcium concentration. It is thought to function in activated platelets by regulating actin filament length. In addition, platelets contain myosin but the actin:myosin ratio is much greater than in skeletal muscle. With platelet activation, nearly all the myosin in platelets becomes associated with actin filaments whereas prior to activation myosin is not so associated. Phosphorylation of two of the four light chains in myosin, catalysed by myosin light chain kinase, coincides with binding of myosin to actin filaments. Ankyrin and spectrin proteins, associated with the submembrane lattice network in red cells, have been identified in other cells including platelets. Actin, associated with the membranes of platelets, may be linked to the membrane by such proteins.

Microtubules form a circular band beneath the platelet membrane, similar to the marginal band found in nucleated erythrocytes. These microtubules are believed to hold the platelet in a disc shape, although microtubule disassembly is not required for platelet shape change, contraction, secretion, irreversible aggregation or clot retraction. The circumferential bands of microtubules do, however, organize into tight rings around centrally concentrated organelles and facilitate the orderly secretion of organelle contents. The microtubules are not themselves contractile but are swept to the center of the activated platelet either directly or indirectly by the actin-myosin contractile system.

Leukocytes

White blood cells consist of three major types: granulocytes, lymphocytes and monocytes. Granulocytes are further divided into three cell types: neutrophils, eosinophils and basophils. The lymphocytes can be divided into two major types: T(thymic)-lymphocytes and B(bone marrow)-lymphocytes. Monocytes mature in the tissues to form a variety of macrophages. Each white cell type has biochemical and structural specializations to subserve different functions. The neutrophil has been a focus for studies on contractile proteins because of its active ameboid movement, its response to chemoattractant molecules, its highly phagocytic nature, and its secretion of granule contents.

Actin is the dominant protein of the contractile protein family in leukocytes and various accessory proteins serve to modulate actin assembly. Profilin has been identified in leukocytes and is thought to stabilize actin in the globular or monomeric configuration impeding nucleation and filament formation. Actin-binding protein which accounts for the nucleation and polymerization of actin and the subsequent cross-linking of actin filaments in leukocytes, is concentrated in the pseudopods and cortical veil of leukocytes; it plays the major role in inducing gelation of cytoplasm.

The special requirements for regulating assembly and disassembly of the actin filaments in non-muscle cells call for a variety of regulator proteins. Two such proteins which have been identified in leukocytes are acumentin and gelsolin. Acumentin binds to the slow-growing end of the filament (pointed end) and blocks addition of actin monomers at that site. Acumentin also facilitates nucleation of actin monomers and in doing so favors formation of short filaments. In contrast, gelsolin binds to the barbed or fast-growing end of the acting filament, but like acumentin, promotes nucleation and short filament formation.

Leukocytes contain myosin which is critical for contraction of the actin network. The myosin content of leukocytes is very low compared to actin. Myosin binds to actin, causes it to move and thereafter dissociates and rebinds at a new site causing further contraction of the actin filament. Thus a relatively small number of myosin molecules can influence a large number of actin molecules in the actin polymer.

FUNCTIONS OF THE CYTOSKELETON

Numerous biological processes, at the cellular level, require the participation of the cytoskeleton. These include mitosis, cell locomotion, cytoplasmic organization, intracellular movement of organelles and maintenance of cell shape. Membrane-related functions, such as endocytosis, exocytosis, membrane flexibility and membrane topology, are also dependent on the proper functioning of the cytoskeleton.

Structural Support and Cell Shape

The cytoskeleton, especially microtubules, plays an important role in determining the distinct geometric characteristics of most eukaryotic cells. This involvement of the various cytoskeleton components is particularly apparent in cell protrusions, such as microvilli and cilia, and in asymmetrical cells.

Cilia are tiny hairlike appendages extending from the surface of many kinds of cells. The arrangements of microtubules in cilia provide not only support but, together with dynein, the molecular basis for the movement of the cilia.

Microvilli are somewhat larger, fingerlike extensions covering the surface of cells requiring a large surface for absorption, such as the epithelial cells of the intestine or the kidneys. Each microvillus contains about 40 actin filaments with the same polarity, organized in a bundle parallel to the length of the microvillus. The actin filaments at the tip of the microvillus are attached to an amorphous region, while at the base, the actin filaments extend into a perpendicular network consisting mainly of actin, called the terminal web. The terminal web also contains myosin, spectrin-like molecules and other actin binding proteins such as fimbrin and villin. A band of antiparallel actin filaments often rings each epithelial cell just basal to the microvilli; contraction of this ring may contribute to brush border motility.

The cytoskeleton not only gives structural support to individual cells, but contributes to the tensile strength of cells associated into tissues. For example, in epithelial cell sheets, neighboring cells are held together mechanically by strong cell junctions, the spot desmosomes. On the cytoplasmic plasmalemma surface, these junctions serve as anchoring sites for keratin-containing intermediate filaments extending through the cell. Since another type of filament extends between the desmosomes, the intermediate filaments form a continuous network of cytoskeleton filaments extending through the epithelial cell sheet and thus serve to give the entire cell sheet its tensile strength.

One of the earliest identified functions of the cytoskeleton was in the maintenance of cell shape; drugs, such as colchicine, vinca alkaloids, and nocodazole, and low temperature, which disrupt cytoplasmic microtubules also cause cells to round up from their substratum. An analogous but normal physiological condition is mitosis where microtubules are disassembled and cells round up. Thus it has been assumed that microtubules are important in maintaining asymmetry.

Actin filaments appear to play a major role in generating asymmetric shape. For example, when embryonic fibroblasts are grown in a hydrated collagen gel they come to resemble bipolar, spindle shaped fibroblasts in situ. The initial steps of filopodia aggregation at opposite ends of the cell, and bipolar pseudopodia extension appear to be actin-mediated, as they are sensitive to cytochalasin D treatment. However, further elongation of the bipolar cell requires microtubules and is disrupted by both nocodazole and taxol.

Neurite formation, both in cultured neurons and in tissues, is also highly dependent on cytoskeleton. Actin is principally localized in the growth cone which is devoid of microtubules and neurofilaments. In turn, extension of the axon appears to be dependent on microtubules which are the first fibers to appear in the lengthening axon. Differentiation of certain neuroblastoma cells, which is recognized by neurite outgrowth, is marked by a dramatic increase in the fraction of polymerized tubulin, as microtubules extend into the newly forming neurites. In other neuroblastoma lines, aggregation of centrioles at the site of the presumptive neurite has been shown to precede neurite formation. Centrioles, or their associated dense material, have been identified as microtubule organizing centers from which cytoplasmic microtubules form, thus their localization may participate in the regulation of neurite formation. Finally neurofilaments invade the axoplasm and their number has been correlated with the diameter of the axon.

There are many other examples where actin filaments and microtubules appear to act coordinately to regulate cell shape. Transformation of the neural plate into the neural groove and, later, closure of the neural tube involves elongation and coincident apical constriction of the neuroepithelial cells. The elongation of these cells is associated with microtubules oriented along the long axis of the cell. A ring of antiparallel actin filaments located at the apex of the cell affects the constriction necessary to curve the tissue sheet into a tube. Several teratogenic agents which interfere with neural tube closure have been reported to disrupt the organization of either the actin filament ring or microtubules.

Intracellular Transport

Transport or movement of intracellular components has been observed in living cells for many years. Familiar examples include muscle contraction, chromsome movement in mitosis, and axonal transport. While the sarcomere arrangement of actin-myosin complexes provides the basis for muscular contraction, other actin-myosin interactions or other proteins generate motive forces in nonmuscle cells (see Jones and Sheetz, this volume).

The familiar arrangement of mitotic chromsomes into the metaphase plate and their separation in anaphase is an important example of an intracellular movement. The mitotic spindle consists of microtubules arising from the spindle pole; some microtubules form the aster, others form the central spindle and some become captured by the kinetochores of the chromosomes. In certain cell types, intermediate filaments form a basket or sphere around the spindle. The molecular basis for chromsomal movement is unclear. Suggested motive forces have included microtubule-dynein complexes, actin-based contraction of filaments, and microtubule assembly-disassembly. While the mechanisms regulating prometaphase chromosome movement remain to be elucidated, anaphase separation of chromosomes is thought to depend, at least in part, on microtubule disassembly.

Axonal transport is an example of intracellular movement that has been better described, although the source of the motive force is still not clear. In axons, the cytoskeleton (as described earlier) is itself, moving slowly centrifugally; in addition it provides the necessary tracks for the rapid movement of organelles and soluble components. These components for axonal transport have been identified on the basis of rate of movement and are summarized in Table 4. Fast axonal transport, 50-400 mm/day, occurs in both the orthograde and retrograde directions and consists primarily of organelles and membranous vesicles and structures. Intact microtubules are necessary for fast transport, as evidenced by the disruptive effects of antimicrotubule agents.

The cytoskeleton itself moves in the slow component of transport. Neurofilaments and microtubules, which often appear to be linked morphologically, move together as the slow component a (SCa) while a second slow component b (SCb) moves at a somewhat faster rate and comprises actin, clathrin, and other metabolic enzymes. Thus, interactions between microfilaments and structures moving as part of the fast component must be transient.

Table 4. Components of Axonal Transport

Component	Rate	Component
Fast		
orthograde	50-400 mm/day	membrane-associated organelles and material, mitochondria, synaptic vesicles
retrograde	200 mm/day	prelysosomal vesicles lysosomal enzymes endocytic vesicles
Slow		
Slow component b (SCb)	2-8 mm/day	microfilaments (actin and associated proteins) spectrin, clathrin, calmodulin
Slow component a (SCa)	0.2-1 mm/day	microtubules (tubulin, MAPs, tau proteins) neurofilaments (triplet proteins) spectrin

While SCa (neurotubules and neurofilaments) provides the scaffold for organelle transport, soluble components (SCb) move by bulk flow within the cytosol. The transport of materials required to maintain the axon and related to its specific transmitter functions appear to be affected within membrane bound macromolecular carrier complexes. Indeed, because of the molecular composition of these complexes, which may contain actin and myosin as well as metabolic enzymes and regulatory proteins, it has been suggested that they may generate the energy and mechanical forces for cytoskeletal movement.

Membrane Properties and Functions

Membrane topology is dependent in part on the integrity of the cytoskeleton. For example, segregation of membrane receptors and receptor internalization can be blocked by the cytochalasins which affect microfilament function. In general, however, antimicrotubule drugs appear to have little effect on membrane receptor movement or endocytosis. Other membrane related events, such as the "ruffling" observed in cultured cells, also appear to involve microfilaments with little if any participation by microtubules or intermediate filaments.

Because the human erythrocyte has no transcellular cytoskeleton, both the characteristic shapes and the mechanical properties of these cells have been ascribed to properties of the plasma membrane. Since the lipid bilayer of human erythrocytes is in the fluid state, the mechanical properties of the membrane must be due to the presence of a membrane skeleton. Despite the vast amount of data available on the structure of the human erythrocyte membrane skeleton and its individual components, the detailed molecular mechanisms underlying the properties of the membrane skeleton still remain to be elucidated. However, it is reasonable to expect that because spectrin constitutes the major part of the membrane skeleton, the physical properties of this molecule somehow play a crucial role in determining the mechanical properties of the human erythrocyte membrane skeleton (see Elgsaeter _et al_., this volume).

For many years it was believed that spectrin was a molecule characteristic of the highly specialized mammalian erythrocytes. However, over the last few years spectrin-like molecules have been found in virtually all cells examined. Often the spectrin molecules are located near the plasma membrane of these cells. It has been speculated therefore that spectrin plays a similar structural role in non-erythroid cells as the molecule does in human erythrocytes, but this still remains to be demonstrated. In addition, the contribution of spectrin to the membrane properties of these cells is confounded by a transcellular cytoskeleton discussed in an earlier section.

Phagocytosis

The human neutrophil is capable of ingesting a variety of particles including bacteria, fungi, red cells and other particles that are properly coated with a bridging molecule like immunoglobulin (Ig). The Fc portion of the Ig molecule fixes to the Fc receptor on the cell membrane and triggers a sequence of steps which eventually leads to internalization of the particle. After contact of the Ig molecule with the receptor, movement of the cytoplasm surrounding the activated receptors, occurs, causing a cup of cytoplasm to protrude over the particle. Eventually the membrane forming the cup fuses so as to internalize the particle in a phagocytic vacuole. As the peripheral cytoplasm advances over the particle, additional Fc receptors are occupied by Ig on the particle perpetuating the movement of cytoplasm until eventually fusion of the membrane and completion of the phagocytic vacuole occur. The movement of cytoplasm is regulated by the formation of actin filaments and their interaction with myosin, a process sensitive to cytochalasins. These events are in turn regulated by accessory proteins such as profilin, acumentin and gelsolin as well as localized changes in free calcium concentration. Discharge of granule contents into the phagocytic vacuole is also thought to be controlled by contractile processes.

Several hypothesis have been developed to explain the integration of the contractile system and to account for its response to stimuli. Microtubules do not appear to be involved directly in phagocytosis, as evidenced by the continued uptake of fluid and particles by cells treated with antimicrotubule drugs. However, there is an abrupt cessation of endocytotic and exocytotic processes as cells enter mitosis, later resumed when cells reassemble an interphase cytoskeleton in early G1.

Motility

It has long been presumed that cell locomotion was primarily a function related to actin-myosin or actin-actin interactions. This view is supported by evidence derived from studies using antiactin and antimicrotubule drugs. For example, the addition of colchicine affects the directed motion of cells but does not abolish amoeboid movement, extension of microvilli or membrane ruffling activity. In contrast, cell motility can be suppressed by addition of either cytochalasin, which prevents actin assembly, or phallodin, which prevents filament depolymerization. Thus, dynamic assembly-disassembly of actin filament appears to play the primary role in cell movement while microtubules may act as vectors.

The translocation of cells depends on the functional forces developed between the cell and a substratum and the ability of the contractile system to push or pull portions of the cell from one point to another. Although there have been longstanding debates about whether cells move by being pushed from a stationary rear end or pulled from a stationary front end, the latter is favored for movement of neutrophils. The behavior of peripheral cytoplasm during cell motility is qualitatively similar to that which occurs during phagocytosis, except that bridging molecules, if any, between the cell and substratum are not required or are less specific than Ig. The advancing portion of the cell forms a broad lamellipodium in which there is a rich reticular mesh of microfilaments. In the presence of a chemoattractant, the cell movement is no longer random but is guided by the chemical gradient, which presumably because of membrane receptor occupancy, causes more directional movement of the cell.

SUMMARY

The brief description, given above, of the structural components, organization and function of the cytoskeleton, points to the varied involvement of the cytoskeleton in normal physiology. Much of what is known about cytoskeleton function has been obtained using drugs or agents which act specifically on components of the cytoskeleton. These selective interactions have been exploited also in the development of therapeutic agents. Finally, toxic agents are being identified that interact with the cytoskeleton either directly or secondarily. It is these two classes of agents which are discussed in Chapter 2.

REFERENCES

THE CYTOSKELETON

Porter, K. (1984) The cytomatrix: a short history of its study, J. Cell Biol. 99:3s-12s.

Weber, K. and Osborn, M. (1982) The cytoskeleton, Natl. Cancer Inst. Monogr. 60:31-46.

MICROFILAMENTS

Ishikawa, H., Bischoff, R. and Holtzer, H. (1969) Formation of arrowhead complexes with heavy meromyosin in a variety of cell types, J. Cell Biol. 43:312-328.

Stossel, T.P. (1984) Contribution of actin to the structure of the cytoplasmic matrix, J. Cell Biol. 99:15s-21s.

Vanderkerckhoeve, J. and Weber, K. (1978) At least six different actins are expressed in a higher mammal: an analysis based on the amino acid sequence of the amino terminal peptide, J. Molec. Biol. 126:783-802.

MICROTUBULES

Dustin, P. (1979) Microtubules, Springer-Verlag, New York.

Roberts, K. and Hyams, J. (1980) Microtubules, Academic Press, New York.

Weisenberg, R.C. (1972) Microtubule formation in vitro in solutions containing low calcium concentrations, Science 177:1104-1105.

INTERMEDIATE FILAMENTS

Lazarides, E. (1982) Intermediate filaments: a chemically heterogeneous, developmentally regulated class of proteins, Ann. Rev. Biochem. 51:219-250.

Steinert, P.M., Jones, J.C.R. and Goldman, R. (1984) Intermediate filaments, J. Cell Biol. 99:22s-27s.

ASSOCIATION BETWEEN COMPONENTS OF THE CYTOSKELETON

Pollard, T.D., Selden, S.C. and Maupin, P. (1984) Interaction of actin filaments with microtubules, J. Cell Biol. 99:33s-37s.

ORGANIZATION OF THE CYTOSKELETON

Cultured Cells

Jockusch, B.M. (1983) Patterns of microfilament organization in animal cells, Molec. Cell Endocrin. 29:1-19.

Olmsted, J.B., Cox, J.V., Asner, C.F., Paryzek, L.M. and Lyons, H.D. (1984) Cellular regulation of microtubule organization, J. Cell Biol. 99:18s-32s.

Steinert, P.M., Jones, J.C.R. and Goldman, R.D. (1984) Intermediate filaments, J. Cell Biol. 99:22s-27s.

Stossel, T.P. (1984) Contribution of actin to the structure of the cytoplasmic matrix, J. Cell Biol. 99:15s-21s.

Erythrocytes

Branton, D., Cohen, C.M. and Tyler, J. (1981) Interaction of cytoskeleton proteins on the human erythrocyte membrane, Cell 24:24-32.

Bennett, V. (1982) The molecular basis for membrane-cytoskeleton association in human erythrocytes, J. Cell Biochem. 18:49-65.

Cohen, C.M. (1983) The molecular organization of the red cell membrane skeleton, Sem. Hematol. 20:141-158.

Neurons

Hoffman, P.N., Griffin, J.W. and Price, D.L. (1984) Control of axonal caliber by neurofilament transport, J. Cell Biol. 99:705-712.

Vallee, R.B., Bloom, G.S. and Theurkauf, W.E. (1984) Microtubule-associated proteins: subunits of the cytomatrix, J. Cell Biol. 99:38s-44s.

Yamada, K.M., Spooner, B.S. and Wessells, N.K. (1970) Axon growth: roles of microfilaments and microtubules, Proc. Natl. Acad. Sci. 66:1206-1212.

Platelets

Lind, S.E. and Stossel, T.P. (1982) The microfilament network of the platelet, Prog. Hemst. Thromb. 6:63-84.

Fox, J.E.B. and Phillips, D.R. (1983) Polymerization and organization of actin filaments within platelets, Sem. Hematol. 20:243-260.

Leukocytes

Southwick, F.S. and Stossel, T.P. (1983) Contractile proteins in leukocyte function, Sem. Hematol. 20:305-321.

Pollard, T.D. (1984) Molecular architecture of the cytoplasmic matrix, in: "White Cell Mechanics. Basic Science and Clinical Aspects," H.J. Meiselman, M.A. Lichtman and P.L. La Celle, eds., pp. 75-86, Alan R. Liss, Inc., New York.

FUNCTION OF THE CYTOSKELETON

Structural Support and Cell Shape

Mooseker, M.S., Bonder, E.M., Conzelman, K.A., Fishkind, J.D.J., Howe, C.L. and Keller III, T.C.S. (1984) Brush border cytoskeleton and integration of cellular functions, J. Cell Biol. 99:104s-112s.

Tomasek, J.J. and, Hay, E.D. (1984) Analysis of the role of microfilaments and microtubules in acquisition of bipolarity and elongation of fibroblasts in hydrated collagen gels, J. Cell Biol. 99:536-549.

Intercellular Transport

Lasek, R.J., Garner, J.A. and Brady, S.T. (1984) Axonal transport of the cytoplasmic matrix, J. Cell Biol. 99:212s-221s.

Lasek, R.J. and Hoffman, P.N. (1976) The neuronal cytoskeleton, axonal transport and axonal growth, in: "Cell Motility," Cold Spring Harbor Symposium, pp. 1021-1049.

Membrane Properties and Functions

Gratzer, W.B. (1981) The red cell membrane and its cytoskeleton, Biochem. J. 198:1-8.

Mangeat, P. and Burridge, K. (1984) Actin-membrane interaction in fibroblasts: what proteins are involved in this association, J. Cell Biol. 99:95s-103s.

Phagocytosis and Motility

Keller, H.U., Hess, M.W. and Cottier, H. (1975) Physiology of chemotaxis and random motility, Sem. Hematol. 12:47-48.

Stossel, T.P. (1975) Phagocytosis: recognition and ingestion, Sem Hematol. 12:83-116.

Silverstein, S.C. (1982) Membrane receptors and the regulation of mononuclear phagocyte effector functions, Adv. Exp. Med. Biol. 155:21-31.

THE CYTOSKELETON AS A TARGET FOR TOXIC AGENTS

Authors: The Editorial Board

Contents:

INTRODUCTION 25

TOXIC AGENTS AS TOOLS FOR STUDYING THE CYTOSKELETON 25

TOXIC AGENTS ACTING ON THE CYTOSKELETON 28

THE NEURONAL CYTOSKELETON 28

    Agents affecting neuronal microtubules 29

    Agents producing neurofibrillary pathology 29

    Agents affecting other targets in the neuronal cytoskeleton 32

NON-NEURONAL CELL CYTOSKELETON 32

SUMMARY 33

REFERENCES 33

THE CYTOSKELETON AS A TARGET FOR TOXIC AGENTS

Tore L.M. Syversen[a], Polly R. Sager[b], Thomas W. Clarkson[c], John B. Cavanagh[d], Anljot Elgsaeter[a], Hans Cato Guldberg[a], Si Duk Lee[e], Marshall A. Lichtman[c], N. Karle Mottet[f], and Joanna B. Olmsted[c]

[a]University of Trondheim, Norway
[b]University of Connecticut, Farmington, Connecticut, USA
[c]University of Rochester, Rochester, New York, USA
[b]Institute of Neurology, London, UK
[e]EPA, Research Triangle Park, North Carolina, USA
[f]University of Washington, Seattle, Washington, USA

INTRODUCTION

In the previous chapter several examples of toxic compounds that have been used as tools to elucidate the structure and function of the cytoskeleton have been described. These examples were presented in the context of cytoskeletal composition and function. In this chapter, we discuss toxic agents that have been used as tools and present a summary of such compounds. Finally, we review a number of compounds for which evidence strongly suggests that the primary effects involve the cytoskeleton.

TOXIC AGENTS AS TOOLS FOR STUDYING THE CYTOSKELETON

Many agents used to alter biochemical events or disrupt cell function were identified on the basis of their toxic action. Claude Bernard (1875) highlighted the importance of using toxins to probe biochemical pathways: "the poison becomes an instrument which dissociates and analyzes the special properties of different living cells; by establishing their mechanisms in causing cell death or changes in cell function we can learn indirectly much about the relation between molecular structure and the physiological process of life". The use of a toxic chemical as a tool requires an understanding of the primary mode of action of the chemical at the cellular level, including the chemistry of the substance and the nature of its interaction with the target macromolecule. A number of substances have been identified that interact selectively with cytoskeleton components. Many of these have been discussed in the preceding chapter and are summarized in Table 1.

Much of what we know of the function of cytoskeleton has been learned from experimental perturbations effected by these probes. Examples include the elucidation of the relative roles of microfilaments and microtubules in axon extension either during development or during regeneration. Addition of cytochalasin D blocks neurite outgrowth if added before extension has begun, while treatment of extending axons, results in cessation of the "ruffling" activity of the growth cone with little immediate effect on the neurite shaft. In contrast, antimicrotubule agents, such as colchicine, leave growth cone activity unaffected but block neurite extension. Continued exposure results in retraction of the cell process, with subsequent formation

Table 1. Selective Probes for Cytoskeletal Components.

Probes	Components
Cytochalasins	microfilament
Phalloidin	microfilament
2,5-Hexanedione	intermediate filaments[a]
Substoichiometric "poisons"; colchicine, colcemid, vinca alkaloids, podophyllotoxin	microtubules
Other disrupting agents benzimidazole carbamates, nocodazole, taxol	microtubules

[a] see Graham et al., this volume for evidence of specific interaction.

of one or more growth cone areas from the cell body. These observations, together with ultrastructural evaluations of treated cells, have clarified the separate roles played by these components of the cytoskeleton in a complex and dynamic biological system. Such insights, of course, require test compounds with well-defined, specific actions.

There are no agents known to selectively disrupt intermediate filaments although the interesting observations of Graham et al., (this volume) suggest that 2,5-hexanedione may act in this way. In cells, vanadate in combination with drugs that disassemble microtubules causes vimentin-type intermediate filaments to collapse around the nucleus; this has been observed both in cell lines and in neuronal precursor cells. In epithelial cells, keratin fiber collapse has been documented after treatment with Ca^{2+} or microtubule and microfilament inhibitors. Disruption by β,β'-iminodipropionitrile (IDPN) of microtubule-neurofilament organization in axons results in the co-localization of certain MAP2 epitopes with the intermediate filaments. The subsequent accumulation of neurofilaments suggests a role for MAP2 in synchronizing the transport of neurofilaments and microtubules, by crosslinking these fibers.

For a small number of antimicrotubule drugs, their interactions with tubulin and microtubules are well understood; the actions of some are summarized in Table 2. The chapters by Wilson, Byard and Gull, and Horwitz et al., in this volume document how these drugs have been used to further our basic understanding of tubulin association into microtubules, and of the role of microtubules in cell function. Other workers have used these same agents to examine regulation of tubulin synthesis, heterogeneity of tubulins, and the genetic alterations in tubulin. Further, by selectively dissociating or stabilizing microtubules, these drugs aid in our understanding of the function of microtubules in cells.

The differences in the mechanism of action of various antimicrotubule agents have been exploited in studies of the regulation of tubulin synthesis as summarized in Table 3. Newly synthesized tubulin was measured in cells treated with colchicine, nocodazole, vinblastine and taxol. In cells incubated with colchicine or nocodazole, microtubules depolymerized and the tubulin monomer pool increased; synthesis in these cells decreased with

Table 2. Microtubule Probes - Interactions with Tubulin and Microtubules.

Agents	Modes of Action
Colchicine and analogues	inhibits assembly by substoichiometric "end poisoning" 1 binding site/dimer
Vinca Alkaloids	at low concentrations (<1 µm): inhibits assembly by substoichiometric "end poisoning", binds at high affinity sites (at microtubule ends) at intermediate concentrations (10-100 µM): direct disassembly of microtubules, self association to form paracrystals, peeling of protofilaments, binds to low affinity sites on microtubule surface at high concentrations (>1 mM): non-specific precipitation
Benzimidazole carbamates, nocodazole	inhibits assembly species specificity in binding affinities competes with colchicine binding
Taxol	stabilizes microtubules to disassembly by cold, colchicine, nocodazole, etc. promotes assembly by reducing critical concentration shifts equilibrium to polymer maximal effect at stoichiometric concentration binds to microtubules

Table 3. Effects of Microtubule Drugs on Tubulin Synthesis.

Microtubule Drug	Microtubules	Monomer Pool	Tubulin Synthesis
Colchicine	disassemble	increase	decrease
Nocodazole	disassemble	increase	decrease
Vinblastine	disassemble	decrease	increase
Taxol	assemble	decrease	increase

time. Vinblastine treatment also caused microtubule disassembly but without increasing free tubulin, since tubulin-vinblastine crystals were formed. Taxol stabilization of microtubules also led to a reduced monomer level in cells. For vinblastine and taxol treated cells, synthesis initially decreased, but after two hours began increasing to control levels or above. Thus synthesis of tubulin appears to follow the cellular levels of free monomer, where increases in free tubulin signal a reduction in translation. In cells treated with drugs such as colchicine, mRNA levels decreased rapidly; the relatively rapid turnover of tubulin mRNA (half-life of about 2 hours) allows the cells to effect the rapid change in message levels.

It is of interest that although colchicine was first identified as a poison of the mitotic spindle in the 1930s; its specific effect was not described until the isolation of tubulin some 40 years later. Taxol, on the other hand, was purified and introduced as an antitumor agent only in 1971. Thus new selective probes continue to be developed which broaden our understanding of microtubule function. Claude Bernard's advice is still as pertinent and compelling today as it was a century ago.

TOXIC CHEMICALS ACTING ON THE CYTOSKELETON

Our understanding of toxicology is to a large extent dependent on our knowledge of organized cellular systems. In most instances, toxicity arises in response to the interaction of the toxic chemical with a specific molecular structure in the cell; this assumes that all toxic responses have a biochemical explanation. Thus, the more we know about the mechanisms of toxicity, the better prepared we are to make rational decisions concerning the safe use of chemicals. The response of biological systems to toxic chemicals may be represented in three phases: disposition, primary reaction and functional changes, as shown in Fig. 1, which illustrates the process from entry of the chemical to the final clinical conditions. The first phase, disposition of the chemical, includes all those processes (distribution, metabolism, excretion, etc.) that affect the delivery of the chemical to its site of action and that depend largely on the physico-chemical properties of the compound. Phase two represents the primary reaction with molecular targets and it is at this point that toxic agents can be used as tools to disrupt normal function. For many toxic substances (T) the interaction efficiency of the substance with the molecular target (M) is related to the dissociation constants of the reaction: $T + M \rightleftharpoons TM$. The maintenance of the primary reaction depends on whether the interaction is dissociable. In some cases, dissociation is not seen because the interaction is via the formation of stable covalent bonds. Subsequent functional changes occur that may be divided into an initial stage of secondary changes, resulting from the reaction of the compound with the primary target, and the end stage of manifest changes, which may result from a cascade of biological responses leading up to clinical signs. The scheme presented in Fig. 1 may be more complex in that a substance may interact with several molecular targets and diverse mechanisms may give rise to secondary changes which are indistinguishable at a clinical level.

This conference was concerned with the molecular components of the cytoskeleton as chemical targets for toxicity. For certain poisons, molecular components of the cytoskeleton are the primary targets. However, changes in the cytoskeleton can also be secondary and indeed may be associated with any of the "functional changes" indicated in Fig. 1. Some papers presented at the conference discussed a number of occupational, environmental and other poisons that affect the cytoskeleton. The neuronal cytoskeleton is the most commonly studied target for such toxic agents and therefore most of the ensuing discussion will deal with agents affecting the nervous system.

THE NEURONAL CYTOSKELETON

Intracellular transport is of prime importance for the proper function of neurons (see previous chapter). Axonal transport depends on an intact neuronal cytoskeleton and impairment of this process causes an accumulation of materials in transit along the axon including neurofilaments. Two types of cytoskeletal poisons have been reported to affect axonal transport. First, there are the microtubule poisons which depolymerize microtubules causing a decrease in fast axonal transport. Second, there are compounds which cause giant swelling of axons by impairing flow characteristics of neurofilaments along the axon; examples of this type are β,β'-iminodipropionitrile (IDPN), 2,5 hexanedione and carbon disulfide.

Figure 1. Phases of toxicity from entry of the chemical to appearance of effects.

	Disposition	Primary Reaction	Effects	
	Fate of chemical	Interaction at target site	Biochemical (B), histochemical (H) physiological (P) changes	Patho-physio-logical changes
E N T R Y	pro-state ↓ Toxic substance	a) T + M ⇌ TM b) T + M → TM	→ (B H P) →	Toxicity
	Distribution, metabolism, excretion binding storage	a) Dissociable complex between T (toxic substance) and M (molecular target) b) covalent binding	Secondary effects (composition, morphology, physiological parameters)	Manifest changes (clinical signs and symptoms syndromes, disease states)

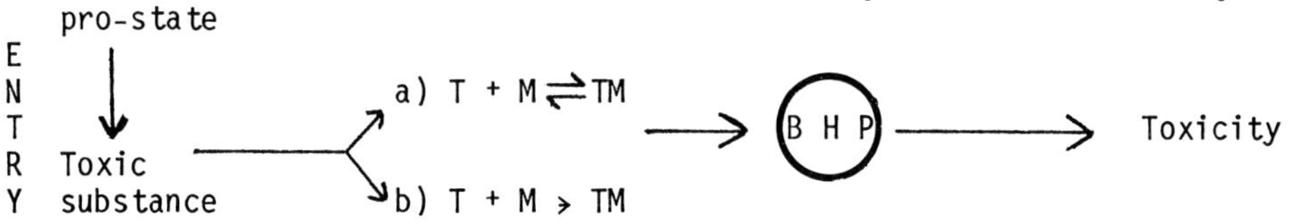

Agents Affecting Neuronal Microtubules

Vinca alkaloids are commonly employed in cancer chemotherapy and peripheral neuropathy develops in more than half the cases. In some patients severe neurologic dysfunctions can limit the use of this agent in therapy.

Neither vinca alkaloids nor colchicine cross the blood-brain or blood-nerve barriers in experimental animals so that not only is it unclear how such a high incidence of nerve damage occurs in humans, but the development of an animal model has been unsuccessful. Local injection of vinca alkaloids and colchicine and their use in cultures of nervous tissue have shown that changes follow their entry into neurons. For example, within 30 min of local injection into rat nerve of 1 mM vincristine, depolymerization of microtubules began and was complete in the injected zone by 2 hours. As tubules disappeared they were replaced by hexagonal paracrystals 30-36 nm in diameter and these grew as very regular structures in the long axes of the nerve. Similar formations were seen in neurons when vincristine was injected into the cerebrospinal fluid of rabbits.

The effect of these changes on peripheral nerve axons was that they underwent wallerian degeneraton from the injection site as if they had been cut. Proximal to this point, organelles accumulated as after nerve transection and neurofilaments began a day or two later to expand the nearby axons. The center of the axons were filled with neurofilaments while mitochondria and other organelles were found in the periphery. Such a dissociation of filaments and tubules is reminiscent of the changes noted in axons in IDPN intoxication.

Agents Producing Neurofibrillary Pathology

Most of the toxic agents that alter neuronal cytoskeleton appear to affect neurofilament distribution. For all compounds investigated so far, very small if any, changes in synthesis or degradation have been observed. Thus, the primary action of such compounds seems to be on intracellular transport of neurofilament protein. These 10 nM intermediate filaments consisting of three major peptides pass along the axon at 0.3-4 mm/day depending on species, age and nerve examined. In adult mammalian peripheral nerve, the neurofilaments often move associated with tubulin at a rate of about 1 mm/day.

Aluminum. As discussed by Shelanski (this volume) local injection of aluminum salts, either into the brain or into cerebrospinal fluid, induced marked accumulations of 10 nm filaments in the perikarya, dendrites and initial segments of large nerve cells. When silver stained, these accumulations superficially resemble the neurofibrillary tangles that are encountered in a number of human disease states, notably Alzheimer's disease, Parkinsonism-Dementia of Guam Islanders, and in senile states. However, the filamentous accumulations are not identical to those in the human disease since they do not show the characteristic ultrastructural paired helical filaments or green dichroism with polarized light after Congo red staining typical of Alzheimer's disease, nor interact with the same intermediate filament antibodies. Nonetheless, this model system is of considerable interest from the point of view of our understanding of the formation and functions of the neuronal cytoskeleton. The biochemical aspects of this experimental system have been discussed by Shelanski (this volume) and, as he points out, there is considerable uncertainty as to how aluminum brings about these cellular events. Clearly the mechanism must be different from that caused by IDPN even though this substance also produces the same general filament accumulation in the perikaryon and proximal segment of the axon.

β,β'-iminodipropionitrile (IDPN). This is an interesting model compound for producing neurofilament accumulation in the proximal region of the axon. The related compound, dimethylaminopropionitrile, has been reported to produce proximal neuropathies in humans, however, the clinical picture differs from that seen in experimental animals.

Intoxication by IDPN caused rats to move in circles and later to develop a hind leg paralysis. The neuropathological changes consisted of a profound swelling of the proximal axons within the first few mm of their course. The swelling contained massive accumulations of 10 nm neurofilaments in which mitochondria and other organelles were trapped. Papasozomenos (this volume) have shown a characteristic reorganization of the cytoskeleton throughout the axon. Normally the microtubules and filaments are evenly arranged across the axon; after IDPN treatment the filaments were distributed peripherally in the axon while the microtubules concentrated in the core of the axon. This remarkable reorganization occurred more completely in the larger axons than in the smaller ones and has been found to take place along the entire length of the sciatic nerve almost simultaneously. Furthermore, the disaggregation of the axonal cytoskeleton was accompanied by evident shrinkage of axons, particularly those of larger diameter. On recovery, the axon slowly returned to its original organization. IDPN inhibited slow axonal transport, and in particular, appeared to selectively inhibit the movement of neurofilament protein, while transport of other constituents of slow axonal transport (e.g. tubulin and actin) was far less affected. Supporting this view is the observation that IDPN had no effect on slow transport in the small fibers of the dorsal motor nucleus of the vagus nerve. These are small diameter fibers which contain small amounts of neurofilament protein.

γ-Diketone (Hexacarbon) Compounds. The interest in the neurotoxicity of hexacarbons arose from an episode of industrial intoxication by methyl-n-butyl ketone. It was later shown that the ultimate toxic chemical was a metabolite common to this compound and to n-hexane, namely 2,5-hexanedione. Clinically the condition manifests itself in man predominantly as a distal motor and sensory neuropathy of slow onset and even slower resolution. Both in man and in experimental animals it has been shown that the axons become swollen with often massive accumulations of 10 nm filaments. Microtubules and other axonal organelles tend to be pushed either to the periphery or to be clustered into groups at the center of the axon. Swellings occur chiefly in larger axons as with IDPN, but in contrast to IDPN, are located principally in the distal half or third of their course. Swollen axons, moreover, are found throughout the nervous system and the filament masses appear to accumulate proximal to the constrictions at the nodes of Ranvier. Since the constrictions at the nodes are less marked in smaller axons, filaments may pass through these more readily than in larger axons. On recovery from intoxication the filament masses move slowly to the preterminal regions of the axons where they undergo degradation.

Despite the swelling along a considerable length of the larger axons, fast axonal transport mechanisms remain normal for a period, but then decline with the onset of physical signs of peripheral neuropathy and evidence of wallerian degeneration. The causes of axonal degeneration are not known, but eventual impoverishment of essential materials for axonal maintenance, due either to obstruction along the length of the axon or some other damage, may play an important part.

Unlike the sequence of events in IDPN intoxication, segregation of microtubules and microfilaments does not occur in the same dramatic and general way in hexacarbon intoxication. However, local application of 2,5-hexanedione to an exposed nerve can create, in the same way as with IDPN, the accumulation of neurofilament masses within axons. There are also other points of similarity, as emphasized by Graham et al. (this volume). By varying the chemical structure of 2,5-hexanedione and incorporating mono- or dimethyl group(s), varying degrees of similarity with the IDPN model system have been demonstrated. These derivatives are capable of producing more proximal axonal swellings. These similar morphological effects do not necessarily imply identical mechanisms of cytoskeletal disruption, but they do suggest that rates of interaction with axonal structures (the faster the rate, the more proximal the swellings) may be playing important roles in determining the localization of the axonal lesion.

Carbon Disulfide. Carbon disulfide poisoning is known to produce a peripheral neuropathy which in animals is morphologically similar to that induced by hexacarbons. Gambetti et al. (this volume) have carried out studies on the dynamics of axonal transport, and the results from studies in the retinotectal pathway are in close agreement with findings using hexacarbon derivatives. The structural differences between the chemical classes to which carbon disulfide and hexacarbons belong, suggest that the mechanisms of action also may be different. However, the resulting neuropathies that these compounds induce appear quite similar both clinically and pathologically.

Acrylamide. Acrylamide causes accumulation of neurofilaments which are first observed in the distal preterminal regions of both sensory and motor axons. As the intoxication proceeds the accumulation develops more proximally. The mechanism of action for acrylamide is not known, and evidence has been presented that the effects on the cytoskeleton might well be of secondary nature (Gold et al., this volume). Most regions of the nervous system show these preterminal swellings regardless of axon length. The presence of such accumulations does not necessarily lead to axonal degeneration, as this always begins in the largest and longest fibers first, and more so in sensory than motor fibers.

Agent Affecting Other Targets in the Neuronal Cytoskeleton

Methylmercury. Methylmercury is a human teratogen. Outbreaks of methylmercury poisoning have revealed that severe damage could be inflicted on the developing nervous system in the fetus, whereas the mother, who was exposed during pregnancy, was only slightly affected. Subsequent animal studies confirmed the greater sensitivity of the developing as compared to the mature central nervous system.

Early experimental work had indicated that methylmercury was a potent mitotic poison. Subsequently, (see Sager and Syversen, this volume) it was shown that methylmercury binds to tubulin and is able to rapidly depolymerize microtubules both in vitro and in vivo. It apppears that in vitro, methylmercury interacts with sulfhydryl groups required for polymerization. Complete inhibition of assembly occurs at equimolar concentrations of methylmercury and tubulin dimers; at higher concentrations, formed microtubules are depolymerized. This led Sager and coworkers to test the hypothesis that the sensitivity of the developing central nervous system was due to methylmercury inhibition of cell division. They went on to show that methylmercury produced mitotic arrest of granule cells of the developing cerebellum in the mouse. These effects were produced at methylmercury concentrations in the brain at or lower than any previously reported effect levels. Of special interest, they noted that at the lowest dose, male mice were much more susceptible than female mice, consistent with findings of an epidemiological study of exposed human infants and children.

Effects of methylmercury on microtubules may have other consequences to the developing central nervous sytem. Autopsy findings on human infants prenatally exposed to methylmercury indicate a picture of disturbed cytoarchitecture of the brain suggesting interference with cell migration.

NON-NEURONAL CELL CYTOSKELETON

Mineral Fibers; Asbestos

Asbestos is a generic term used to describe chain silicates, which occur naturally in a fibrous form; chrysotile and amphibole are the two main types. Chrysotile is a hydrated magnesium silicate that forms a soft flexible type of fiber. Chrysotile asbestos has proven to be immensely useful in industry because of its resistance to abrasion and heat, and its high tensile strength and flexibility. Because of its small diameter, it has properties which cannot yet be achieved by manmade fibers, but which in turn increase its deposition in lung tissue. As a result humans are exposed to aerosols of asbestos fibers which produce a variety of pathological reactions including (1) asbestosis - a fibrosing lesion of the lung, (2) pleural placques, and (3) bronchogenic carcinomas and mesotheliomas of the pleural and peritoneal linings.

The inhalation of asbestos fibers and the attributes of fiber size and shape affect the production and location of the pathological lesions associated with it. Some asbestos fibers traverse the epithelial cells to the lung interstitium. Brody et al., (this volume) have gathered evidence that actin-containing microfilaments play a role in the intracellular translocation of asbestos. Following inhalation of asbestos, it was shown that the fibers were associated with microfilaments in alveolar epithelial cells. In cultured fibroblasts, intracellular asbestos induced polymerization of actin filaments and was transported to a perinuclear location; in dividing cells these fibers produced chromosome damage and inhibition of cytokinesis. It has been speculated that the perinuclear asbestos fibers damage or mechanically impede cytoskeletal elements essential for normal chromosomal segregation and cell division.

SUMMARY

The two preceding introductory chapters were included to provide background of the current state of knowledge about the cytoskeleton (Chapter 1) and the known effects of toxic agents on cytoskeletal components (this chapter). For toxicologists interested in the contents of this volume, we hope to provide in Chapter 1, a succinct resume of the functional roll of cytoskeletal structures; for the cell biologist, physiologist or biochemist we provide in Chapter 2, oppportunities for learning more about the cytoskeleton as a target for the deleterious effects of toxic agents on cells, tissues, and organs.

The succeeding chapters contain new perspectives on the cytoskeleton as a target for toxic effects. The intention of this volume is to encourage and stimulate further interactions between cell biologists and toxicologists to examine more thoroughly the cytoskeleton as a site of toxic actions.

REFERENCES

TOXIC AGENTS FOR STUDYING THE CYTOSKELETON

Ben-Ze'ev, A., Farmer S.R. and Penman S. (1979) Mechanisms regulating tubulin synthesis in cultured mammalian cells, Cell 17:319-325.

Bernard, C. (1875) la Science Experimentale, p. 237, Bailliere, Paris.

Cleveland, D.W., Lopata, M.A., Sherline, P. and Kirschner, M.W. (1981) Unpolymerized tubulin modulates the level of tubulin mRNAs, Cell 25:537-546.

Yamanda, K.M. Spooner, B.S. and Wessells, N.K. (1970) Axon Growth: roles of microtubulin and microfilaments, Proc. Natl. Acad. Sci. 66:1206-1212.

TOXIC CHEMICALS ACTING ON THE CYTOSKELETON

Aldridge, W.N., (1981), Mechanisims of toxicity. New concepts are required in toxicology, Trends Pharmacol. Sci. 2:228-231.

THE NEURONAL CYTOSKELETON AGENTS AFFECTING NEURONAL MICROTUBULES

Roderiguez Echandia, E.L., Ramirez, B.U. and Fernandez, H.L. (1973) Studies on the mechanism of inhibition of axoplasmic transport of neuronal organelles, J. Neurocytol. 2:149-156.

Schlaepfer, W.W. (1971) Vincristine induced axonal alterations in rat peripheral nerve, J. Neuropathol. Exper. Neurol. 30:488-505.

Shelanski, M.L. and Wisniewski, H. (1969) Neurofibrillary degeneration induced by vincrestine neuropathy, Arch. Neurol. 20:199-206.

Watkins, S.M. and Griffin, J.P. (1978) High incidence of vincristine induced neuropathy in lymphomas, Brit. Med. J. 1:610-612.

AGENTS PRODUCING NEUROFIBRILLARY PATHOLOGY

Gajdusek, D.C. (1985) Hypothesis: interference with axonal transport of neurofilament as a common pathogenic mechanism in certain diseases of the central nervous system, N. Eng. J. Med. 312:714-719.

Aluminum

Wisniewski, H.M., Sturman, J.A. and Shek, J.W. (1980) Aluminium chloride induced neurofibrillary changes induced in the developing rabbit: a chronic animal model, Ann. Neurol. 8:479-490.

IDPN

Clark, A.W., Griffin, J.W. and Price D.L. (1980) The axonal poathology in chronic IDPN intoxication, J. Neuropathol. Exper. Neurol. 39:42-55.

Hoffman, P.N., Clark, L.C., Carroll, P.T. and Price D.L. (1978) Slow axonal transport of neurofilament proteins: impairment by β,β'-iminodipropionitrile administration, Science 202:633-635.

Pestronk, A., Keoyh, J.P. and Griffin, J.W. (1980) Dimethylaminoproprontrile, in: "Experimental and Clinical Neurotoxicology," P.S. Spencer and H.H. Schvamburg, eds., pp 422-429, Williams and Wilkins, Baltimore.

γ-Diketone (Hexacarbon) Compounds

Cavanagh, J.B. (1982) The pattern of recovery of axons following 2,5 hexanediol intoxication: a question of rheology, Neuropathol. Neuropiol. 8:19-34.

Spencer, P.S., Schaumburg, H.H., Sabri, N.I. and Veronesia, B. (1980) The changing view of hexacarbon neuropathy, CRC Crit. Toxicol. 7:279-356.

Carbon Disulfide

Szendzikowski, S., Stetkiewicz, J. Wronska-Nofer, T. and Zdrajkowska, I. (1973) Structural aspects of experimental carbon disulphide neuropathy. I. Development of neurohistological changes in chronically intoxicated rats, Internat. Arch. Arbeitsmed. 31:135-149.

SESSION 1. MICROTUBULES AND MICROTUBULE INHIBITORS

    Chairpersons: J.B. Olmsted,
                           University of Rochester

                      M.W. Anders
                      University of Rochester School of Medicine

    Rapporteurs: P.R. Sager
                           University of Connecticut Health Center

                      T.L.M. Syversen
                      University of Trondheim School of Medicine

MICROTUBULES AS TARGETS FOR DRUG AND TOXIC CHEMICAL ACTION: THE MECHANISMS OF ACTION OF COLCHICINE AND VINBLASTINE

Leslie Wilson

Department of Biological Sciences
University of California
Santa Barbara, California, USA

INTRODUCTION

Colchicine (Fig. 1), the earliest chemical substance discovered that acts on microtubules, was initially described as a spindle poison in the 1930s because it produced strikingly disorganized mitotic spindles and chromosome patterns in treated cells (reviewed by Dustin, 1978). It was many years before it was established that the mechanism of action of colchicine involved disruption of microtubules (see Borisy and Taylor, 1967a,b; Wilson and Friedkin, 1967). Other microtubule-disruptive substances originally identified as spindle poisons include griseofulvin (see Deysson, 1964), podophyllotoxin, a plant alkaloid that binds to tubulin in the vicinity of the colchicine binding site (see Wilson, 1975), and a clinically important group of alkaloids commonly known as the "vinca alkaloids." These drugs were discovered in 1957 in extracts of the plant Catharanthus rosea, originally called Vinca rosea (Cutts et al., 1957; Noble et al., 1958). Three vinca alkaloids, vinblastine, vincristine (Fig. 1), and vindesine, are currently used for the treatment of several forms of cancer (Gerzon, 1980).

More recently, the list of chemical substances known to disrupt microtubule structure and function in cells has expanded greatly, to include more than 20 distinct chemical classes of agents. Several of the agents are discussed in the proceedings of this conference, including the benzimidazole carbamates (Byard and Gull, this volume), the plant product, taxol (Horwitz et al., this volume) and methylmercury (Sager and Syverson, this volume). Other microtubule-disrupting substances, not discussed at this conference, include maytansine (Remillard et al., 1975), chlorpromazine (Cann and Hinman, 1975), cytochalasin A (Himes and Houston, 1976), diphenylhydantoin (MacKinney et al., 1978), stypoldione (O'Brien et al., 1983), heavy metals such as aluminum (Bonhaus et al., 1980), and RNA, DNA, and other polyanionic substances (Bryan et al., 1975; Corces et al., 1980). The foregoing list is far from complete, and it has become clear that the large numbers of chemical substances that disrupt microtubule function in cells do so by a wide variety of mechanisms.

It is convenient to group the chemical substances that disrupt microtubules into two broad categories. One category includes the drugs that interact with distinct and specific binding sites on tubulin or on the microtubule surface lattice. Not surprisingly, a broad range of potencies

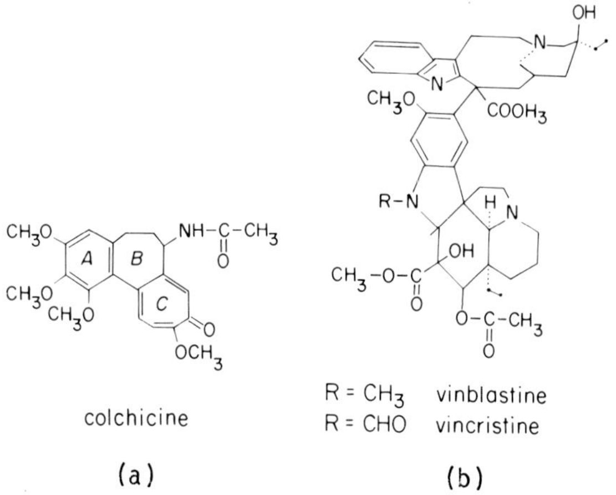

Fig. 1. Chemical structures of colchicine, vinblastine, and vincristine.

has been observed for the large number of substances that interact with specific sites on the tubulin molecule. For example, colchicine, vinblastine, and podophyllotoxin exhibit relatively high affinities for mammalian tubulin (see Wilson, 1975), while diphenylhydantoin, griseofulvin, and a number of the benzimidazole carbamates have relatively low affinities for mammalian tubulin.

Interestingly, several substances bind to tubulins from different sources with substantially different affinities. For example, methyl N-(benzimidazol-2-yl) carbamate, which binds weakly to mammalian tubulin, binds strongly to tubulin from the yeast Saccharomyces cerevisiae, while colchicine, which binds strongly to mammalian tubulin, binds weakly to yeast tubulin (Kilmartin, 1981). The weak binding of several of the benzimidazole carbamates to mammalian cell tubulin, as compared to fungal cell tubulin, is considered to form the basis for the selective fungicidal activity of these compounds. The second major category of substances interacting with tubulin or microtubules consists of those agents that interact with tubulin with little or no selectivity. These substances include sulfhydryl reactive agents, polyanionic substances such as RNA and DNA, and cationic substances such as the heavy metals.

TOXICOLOGICAL EFFECTS OF MICROTUBULE-DISRUPTING SUBSTANCES

Due to increased knowledge regarding the widespread involvement of microtubules in many biological processes and the realization that a large number of chemical substances can disrupt microtubules by a variety of mechanisms, it is becoming clear that considerably more attention need be given to the toxicological evaluation of microtubule-disruptive substances. The most obvious and best studied toxic effects produced by chemical substances that impair microtubule function are associated with inhibition of cell growth. Central nervous system toxicities comprise another obvious toxicity category. Neurons, for example, contain large numbers of microtubules that are required for axonal transport, and many substances that disrupt microtubules have been shown to affect axonal transport. Unfortunately, the toxicological effects in the central nervous system due to chronic exposure to microtubule-disrupting substances have not been sufficiently evaluated.

Much less well studied are the mutagenic and teratogenic effects of microtubule-disrupting agents. Of notable concern are the possible mutagenic or teratogenic effects of weakly acting substances. For example, a slight perturbation of the mitotic spindle during anaphase chromosome movement by weakly acting microtubule inhibitors could result in chromosomal nondisjunction in otherwise perfectly functioning spindles: an action which could produce both mutagenic and teratogenic effects. Such weakly acting substances might be considered safe for human exposure because they do not produce detectable inhibition of cell growth at human exposure levels.

MECHANISMS OF ACTION OF COLCHICINE AND VINBLASTINE: TWO WELL-STUDIED MICROTUBULE-DISRUPTIVE SUBSTANCES

My laboratory group has been interested in the biochemical mechanisms of action of substances that affect microtubule function. Two of the most thoroughly studied and understood substances that disrupt microtubules are colchicine and vinblastine, although many specific aspects of the mechanisms of these two substances remain to be elucidated. I will describe in overview form my own present understanding of the mechanisms of action of these two compounds, drawing in large part from work done in my own laboratory. My hope is that the discussion will serve to illustrate the fact that chemical agents can interfere with microtubule structure and function in many ways, and that microtubules are potentially toxicological targets of a wide variety of chemical substances.

Colchicine

The binding reaction between colchicine and tubulin has been studied extensively since 1967 (Borisy and Taylor, 1967a,b; Wilson and Friedkin, 1967) using tubulins from a wide range of species. There is a single, relatively strong binding site for colchicine on the tubulin dimer, with an affinity that differs substantially depending on the species source of the tubulin, and to some extent, the method of measurement (discussed in Wilson et al., 1984). The binding reaction between colchicine and tubulin is unusual in several respects, including a very slow association rate and the likelihood of conformational changes occurring in both the tubulin and the colchicine (Garland, 1978; Detrich et al., 1981).

Interestingly, the affinity of colchicine for tubulin differs even for tubulin from the same species (Wilson et al., 1984). Shown in Fig. 2 are Van't Hoff plots for the binding of colchicine to purified tubulins from unfertilized eggs and from the sperm flagella outer doublet microtubules of the sea urchin Strongylocentrotus purpuratus, from which the thermodynamic parameters governing the binding of colchicine to tubulin were derived. The data demonstrate that significant chemical differences exist in the tubulins from two functionally distinct microtubule systems of the same species and that the differences are expressed at the level of the colchicine binding site on the native tubulin dimer. Perhaps the colchicine binding site is an important regulatory site on the tubulin molecule, and colchicine is mimicking a natural regulatory molecule involved in the control of microtubule assembly or function.

Colchicine inhibits the polymerization of microtubule protein into microtubules in vitro and in vivo (reviewed in Dustin, 1978). The early observation of Olmsted and Borisy (1973) that inhibition of microtubule polymerization in vitro in crude brain extracts occurred at colchicine concentrations well below the concentration of soluble tubulin in the extract suggested that the drug was not stoichiometrically inactivating the tubulin. An additional important early observation was that colchicine does not bind to microtubule surfaces along the lengths of the microtubules

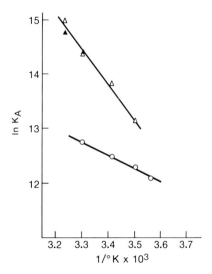

Fig. 2. Van't Hoff analysis of colchicine binding to sea urchin tubulins. The binding of tritium-labeled colchicine to purified tubulin obtained from unfertilized eggs and from sperm tail outer doublet microtubules was ascertained at a variety of temperatures by Scatchard analysis, and the data plotted in the form of Van't Hoff plots to obtain the thermodynamic parameters describing the binding reactions (see Wilson et al., 1984, for experimental details). △, outer doublet tubulin solubilized with the French pressure cell; ▲, outer doublet tubulin purified by sonication and two cycles of in vitro microtubule assembly and disassembly; ○, purified egg tubulin. Thermodynamic parameters for egg tubulin were: ΔH_o, 4.6 kcal/mole; ΔS_o, 40.4 entropy units; and $\Delta G_{13°C}$, -6.9 kcal/mol. Thermodynamic parameters for outer doublet tubulin were: ΔH_o, 13.6 kcal/mole; ΔS_o, 73.4 entropy units; and $\Delta G_{13°C}$, -7.4 kcal/mole.

(Wilson and Meza, 1973). The initial observations were confirmed by Margolis and Wilson (1977) and Margolis et al. (1980) who further found that the basis of the substoichiometric poisoning phenomenon was the formation of tubulin-colchicine complexes which add to microtubule ends and inhibit growth (schematically depicted in Fig. 3).

A typical example of an experiment that begins with a solution of unassembled bovine brain microtubule protein and that demonstrates the substoichiometric nature of the poisoning mechanism is shown in Fig. 4. In this experiment, previously formed colchicine-tubulin complexes were added to the microtubule protein solution at a ratio of 2.6 moles of colchicine-tubulin complex to 100 moles of uncomplexed tubulin; assembly was monitored by light scattering at 350 nm. The assembly rate in the drug-containing mixture was inhibited by 41 percent as compared with assembly rates in controls. Similar results were obtained when colchicine was added to preformed microtubules at steady state. With the 3-cycle-purified bovine-brain microtubules used in this work, the microtubules undergo a continuous net addition of tubulin at one end of each microtubule, called the net assembly- or A-end, and a precisely equal net loss of tubulin at the opposite end of each microtubule, called the D-end (Margolis and Wilson, 1981; see schematic diagram, Fig. 5). Colchicine inhibits the net uptake of tubulin at the A-end (Fig. 6) and, therefore, net tubulin flux from one end of the microtubule to the other, when added to the steady-state microtubule suspension. The initial rate of tubulin uptake can be determined for a

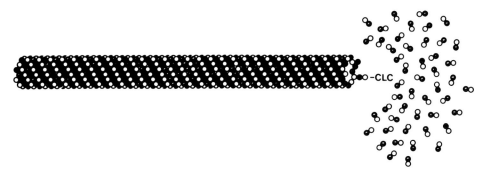

Fig. 3. A schematic representation of substoichiometric inhibition of net tubulin addition to microtubule ends by colchicine-tubulin complexes. Inhibition of net tubulin addition is obtained with small numbers of colchicine-tubulin complexes bound at the microtubule A-ends by the capacity of the colchicine-tubulin complexes to inhibit both tubulin addition and loss rates. Colchicine-tubulin complexes also affect tubulin addition and loss at microtubule D-ends (Bergen and Borisy, 1983; Farrell and Wilson, 1984), but apparently not when added in low concentrations to steady-state bovine brain microtubules (see text).

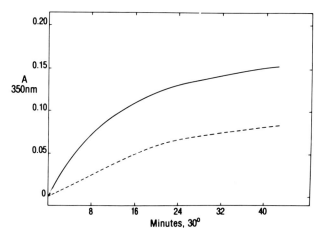

Fig. 4. The effect of added colchicine-tubulin complex on microtubule polymerization. Bovine brain microtubule protein, 1 mg/ml of total protein (0.75 mg/ml tubulin), was polymerized at 30°C in the absence (——) and presence (– – –) of preformed colchicine-tubulin complex (ratio of tubulin in the form of colchicine-tubulin complex to uncomplexed tubulin, 2.6:100 and analyzed by light scattering at 350 nm, as described in Margolis and Wilson (1977).

range of colchicine concentrations, and the results can be plotted in the form of a double-reciprocal plot to obtain an apparent inhibition constant (Fig. 7). Half-maximal inhibition of net tubulin uptake at the A-ends of the steady-state microtubules, at the conditions used, occurred at a colchicine concentration of 0.133 μM; at 50 percent inhibition of assembly, only 2.1 percent of the soluble tubulin was bound by the drug (see Margolis et al., 1980).

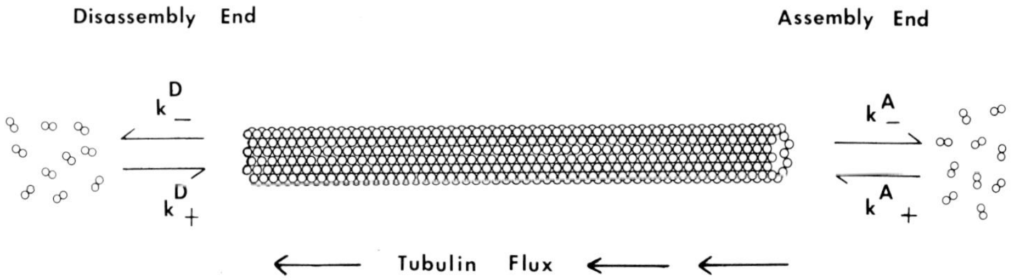

Fig. 5. Steady-state flux of tubulin. Equilibrium reactions exist at both ends of the MAP-containing bovine-brain microtubules used in this study. At steady state in the presence of GTP and a GTP regenerating system in vitro, the critical subunit concentration for subunit growth at the assembly ends is lower than the critical subunit concentration for growth at disassembly ends, giving rise to subunit flux (precisely balanced net addition of tubulin at A-ends and net loss of tubulin at D-ends). Microtubule flux has not been demonstrated to exist in vivo (see Margolis and Wilson, 1981). Under the conditions used in our experiments with reassembled bovine brain microtubule proteins, the steady-state net growth rate at the A-ends, and loss rate at the D-ends (i.e., flux rate) is approximately 1 μm/hr.

The precise mechanism by which colchicine-tubulin complexes inhibit the polymerization reaction at microtubule ends, however, has remained highly controversial. This is due certainly to the complex nature of the mechanism, but also in part, to the use of different preparations of microtubule proteins in different laboratories and to differences in the conditions employed to study the mechanism of inhibition (discussed in Farrell and Wilson, 1984; see also Lambeir and Engelborghs, 1980; Deery and Weisenberg, 1981; Bergen and Borisy, 1983). For example, Margolis et al. (1980) developed a capping model to describe the inhibition of assembly by colchicine, based on experiments in which radioactive colchicine was added to suspensions of preassembled steady-state bovine brain microtubules in vitro. The idea was that colchicine-tubulin complexes bound tightly to the net assembly ends of steady-state microtubules, and that when bound at the ends, the complexes prevented both the net addition of tubulin as well as the net loss of tubulin (a kinetic "cap"). For example, at conditions of half-maximal inhibition of net tubulin addition to steady-state bovine brain microtubules, an average of 0.48 molecules of colchicine were bound at the net assembly end (Margolis et al., 1980). The data suggest that a single colchicine-tubulin complex bound at the assembly end of a microtubule, under steady-state conditions, reduces the net tubulin addition rate to nearly undetectable levels.

In a different kind of experiment, a high colchicine concentration (100 μM) was added to microtubules containing tritium-labeled GTP at their A-ends. The microtubules were then diluted five-fold into a buffer containing 100 μM colchicine, and the rate of loss of A-end label was quantitated and compared to the rate of loss of A-end label from control microtubules. As shown in Fig. 8, the colchicine-treated microtubules lost A-end label much more slowly upon dilution than the control microtubules, indicating that colchicine-tubulin complexes bound to the A-ends of the microtubules kinetically slowed ("capped") the A-ends.

However, a different mechanism was described by Sternlicht and Ringel (1979), for microtubule protein that spontaneously polymerized in the presence of colchicine-tubulin complexes. These investigators found that

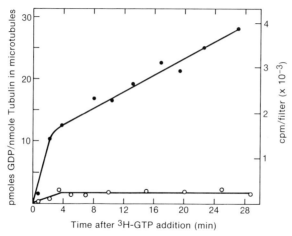

Fig. 6. Inhibition by colchicine of radioactive guanine nucleotide incorporation at the A-ends of steady-state microtubules. A solution of purified bovine brain microtubule protein was assembled to steady state at 30°C for 2.5 hr and then split into two portions. One portion was incubated with 100 μM colchicine for 10 min. At zero time for the experiment, a pulse of tritium-labeled GTP was added to each portion, and incubation continued. At the indicated times, aliquots were removed and assayed for labeled nucleotide incorporation after stabilizing the microtubules and collecting them on glass fiber filter discs (described in detail in Wilson et al., 1982a). An initial burst of labeled nucleotide occurred during the first several minutes, followed by a slower and linear incorporation rate. Flux rates are routinely determined after the initial burst rate has subsided. Control (●); colchicine-treated (○). Colchicine inhibited the net tubulin addition rate in this experiment essentially completely.

large numbers of colchicine molecules became incorporated into the microtubule polymers during assembly, thus forming copolymers; they proposed that colchicine inhibited polymerization by an end-conserving copolymerization mechanism. Farrell and Wilson (1984) have confirmed the results of Sternlicht and Ringel (1979). Using a microtubule-dilution procedure and a double-label protocol that permitted simultaneous measurement of tubulin loss rates from both the A- and D-ends of the microtubules, Farrell and Wilson (1984) also observed that tubulin dissociation from the copolymers was slowed at both ends of the copolymers as compared to control microtubules. Further, and most interestingly, the kinetic slowing of tubulin loss at the A-ends of the microtubules was far more pronounced than the effect at the D-ends.

For example, at a tubulin-colchicine complex to tubulin mole fraction in solution of 0.02, the apparent dissociation rate constant at the copolymer A-ends was 3/sec, as compared with a dissociation rate constant in control microtubules at the A-ends of approximately 21/sec (85 percent inhibition), while the dissociation rate constant at the D-ends of copolymers was essentially unaffected, at 64/sec, as compared with control values of 60/sec. At a tubulin-colchicine complex to tubulin mole fraction of 0.06, the dissociation rate constant at A-ends was approximately 0.05/sec (93 percent inhibition), while the dissociation rate constant at D-ends was approximately 29/sec (52 percent inhibition; Farrell and Wilson, 1984). Thus, colchicine appears to be able to stabilize opposite microtubule ends differentially, and at low colchicine-tubulin complex to tubulin ratios, it greatly augments the kinetic differences between the two ends.

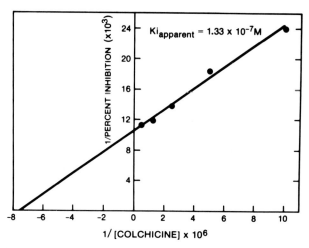

Fig. 7. Inhibition of microtubule assembly at steady state by colchicine (double-reciprocal plot). The extent of tubulin addition to A-ends of steady-state bovine brain microtubules *in vitro* was determined in control samples and in the presence of different concentrations of colchicine (between 0.1 and 2 μM) at 30°C 90 min. after colchicine addition (30 min. after addition of the tritium-labeled GTP pulse), as described in detail in Margolis et al. (1980). Guanine nucleotide incorporation was determined after collecting the microtubules through 50 percent sucrose cushions.

Although the mechanism of colchicine action is still not fully understood, it seems likely that colchicine-tubulin complexes slow potently A-end tubulin addition and loss rates, under both steady-state and nonsteady-state conditions, to an extent, that at high colchicine concentration, all but eliminates tubulin addition and loss at steady state, i.e., caps the microtubule A-end. D-ends, on the other hand, are affected at relatively higher colchicine-tubulin complex concentrations compared to the A-ends, and the kinetic slowing of tubulin loss and addition at the D-ends is much less complete than at the A-ends. It is reasonable to propose that copolymers form by predominant D-end addition of tubulin and colchicine-tubulin complexes when microtubule protein is assembled in the presence of colchicine- dimer complexes; A-ends remain kinetically slow, with the polymerization reaction being driven at D ends by the high tubulin concentration present at the beginning of the polymerization reaction. Addition of colchicine to steady-state microtubules does not result in copolymer formation because the steady-state free tubulin concentration is insufficiently high to drive copolymer addition at the D-ends.

Vinca Alkaloids: Vinblastine

The vinca alkaloids are potent inhibitors of cell growth and appear to exert their antiproliferative effects predominantly by disruption of microtubules. The binding reaction between vinblastine and tubulin has been studied in some detail (Owellen et al., 1972; Bryan, 1972; Wilson, 1975; Lee et al., 1975; Bhattacharyya and Wolfe, 1976; Himes et al., 1976; Wilson et al., 1978; Na and Timasheff, 1980a,b; Wilson et al., 1982b). Binding of vinblastine to tubulin dimers occurred very rapidly, was readily reversible, and was relatively independent of temperature between 0°C and 37°C. It seemed clear that there were two specific vinblastine binding sites per tubulin dimer, but there has been disagreement about the magnitude of the binding affinities for the sites. Values have ranged from a high of

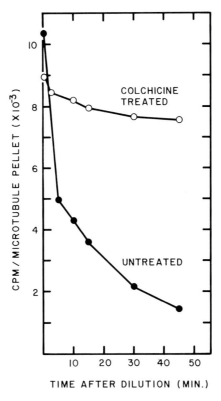

Loss of Assembly-End Label

Fig. 8. Kinetic "capping" of microtubule ends by colchicine. Bovine brain microtubule protein (2.45 mg/ml of total protein) was assembled to steady state at 30°C in the presence of a GTP regenerating system and pulsed for one hr. with tritium-labeled GTP to label the A-ends. The microtubule suspension was then split into two portions and incubation of one portion was continued for an addition 15 min with 100 µM colchicine in the presence of an unlabeled GTP chase (○). The drug-treated microtubule suspension was diluted five-fold into buffer containing 100 µM colchicine and 2.5 mM GTP and incubated for the times indicated in the figure. Drug-free microtubules were similarly diluted, but into buffer containing only the unlabeled GTP (●). Remaining tritium-labeled nucleotide in the microtubules after dilution was determined after collection of the microtubules through 50 percent sucrose cushions. Control microtubules, which initially were 8.0 µm in length and were labeled for 0.8 µm at the A-ends, lost 85.9 percent of the assembly-end label. The colchicine-treated microtubules lost 15.1 percent of the label, or 0.12 µm of length, 45 min. after dilution. The results demonstrate that an A-end block by colchicine greatly slows the rate of tubulin loss from this end. It is important to point out that this experiment was initially represented in an incorrect fashion (Fig. 1, Margolis et al., 1980). In the initial version, the specific activity of label remaining at the A-ends rather than total activity was reported. This error led to an overstatement of the degree of label retention in the presence of colchicine.

approximately 8×10^6 liters/mole at 37°C (Owellen et al., 1972; Bhattacharyya and Wolfe, 1976) to a low of 2.3×10^4 liters/mole at 25°C (Lee et al., 1975). Vinblastine induces a self-association of tubulin in vitro which has been studied extensively by Na and Timasheff (1980a,b). These investigators observed at the vinblastine-induced self-association of tubulin can be described as an indefinite isodesmic self-association reaction, in which tubulin subunits add indefinitely to a growing polymer chain with identical association constants for each step. On the basis of their studies, they have suggested that cooperativity between vinblastine binding to tubulin and tubulin self-association could account, at least in part, for the different vinblastine-tubulin binding constants that have been observed.

Interestingly, vinblastine exerts a number of mechanistically distinct actions on tubulin and on microtubules, depending on the conditions and vinblastine concentrations used. An unusual property of vinblastine is its ability to cause tubulin to assume a variety of polymorphic forms. At low concentrations, below approximately 1 μM both in cells and in vitro, vinblastine inhibits the polymerization of microtubule protein into microtubules. No unusual polymorphic effects of vinblastine on tubulin or on microtubules have been observed below 1 μM vinblastine. At intermediate concentrations, between 10-100 μM, vinblastine has the unusual property of causing cytoplasmic tubulin in a variety of cells to rearrange to form birefringent uniaxial crystals composed of tubulin complexed with vinblastine (Schochet et al., 1968; Bensch and Malawista, 1969; Bryan, 1971, 1972; Fujiwara and Tilney, 1975). Suspensions of highly pure crystals can be isolated from unfertilized sea urchin eggs (Fig. 9), and such crystal preparations have been very useful in studying the binding of vinblastine to tubulin (Bryan, 1972; Wilson et al., 1978). Also at intermediate concentrations > 2-4 μM in vitro, addition of vinblastine to pre-formed brain microtubules causes peeling of protofilament arrays from both microtubule ends as shown in Fig. 10.

At still higher concentrations, >1 mM, vinblastine causes tubulin to precipitate from solution (Marantz et al., 1969; Olmsted et al., 1970; Wilson et al., 1970; Mori and Kurokawa, 1979). This effect of vinblastine is not restricted to tubulin, but occurs also with a number of other proteins including the two other major cytoskeletal proteins: actin (Wilson et al., 1970) and neurofilament protein (Mori and Kurokawa, 1979). Vinblastine also can cause the aggregation of polyribosomes in eukaryotic cells (Krishan and Hsu, 1969) and also in the prokaryotic organism, Escherichia coli (Kingsbury and Voelz, 1969), which is considered to contain no microtubules. This activity of vinblastine appears to be nonspecific; due to the strong cationic character of vinblastine, it may result from the binding of the drug to large numbers of nonspecific ionic sites on the surface of the various proteins. In support of this possibility, we have observed that vinblastine-tubulin paracrystals from S. purpuratus eggs bound approximately 7 moles of vinblastine per mole of tubulin at 1 mM vinblastine, thus indicating the presence of a large number of low affinity binding sites on the crystal tubulin (Wilson et al., 1978). It is possible that the effects of metal ions on tubulin and microtubule assembly may, in part, be mediated by nonspecific interactions with anionic sites on the protein surfaces.

Like colchicine, vinblastine has the capacity to inhibit microtubule polymerization in vitro at concentrations that are considerably below the concentration of free tubulin. For example, 50 percent inhibition of net tubulin addition to steady-state bovine brain microtubules in vitro occurs at 0.14 μM vinblastine, at a ratio of vinblastine-bound tubulin to uncomplexed tubulin of approximately 2:100 (Wilson et al., 1982b). As with colchicine, the binding of vinblastine to steady-state bovine brain

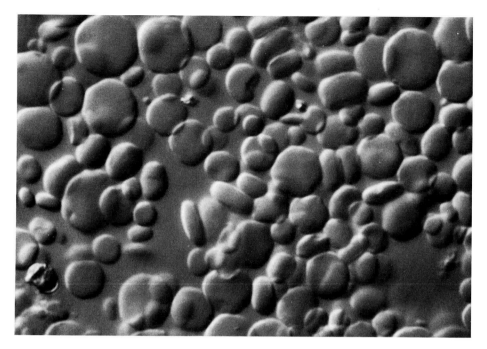

Fig. 9. Vinblastine-tubulin crystals. Vinblastine-tubulin crystals were produced in unfertilized sea urchin oocytes by incubation of the oocytes with 100 μM vinblastine for 36 hr. and were isolated as described by Wilson et al. (1978); vinblastine-tubulin crystals were suspended in stabilizing buffer (Nomarski optics, original magnification, 1325x). Light micrograph by Dr. Paul Matsudaira.

microtubules has been measured over a wide concentration range. Interestingly, two affinity classes of sites have been detected (Fig. 11). At low concentrations, a class of sites with a high affinity and low capacity was detected (16-17 sites per microtubule with a dissociation equilibrium constant of 1.9 μM). The most likely location of these sites is at one or both ends of the microtubule (Wilson et al., 1982b). At 0.14 μM, the vinblastine concentration that produces half-maximal inhibition of tubulin addition at the net assembly ends of the microtubules, approximately 1.2 molecules of vinblastine were bound per microtubule.

The mechanism of inhibition of microtubule polymerization by vinblastine differs in several significant respects from that of colchicine. In contrast to colchicine, vinblastine can bind rapidly and reversibly to the high affinity binding sites at microtubule ends in the absence of soluble tubulin (Wilson et al., 1982b). Thus, with vinblastine, the poisoning reaction is not mediated by addition of drug-modified tubulin molecules to microtubule ends. The kinetically rapid binding of vinblastine to a limited number of high affinity binding sites at the net assembly end of a microtubule appears to be all that is required to reduce the rate of further tubulin addition. If an equal distribution of high affinity binding sites exists at the two ends (i.e., approximately 8 per end), then at 50 percent inhibition of assembly, there would be one molecule of vinblastine at the net assembly end for every two microtubules in suspension. If this distribution were correct, the binding of a single vinblastine molecule to the A-end of a microtubule would be able to reduce the net rate of tubulin addition to below detectable levels. If all of the high affinity binding sites were located at the net assembly ends, the binding of only two vinblastine molecules per microtubule would be sufficient to block assembly.

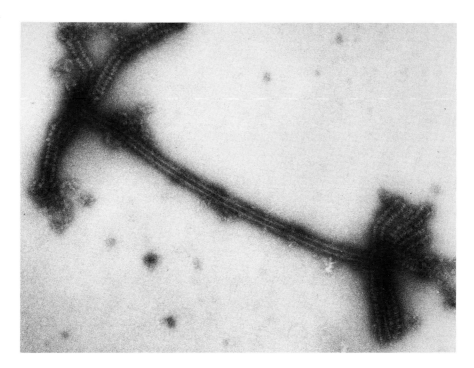

Fig. 10. Peeling of protofilament arrays at microtubule ends by vinblastine. A steady-state bovine brain microtubule that had begun to depolymerize in the form of spirals at both ends after 4 min. incubation in 190 µM vinblastine is shown. Six spirals can be seen emanating from one end of the microtubule, and five spirals from the opposite end. No spirals are observed along the length of the polymer. The microtubule suspension (1.7 mg/ml of total protein) was assembled to steady state at 30°C in the presence of 0.2 mM GTP and a GTP regenerating system (see Wilson et al., 1982b). Vinblastine (final concentration, 190 µM) was added and 4 min. later a 10 µl aliquot of the suspension was placed on 10 µl of 40 percent buffered sucrose on a parlodion-carbon coated electron microscope grid and negatively stained with cytochrome c and uranyl acetate. (Original magnification, 106,000x.) Electron micrograph by Dr. Mary Ann Jordan.

It seems unlikely that the interaction of vinblastine with the high affinity class of sites is responsible for the direct depolymerization of microtubules and the protofilament peeling reaction that occurs at intermediate vinblastine concentrations. The high affinity sites become saturated at approximately 1-2 µM drug, and depolymerization of microtubules and the peeling of protofilament strands does not occur below approximately 1-3 µM drug. Apparently, low affinity binding to a large number of sites located on the microtubule surface occurs in the same vinblastine concentration range as the protofilament peeling reaction; binding and the degree of peeling increase in parallel with one another (Wilson et al., 1982b). These results indicate that interaction of vinblastine with this class (or classes) of sites is associated with, and possibly responsible for, the ability of the drug to depolymerize microtubules directly. A likely possibility is that the low affinity binding of vinblastine to microtubule surfaces represents the binding of vinblastine and its specific binding sites on tubulin, but the sites have been modified in such a way as to reduce considerably the affinity for vinblastine. Recent evidence from

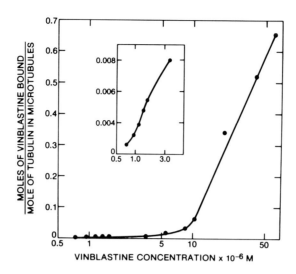

Fig. 11. Binding of vinblastine to steady-state microtubules: extended concentration range. Portions of a bovine brain microtubule suspension at steady state (2.8 mg/ml of total protein, mean microtubule length, 1.09 μm) were incubated with different concentrations of radioactive vinblastine for 30 min. at 30°C. Vinblastine binding was determined after collection of the microtubules through 50 percent sucrose cushions, as described in Wilson et al. 1982b). V = the moles of vinblastine bound/mole of tubulin in the pellet of drug-treated microtubules. The inset shows the low vinblastine concentration range expanded.

our laboratory indicates that the low affinity sites are located at the surface of the microtubule along its length (Jordan et al., 1983).

Thus, the vinblastine binding sites on tubulin can display different affinities depending on their locations in the microtubule surface lattice. The low affinity sites may be sterically blocked or, alternatively, the tubulin may undergo a conformational change during assembly that reduces the affinity for vinblastine. This latter possibility is consistent with the data of Na and Timasheff discussed previously (1980a,b), which indicated that tubulin self-association could be associated with altered vinblastine binding affinity.

SUMMARY

In summary, the foregoing discussion was intended to provide a mechanistic overview of how two well-known microtubule-disrupting drugs act. A broad range of mechanisms are possible in addition to the end poisoning mechanisms of colchicine and vinblastine. An important concept that emerges from consideration of the mechanisms of presently known microtubule-disruptive agents is that it is possible to interfere with microtubule structure and function in cells by a wide number of different kinds of chemical substances by multiple types of mechanisms. It seems clear that microtubules are prime targets for the action of toxicological substances, and that the toxicological consequences of microtubule-disrupting agents can be far reaching because of the many cell processes dependent upon proper microtubule function.

ACKNOWLEDGMENTS

The work described from my laboratory has been supported by U.S. Public Health Service Grants NS13560 from the Institute of Neurological, Communicative Diseases, and Stroke (L.W.) and CA36389 from the National Cancer Institute (L.W.).

REFERENCES

Bensch, K.G. and Malawista, S.E., 1969, Microtubular crystals in mammalian cells, J. Cell Biol., 40:95-107.

Bergen, L.G. and Borisy, G.G., 1983, Tubulin-colchicine complex inhibits microtubule elongation at both plus and minus ends, J. Biol. Chem., 258:4190-4194.

Bhattacharyya, B. and Wolfe, J., 1976, Tubulin aggregation and disaggregation: Mediation by two distinct vinblastine binding sites, Proc. Natl. Acad. Sci., 73:2375-2378.

Bonhaus, D.W., McCormack, K.M., Mayor, G.H., Mattson, J.C. and Hook, J.B., 1980, The effects of aluminum on microtubular integrity using in vitro and in vivo models, Toxicol. Let., 6:141-147.

Borisy, G.G. and Taylor, E.W., 1967a, The mechanism of action of colchicine. Binding of colchicine-^{3}H to cellular protein, J. Cell Biol., 34:525-533.

Borisy, G.G. and Taylor, E.W., 1967b, The mechanism of action of colchicine. Colchicine binding to sea urchin eggs and the mitotic apparatus, J. Cell Biol., 34:535-548.

Bryan, J., 1971, Vinblastine and microtubules I. Induction and isolation of crystals from sea urchin oocytes, Exp. Cell Res., 66:129-136.

Bryan, J., 1972, Definition of three classes of binding sites in isolated microtubule crystals, Biochem., 11:2611-2616.

Bryan, J., Nagle, B.W. and Meza, I., 1975, Inhibition of tubulin assembly by RNA and other polyanions. A mechanism, in: "Microtubules and Microtubule Inhibitors," M. Borgers and M. de Brabander, eds., pp. 91-101, North-Holland, Amsterdam.

Cann, J.R. and Hinman, N.D., 1975, Interactions of chlorpromazine with brain microtubule subunit protein, Mol. Pharmacol., 11:256-267.

Corces, V.G., Manso, R., De La Torre, J. and Avila, J., 1980, Effects of DNA on microtubule assembly, Eur. J. Biochem., 105:7-16.

Cutts, J.H., Beer, C.T. and Noble, R.L., 1957, Effects on hematopoiesis in rats of extracts of Vinca rosea, Rev. Canad. de biol., 16:487.

Deery, W.J. and Weisenberg, R.C., 1981, Kinetic and steady-state analysis of microtubules in the presence of colchicine, Biochem., 20:2316-2324.

Detrich, H.W., III, Williams, R.C., Jr., MacDonald, T.L., Wilson, L. and Puett, D., 1981, Changes in the circular dichroic spectrum of colchicine associated with its binding to tubulin, Biochem., 20:5999-6005.

Deysson, G., 1964, Sur les proprietes antimitotiques de la griseofulvine, Ann. Pharm. Fr., 22:17-25.

Dustin, P., 1978, "Microtubules," Springer-Verlag, Berlin.

Farrell, K.W. and Wilson, L., 1984, The differential kinetic stabilization of opposite microtubule ends by tubulin-colchicine complexes, Biochem., 23:3741-3748.

Fujiwara, K. and Tilney, L.G., 1975, Substructural analysis of the microtubule and its polymorphic forms, Ann. N.Y. Acad. Sci., 253:27-50.

Garland, D.L., 1978, Kinetics and mechanism of colchicine binding to tubulin: evidence for ligand-induced conformational change, Biochem., 17:4266-4272.

Gerzon, K., 1980, Dimeric catharanthus alkaloids, in: "Anticancer Agents Based on Natural Product Models," J.M. Cassaday and J.D. Douros, eds., pp. 271-317, Academic Press, New York.

Himes, R.H. and Houston, L.L., 1976, The action of cytochalasin A on the in vitro polymerization of brain tubulin and muscle G-actin, J. Supramol. Struct., 5:81-90.
Himes, R.H., Kersey, R.N., Heller-Bettinger, I. and Samson, F.E., 1976, Action of the Vinca alkaloids vincristine, vinblastine, and desacetyl vinblastine amide on microtubules in vitro, Cancer Res., 36:3798-3802.
Jordan, M.A., Himes, R.H., Margolis, R.L. and Wilson, L., 1983, Interaction of vinblastine with steady-state microtubules in vitro at concentrations that produce spiral protofilament arrays, J. Cell Biol., 97:214a (Abstract).
Kilmartin, J., 1981, Purification of yeast tubulin by self-assembly in vitro, Biochem., 20:3629-3633.
Kingsbury, E.W. and Volez, H., 1969, Induction of helical arrays of ribosomes by vinblastine sulfate in Escherichia coli, Science, 166:768-769.
Krishan, A. and Hsu, D., 1969, Observations on the association of helical polyribosomes and filaments with vincristine-induced crystals in Earle's L-cell fibroblasts, J. Cell Biol., 43:533-563.
Lambeir, A. and Engelborghs, Y., 1980, A quantitative analysis of tubulin-colchicine binding to microtubules, Eur. J. Biochem., 109:619-624.
Lee, J.C., Harrison, D. and Timasheff, S.N., 1975, Interaction of vinblastine with calf brain microtubule protein, J. Biol. Chem., 250:9276-9282.
MacKinney, A.A., Vyas, R.S. and Walker, D., 1978, Hydantoin drugs inhibit polymerization of pure microtubular protein, J. Pharmacol. Exper. Therap., 204:189-194.
Marantz, R., Ventilla, M. and Shelanski, M., 1969, Vinblastine-induced precipitation of microtubule protein, Science, 165:498-499.
Margolis, R.L. and Wilson, L., 1977, Addition of colchicine-tubulin complex to microtubule ends: The mechanism of substoichiometric colchicine poisoning, Proc. Natl. Acad. Sci., 74:3466-3470.
Margolis, R.L. and Wilson, L., 1981, Microtubule treadmills-possible molecular machinery, Nature (London), 293:705-711.
Margolis, R.L., Rauch, C.T. and Wilson, L., 1980, Mechanism of colchicine-dimer addition to microtubule ends: Implications for the microtubule polymerization mechanism, Biochemistry, 19:5550-5557.
Mori, H. and Kurokawa, M., 1979, Purification of neurofilaments and their interaction with vinblastine sulfate, Cell Struct. Funct., 4:163-167.
Na, G.C. and Timasheff, S.N., 1980a, Stoichiometry of the vinblastine-induced self-association of calf brain tubulin, Biochem., 19:1347-1354.
Na, G.C. and Timasheff, S.N., 1980b, Thermodynamic linkage between tubulin self-association and the binding of vinblastine, Biochem., 19:1355-1369.
Noble, R.L., Beer, C.T. and Cutts, J.H., 1958, Role of chance observations in chemotherapy: Vinca rosea, Ann. N.Y. Acad. Sci., 76:882-894.
O'Brien, E.T., Jacobs, R.S. and Wilson, L., 1983, Inhibition of bovine brain microtubule assembly in vitro by stypoldione, Mol. Pharmacol., 24:493-499.
Olmsted, J.B. and Borisy, G.G., 1973, Characterization of microtubule assembly in porcine brain extracts by viscometry, Biochem., 12:4282-4289.
Olmsted, J.B., Carlson, K., Klebe, R., Ruddle, F. and Rosenbaum, J., 1970, Isolation of microtubule protein from cultured mouse neuroblastoma cells, Proc. Natl. Acad. Sci., 65:129-136.
Owellen, R.J., Owens, A.H., Jr. and Donigian, D.W., 1972, The binding of vincristine, vinblastine and colchicine to tubulin, Biochem. Biophys. Res. Commun., 47:685-691.
Remillard, S., Rebhun, L.I., Howie, G.A. and Kupchan, S.M., 1975, Antimitotic activity of the potent tumor inhibitor maytansine, Science, 189:1002-1005.
Schochet, S.S., Jr., Lambert, P.W. and Earle, K.M., 1968, Neuronal changes induced by intrathecal vincristine sulfate, J. Neuropathol. Exp. Neurol., 27:645-658.

Sternlicht, H. and Ringel, I., 1979, Colchicine inhibition of microtubule assembly via copolymer formation, J. Biol. Chem., 254:10540-10550.

Wilson, L., 1975, Microtubules as drug receptors: Pharmacological properties of microtubule protein, Ann. N.Y. Acad. Sci., 253:213-231.

Wilson, L. and Friedkin, M., 1967, The biochemical events of mitosis. II. The in vivo and in vitro binding of colchicine in grasshopper embryos and its possible relation to inhibition of mitosis, Biochem., 6:3126-3135.

Wilson, L. and Meza, I., 1973, The mechanism of action of colchicine. Colchicine binding properties of sea urchin sperm tail outer doublet tubulin, J. Cell Biol., 58:709-719.

Wilson, L., Bryan, J., Ruby, A. and Mazia, D., 1970, Precipitation of proteins by vinblastine and calcium ions, Proc. Nat. Acad. Sci., 66:807-814.

Wilson, L., Morse, A.N.C. and Bryan, J., 1978, Characterization of acetyl-^{3}H-labeled vinblastine binding to vinblastine-tubulin crystals, J. Mol. Biol., 121:255-268.

Wilson, L., Snyder, K.B., Thompson, W.C. and Margolis, R.L., 1982a, A rapid filtration assay for analysis of microtubule assembly, disassembly and steady-state tubulin, Methods in Cell Biol., 24:159-169.

Wilson, L., Jordan, M.A., Morse, A. and Margolis, R.L., 1982b, Interaction of vinblastine with steady-state microtubules in vitro, J. Mol. Biol., 159:125-149.

Wilson, L., Miller, H.P., Pfeffer, T.A., Sullivan, K.F. and Detrich, H.W., III, 1984, Colchicine binding activity distinguishes sea urchin egg and outer doublet tubulins, J. Cell Biol., 99:37-41.

TAXOL: A PROBE FOR STUDYING THE STRUCTURE AND FUNCTION OF MICROTUBULES

Susan B. Horwitz, Peter B. Schiff, Jerome Parness,
James J. Manfredi, Wilfredo Mellado, and Samar N. Roy

Department of Molecular Pharmacology and Cell Biology
Albert Einstein College of Medicine
Bronx, New York, USA

INTRODUCTION

Plant alkaloids, particularly colchicine and the vinca alkaloids, have played a major role in furthering our understanding of the complex mechanisms involved in the polymerization and depolymerization of microtubules. This has been true in observing cells throughout the cell cycle and in studies done in cell-free systems with purified components. In addition, the vinca alkaloids are important cancer chemotherapeutic drugs used in the treatment of human malignancies.

Taxol, also a plant product, is a new addition to this group of compounds. It was isolated in 1971 from Taxus brevifolia and was shown to have antileukemic activity in experimental murine systems (Wani et al., 1971). Taxol (Fig. 1), a taxane derivative with an oxetan ring, has an unusual chemical structure whose biological activity had not previously been ascertained. The following review will discuss the studies that have been done in our laboratory to define the mechanism of action of this drug.

EFFECTS OF TAXOL IN CELLS

Taxol is a potent inhibitor of cell replication. At concentrations of drug which completely inhibit cell division, cells remain viable and capable of synthesizing macromolecules. When HeLa cells were incubated with a low concentration (0.25 µM) of taxol and microfluorometry was used to examine the effect of the drug on the distribution of DNA, it became clear that within 18 hours essentially all cells had a tetraploid DNA content (Fig. 2) (Schiff and Horwitz, 1980). Examination of these cells indicated that approximately 70 percent had lost their nuclear membranes and contained condensed chromosomes. If taxol was added to synchronized HeLa cells at the beginning of the S phase of the cell cycle, the cells proceeded through S phase normally, but accumulated in the G_2 and M phases of the cell cycle. The reversibility of these effects was dependent on the cell type; for example, in the murine macrophage-like cell line J774.2, cells replicated after the removal of taxol from the medium, whereas in HeLa cells, the effects of taxol were irreversible (Horwitz, S.B., personal communication).

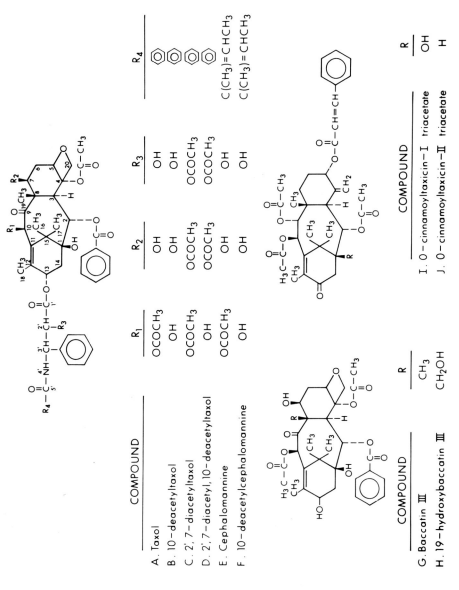

Fig. 1. Structure of taxol and related taxanes.

It became clear that taxol was an antimitotic agent, but it was unexpected to find that interphase cells which were incubated with taxol maintained their cytoplasmic microtubules and developed prominent bundles of microtubules. Such treatment stabilized the cellular microtubules to depolymerization by cold and antimitotic agents such as steganacin and colchicine (Schiff et al., 1979; Schiff and Horwitz, 1980; Crossin and Carney, 1981; Thompson et al., 1981). Electron micrographs of HeLa cells treated with 10 µM taxol for 20 hours revealed the presence of microtubule bundles and indicated an association between the microtubules and the rough endoplasmic reticulum (Fig. 3). The presence of bundles of microtubules is characteristic of the cytoskeleton of taxol-treated cells and has been observed in a number of cell lines (Schiff and Horwitz, 1980; Albertini and Clark, 1981; DeBrabander et al., 1981; Brenner and Brinkley, 1982; DeBrabander, 1982; Manfredi et al., 1982; Tokunaka et al., 1983). These bundles tend not to be associated with the microtubule organizing center; in taxol-treated PtK$_2$ cells essentially none of the microtubule bundles were associated with the centrosome (DeBrabander et al., 1981). Similar observations were made when J774.2 cells were treated with a high concentration (30 µM) of taxol for 30 min (Manfredi and Horwitz, 1984). The microtubule bundles in these cells were stable to extraction by 1 percent Triton X-100 even when the extraction was done in the presence of calcium-containing buffers (Manfredi and Horwitz, 1984). When organotypic mouse spinal cord-ganglion cultures were exposed to 1 µM taxol for 6 days, unusually numerous microtubules and a variety of arrays of microtubules in dorsal root ganglion neurons and supporting cells were seen (Fig. 4) (Masurovsky et al., 1981; 1982; 1983). Ordered microtubular arrays occurred along the endoplasmic reticulum cisternae; at times, short linear elements extended between them. These results suggest that the endoplasmic reticulum may be involved in the organization of microtubules in taxol-treated cells, possibly by serving as an attachment site for microtubules. The presence of taxol at sites of tubulin synthesis could result in polymer formation at rough endoplasmic reticulum loci. Tubulin has been found to be synthesized on both membrane-bound and free polyribosomes prepared from brain (Soifer and Czosnek, 1980).

The specific binding of [^{3}H]taxol to J774.2 cells saturated in a concentration dependent manner with maximal binding occurring at 21 pmol bound mg total cellular protein (Fig. 5) (Manfredi et al., 1982). Unlabeled taxol displaced tritiated taxol and the binding was completely reversible even after a 90 minute incubation with [^{3}H]taxol. Pretreatment of cells with colchicine or vinblastine resulted in the inhibition of [^{3}H]taxol binding. When such cells were examined by immunofluorescence using antibodies against tubulin, it was evident that the pretreatment had completely depolymerized the microtubule cytoskeleton. The cellular receptor for taxol was lost, and saturable binding of [^{3}H]taxol was no longer observed. Saturable binding also was not seen in mature mammalian red blood cells (Manfredi and Horwitz, 1984) which do not contain microtubules. Further evidence that the cellular receptor for taxol is the polymer form of tubulin comes from experiments in which the binding of [^{3}H]taxol to detergent extracted cytoskeletons, which contain only microtubules and no unassembled tubulin dimer, was observed. However, when the detergent extraction was done in the presence of calcium, specific binding of [^{3}H]taxol was not observed, since the microtubules had been depolymerized and were not present in the cytoskeleton (Manfredi and Horwitz, 1984).

Although taxol binds to detergent extracted cytoskeletons, microtubule bundles are never observed in these preparations. The depletion of ATP in J774.2 cells by treatment with sodium azide does not influence the binding of [^{3}H]taxol to cells; however, microtubule bundles are not formed in these cells (Manfredi et al., 1982). The process by which taxol induces the

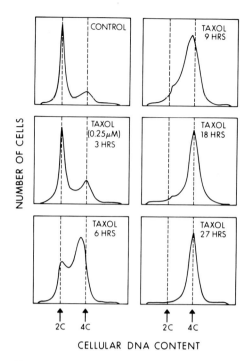

Fig. 2. Flow microfluorometry of the DNA content of HeLa cells in the absence and presence of taxol. Cells growing exponentially were diluted to 3.2×10^5/ml at the start of the experiment. Approximately 6×10^4 cells were analyzed for DNA content at each time point. The histograms depict 100-channel analyses of cellular DNA content. The arrows indicate the modal positions of cells having diploid (2C) and tetraploid (4C) DNA contents. The full ordinate scale indicates a cell count of approximately 4×10^5 cells per channel. In control cultures, the proportion of cells with various DNA contents did not vary signficantly during the time course of the experiment. (From Schiff and Horwitz, 1980, reproduced by permission of National Academy of Sciences, U.S.A.)

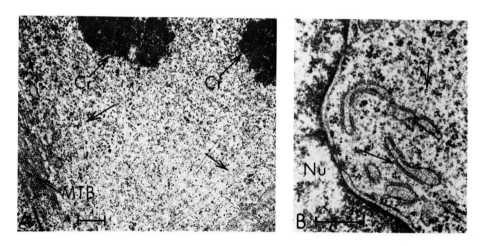

Fig. 3. Electron micrographs of thin sections of HeLa cells treated with 10 μM taxol for 20 hr. NU, nucleus; arrows point to microtubules in cross section. Scale bars: 0.5 m. (From Schiff and Horwitz, 1980, reproduced by permission of National Academy of Sciences, U.S.A.)

formation of microtubule bundles is not known; however, an intact cell with normal ATP levels is necessary for the cytoskeletal reorganization that is required for the formation of microtubule bundles.

EFFECTS OF TAXOL IN VITRO

Taxol also has unique effects on the assembly of microtubules in vitro. The drug has the capacity to shift the equilibrium between the tubulin dimer and polymer in favor of the microtubule. When microtubule assembly was monitored by turbidity, sedimentation or electron microscopy, in the presence of taxol, it became clear that the drug altered the kinetics of assembly. The net result was an enhancement of both the rate of microtubule assembly and the yield of microtubules. Under our normal assembly conditions, there is a three- to four-minute lag time that is eliminated in the presence of 5 μM taxol (Fig. 6) (Schiff et al., 1979). The usual sigmoid progress curve, as followed by turbidity measurements, becomes hyperbolic.

Microtubules assembled in the presence of taxol, or to which taxol was added at steady-state, resist depolymerization by either 4 mM calcium or cold (4°C). In addition, treadmilling of in vitro microtubules was reduced in the presence of taxol (Schiff and Horwitz, 1981a; Kumar, 1981; Caplow and Zeeberg, 1982). Maximal effects of taxol were observed when the concentration of drug was stoichiometric with the tubulin dimer concentration. These data suggest that there is a taxol binding site on the microtubule that is distinct from the exchangeable GTP binding site and the binding sites for colchicine, podophyllotoxin and vinblastine which are present on the tubulin dimer (Schiff and Horwitz, 1981a; Kumar, 1981). The preparation and use of [^{3}H]taxol have made it clear that, as in cells, the drug bound specifically and reversibly to assembled microtubules in vitro and the stoichiometry of binding approached one mole of taxol bound per mole of tubulin dimer (Parness and Horwitz, 1981). Studies by Carlier and Pantaloni (1983) suggest that taxol bound to the tubulin-colchicine complex with an affinity 10-fold lower than its affinity for dimers that were present in polymerized microtubules. Although antimitotic drugs such as vinblastine do not bind at the taxol site, they can inhibit taxol binding if added at the time of assembly. This probably is due to an inhibition of assembly by vinblastine that results in a decrease in the number of microtubules, the target for taxol binding (Horwitz et al., 1982).

Taxol decreased the critical concentration of microtubule protein (tubulin plus microtubule-associated proteins) required for microtubule assembly (Schiff et al., 1979). In the presence of 5 μM taxol, the critical concentration decreased from 0.2 mg of microtubule protein per ml to less than 0.01 mg per ml. In addition to the decrease in the critical concentration of protein and the lag time for assembly, it must be noted that microtubules polymerized in the presence of taxol were shorter than microtubules formed in the absence of drug (Schiff and Horwitz, 1981b). The average length of control microtubules was 4.1 ± 2.0 μm, whereas microtubules polymerized in the presence of 10 μM taxol had an average length of 1.5 ± 0.7 μm. Sedimentation experiments indicated that even though the microtubules assembled were shorter, the drug increased the yield of the microtubule reaction and a greater proportion of the microtubule protein was in the form of polymer. If the average length of the microtubules and the quantity of protein in the polymer form were measured after assembly in the presence of 10 μM taxol, the drug must have been responsible for an approximate 4-fold increase in the number of nucleation events at the start of microtubule assembly.

The capacity of the drug to enhance the assembly of microtubules and shift the equilibrium toward the polymer can be appreciated by considering

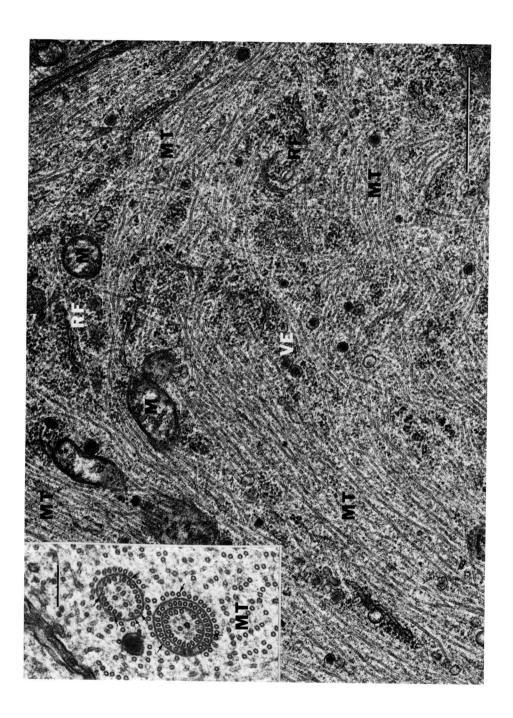

Fig. 4. Electron micrograph illustrating an unusual abundance of microtubules (MT) coursing through the cytoplasm near an exiting process of a neuron in a 13-day fetal mouse dorsal-root ganglion explant exposed to taxol (1 µM) for 6 days (+ nerve growth factor), after an initial period of development for more than 2 weeks in control culture medium. The MTs appear in various orientations interspersed with foci of vesicles (VE), mitochondria (M), and ribosomal formations (RF). X 40,000. Scale bar: 1 µm. Inset: Transverse section through concentric ordered arrays of MTs alternating with layers of macromolecular material in a portion of a neuritic extension near the soma. Connections between some MTs and these nonmembranous lamellae appear at various points in these complexes (e.g., arrows). Some nearby MTs appear to be deployed in various linear and other groupings. X 80,000. Scale bar: 0.2 µm. From Masurovsky et al. (1983), courtesy of IBRO and Pergamon Press.

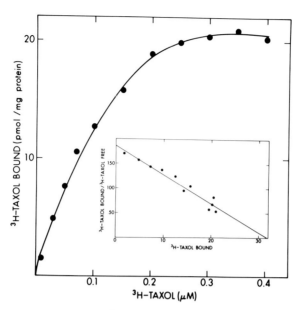

Fig. 5. Specific binding of [^{3}H]taxol to J774.2 cells. Confluent 35 mm dishes of J774.2 cells were incubated for 60 min at 37° C in 2 ml of complete medium containing various concentrations of [^{3}H]taxol. Cells were washed five times and lysed in 1 ml of 0.1 N NaOH. Radioactivity was determined by liquid scintillation counting and protein concentrations were determined. Specific binding is calculated as the difference between binding of [^{3}H]taxol in the presence and absence of a 100-fold excess of unlabeled taxol. Inset: The specific binding data have been plotted in the appropriate Scatchard analysis form. Scatchard analysis reveals a single class of specific binding sites for [^{3}H]taxol in J774.2 cells. The line was determined by a least squares linear regression. Units on the abscissa are pmol [^{3}H]taxol/mg protein. Units on the ordinate are pmol [^{3}H]taxol/mg protein/μM.

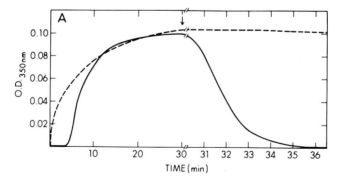

Fig. 6. Effect of taxol on the kinetics of calf brain microtubule polymerization and depolymerization. Microtubule polymerization was measured by turbidity at a final volume of 1.0 ml at 37°C. The assay mixture contained 1 mM EGTA, 0.5 mM $MgCl_2$, 1 mM GTP, 0.1 M 2-[N-morpholino]ethane sulphonic acid (MES) at pH 6.6 and 1 mg/ml tubulin. Control (——). 10 μM taxol (---). The arrow (↓) indicates a shift in temperature from 37°C to 8°C.

the conditions under which taxol assembles microtubules. Taxol assembled tubulin in the absence of microtubule-associated proteins, exogenously added GTP, organic buffer and even at low temperatures (Kumar, 1981; Schiff and Horwitz, 1981a; Hamel et al., 1981; Thompson et al., 1981). The drug, in the absence of GTP, induced the assembly of calcium stable microtubules from flagellar tubulin solubilized from sea urchin sperm tail outer doublets by sonication and dramatically reduced the critical concentration of protein required for polymerization (Parness et al., 1983). The enhancement of both the rate and yield of microtubule assembly reactions by taxol makes this drug distinct from all other antimitotic agents whose mechanism of action has been analyzed.

STRUCTURE-ACTIVITY RELATIONSHIPS

In addition to taxol, other taxanes have been isolated from Taxus species (Fig. 1) (Della Casa de Marcano and Halsall, 1975; Powell et al., 1979; Miller et al., 1981; McLaughlin et al., 1981). Among these, 10-deacetyltaxol (B) cephalomannine (E) and 10-deacetylcephalomannine (F) were reported to be cytotoxic. These natural products plus baccatin III (G) and 19-hydroxybaccatin III (H) and the semisynthetic derivatives, 2'-7 diacetyltaxol (C), 2'-7-diacetyl,10-deacetyltaxol (D), o-cinnamoyltaxicin-I-triacetate (I) and o-cinnamoyltaxicin-II triacetate (J) have been examined in our laboratory to acquire information on the structural components of taxol that are required for biological activity (Parness et al., 1982). Three parameters of drug activity have been determined. The ability of each compound to: 1) induce microtubule assembly in vitro in the absence of exogenous GTP, a process dependent on taxol, 2) inhibit [^{3}H]taxol binding to microtubules, and 3) decrease the growth of J774.2 cells in culture. It became clear that an intact taxane ring alone was inactive and esterification at C-13 appeared necessary for any of the activities of taxol. Small changes in the ring structure, such as loss of the acetyl group at C-10 as in 10-deacetyltaxol, or small alterations on the side chain, such as a change of the N-acyl substituent as in cephalomannine, had little effect on the activities of taxol.

The semisynthetic derivative of taxol, 2',7-diacetyltaxol, although cytotoxic to cells, did not promote microtubule assembly in vitro. To dissect the roles of the acetyl groups at C-7 and C-2', 2'-acetyltaxol and 7-acetyltaxol have been synthesized (Mellado et al., 1984). Whereas 2'-acetyltaxol inhibited the growth of cells in culture, this compound did not induce microtubule assembly in vitro. The cytotoxic activity of 2'-acetyltaxol suggests that this compound may be converted intracellularly to either taxol or an unknown taxol metabolite. The 7-acetyltaxol, however, was very similar to taxol in both its effects on the replication of J774.2 cells and on in vitro microtubule polymerization. Recent studies in which the effects of taxol and 7-acetyltaxol on the assembly of microtubules were measured, also indicated that these two compounds have similar activity (Lataste et al., 1984). The results indicating that neither 2'-acetyltaxol nor 2',7-diacetyltaxol is active in vitro, suggests that the C-2' hydroxyl group on the ester side chain either was involved in a specific interaction with microtubules, altered the conformation of the drug to an inactive form, or inhibited its interaction with microtubules by steric hindrance.

Taxol has extremely limited solubility in aqueous solvents and this characteristic has delayed its pharmacological formulation and introduction into clinical studies. Our studies demonstrating that the properties of 7-acetyltaxol and taxol were similar, indicate that a free hydroxyl group at C-7 is not required for in vitro activity and that this position is available for structural modifications. Although 7-acetyltaxol has no advantage over taxol in terms of solubility, the introduction of charged or hydrophilic groups at C-7 could improve solubility while maintaining cytotoxic activity.

TAXOL RESISTANT CELLS

Taxol-resistant cells have been described by Warr et al. (1982), Gupta (1983) and Cabral et al. (1981). In the latter case, a 2- to 3-fold taxol-resistant Chinese hamster ovary cell line that is temperature sensitive and has an altered α-tubulin has been described. These investigators have also studied a cell line that requires the continuous presence of taxol for normal cell division to proceed (Cabral, 1983; Cabral et al., 1983).

In our laboratory, a drug-resistant cell line was developed over a one-year period by growing J774.2 cells in the presence of stepwise increases in the concentration of taxol (Roy and Horwitz, 1985). The concentration was raised to 50 μM and cells were maintained in this concentration of taxol. This subline demonstrated an 800-fold resistance to the drug and also a cross-resistance to vinblastine (43-fold) and to colchicine (58-fold). In addition, an approximate 4-fold increase in sensitivity to bleomycin was detected. The taxol-resistant cell line demonstrated a major reduction, approximately 90 percent, in steady-state accumulation of taxol compared to the parental cell line. The plasma membranes of the resistant cells included a phosphoglycoprotein with an approximate molecular weight of 135,000 daltons that was essentially not detected in the taxol-sensitive cells. The presence of the 135,000 dalton phosphoglycoprotein in the plasma membranes of taxol resistant cells correlated very well with resistance to the drug. Resistance to the growth inhibitory properties of taxol was extremely dependent on the presence of the drug. When cells were grown in the absence of drug for 30 days there was a 90 percent increase in their sensitivity to taxol and a 90 percent reduction in the content of the 135,000 dalton membrane phosphoglycoprotein, as analyzed by gel electrophoresis. However, resistance did not appear to correlate closely with the steady-state accumulation of taxol in cells. In cells that had not been exposed to the drug for one month, the steady-state association of the drug with the cells remained reduced by approximately 80 percent.

Analysis of tubulin from the resistant cells has not indicated any major alterations as compared with tubulin in the drug-sensitive cells. The taxol-resistant cell line may be related to a category of resistant cells that have been described as demonstrating pleiotropic or multi-drug resistance (Carlsen et al., 1977; Riordon and Ling, 1979; Biedler and Peterson, 1981; Beck et al., 1984). Such cells are cross-resistant to drugs that are structurally unrelated, have altered permeability properties, contain high molecular weight glycoproteins in their plasma membranes and often possess alterations in their karyotypes, such as the inclusion of double minute chromosomes or homogeneously staining regions (Biedler and Spengler, 1976; Grund et al., 1983; Kuo et al., 1982; Robertson et al., 1984). Although the taxol-sensitive J774.2 cell line did not have any double minute chromosomes, double minute chromosomes were clearly visible in the taxol-resistant cell line. The existence of the double minute chromosomes, the membrane glycoprotein and the expression of drug resistance were all closely dependent on the presence of taxol in the growth medium and may be associated with gene amplification in the drug resistant cell lines. The presence of amplified DNA sequences in Chinese hamster cells and their correlation with multi-drug resistance has been shown recently by Roninson et al. (1984). The function of the 135,000 phosphoglycoprotein in the membranes of taxol-resistant J774 cells is not clear and is being investigated.

SUMMARY

Taxol is an antimitotic agent that binds to microtubules and stabilizes microtubules against depolymerization, both in cells and <u>in vitro</u>. The drug promotes the rate and extent of microtubule assembly. Although the mechanism of action of this potential antitumor agent has not been delineated on a molecular level, the drug may bind to small tubulin oligomers strengthening the interaction between tubulin dimers. The oligomers become resistant to depolymerization, act as nucleating sites and shift the equilibrium in favor of polymer formation.

Information on the effects of chemical alterations of the taxol molecule on biological activity will improve our understanding of the interaction of the drug with its cellular target. The use of a series of taxol-resistant cells should also be of value in probing the unique mechanism of action of this drug.

The specificity of taxol for the tubulin-microtubule system has made the drug useful in a variety of situations. For example, the drug has aided in the identification of mitosis-specific microtubule associated proteins (Zieve and Solomon, 1982) and in the isolation of microtubule associated proteins (Vallee, 1982). The role of taxol as a tool for studying the regulation of microtubule assembly and organization in cells is being explored in a number of different laboratories and the drug is proving to be an important addition to the pharmacological agents presently available.

ACKNOWLEDGMENTS

The authors are grateful to Dr. E. Masurovsky for valuable discussions. Research that originated in the authors' laboratory was supported by USPHS Grants CA 17514 and GM 29042 and by an American Cancer Society Grant CH-86.

REFERENCES

Albertini, D.F. and Clark, J.I., 1981, Visualization of assembled and disassembled microtubule protein by double label fluorescence microscopy, Cell Biol. Int. Rep., 5:387-397.

Beck, W.T., Cirtain, M.C. and Lefko, J.L., 1984, Energy-dependent reduced drug binding as a mechanism of Vinca alkaloid resistance in human leukemic lymphoblasts, Mol. Pharmacol. 24:485-492.

Biedler, J.L. and Peterson, R.H.F., 1981, Altered plasma membrane glycoconjugates of Chinese hamster cells with acquired resistance to actinomycin D, daunorubicin, and vincristine, in: "Molecular Actions and Targets for Cancer Chemotherapeutic Agents," A.C. Sartorelli, J.S. Lazo and J.R. Bertino, eds., Bristol-Myers Cancer Symposium, Vol. 2, pp. 453-482, Academic Press, New York.

Biedler, J.L. and Spengler, B.A., 1976, Metaphase chromosome anomaly: Association with drug resistance and cell specific products, Science 191:185-187.

Brenner, S.L. and Brinkley, B.R., 1982, Tubulin assembly sites and the organization of microtubule arrays in mammalian cells, Cold Spring Harb. Symp. Quant. Biol., 46:241-254.

Cabral, F.R., 1983, Isolation of Chinese hamster ovary cell mutants requiring the continuous presence of taxol for cell division, J. Cell. Biol., 97:22-29.

Cabral, F.R., Abraham, I. and Gottesman, M.M., 1981, Isolation of a taxol-resistant Chinese hamster ovary cell mutant that has an alteration in α-tubulin, Proc. Natl. Acad. Sci. USA, 78:4388-4391.

Cabral, F., Wible, L., Brenner, S. and Brinkley, B.R., 1983, Taxol-requiring mutant of Chinese hamster ovary cells with impaired mitotic spindle assembly, J. Cell Biol., 97:30-39.

Caplow, M. and Zeeberg, B., 1982, Dynamic properties of microtubules at steady state in the presence of taxol, Eur. J. Biochem., 127:319-324.

Carlier, M.-F. and Pantaloni, D., 1983, Taxol effect on tubulin polymerization and associated guanosine 5'-triphosphate hydrolysis, Biochemistry, 22:4814-4822.

Carlsen, S.A., Till, J.E. and Ling, V., 1977, Modulation of drug permeability in Chinese hamster ovary cells: Possible role for phosphorylation of surface glycoproteins, Biochim. Biophys. Acta, 467:238-250.

Crossin, K.L. and Carney, D.H., 1981, Microtubule stabilization by taxol inhibits initiation of DNA synthesis by thrombin and by epidermal growth factor, Cell, 27:341-350.

DeBrabander, M., 1982, A model for the microtubule organizing activity of the centrosomes and kinetochores in mammalian cells. Cell Biol. Int. Rep. 6:901-915.

DeBrabander, M., Geuens, G., Nuydens, R., Willerbrords, R. and DeMey, J., 1981, Taxol induces the assembly of free microtubules in living cells and blocks the organizing capacity of the centrosomes and kinetochores, Proc. Natl. Acad. Sci. USA, 78:5608-5612.

Della Casa de Marcano, D.P. and Halsall, T.G., 1975, Structures of some taxane diterpenoids, Baccatins-II, -IV, -VI, and -VII and 1-dehydroxybaccatin-IV, possessing an oxetan ring, J. Chem. Soc., Chem. Commun., 365-366.

Grund, S.H., Patil, S.R., Shah, H.O., Pauw, P.G. and Stadler, J.K., 1983, Correlation of unstable multidrug cross-resistance in Chinese hamster ovary cells with a homogeneously staining region on chromosome 1, Mol. Cell. Biol., 3:1634-1647.

Gupta, R.S., 1983, Taxol resistant mutants of Chinese hamster ovary cells: Genetic, biochemical and cross-resistant studies, J. Cell. Physiol., 114:137-144.

Hamel, E., Del Campo, A.A., Lowe, M.C. and Lin, C.M., 1981, Interactions of taxol, microtubule-associated proteins and guanine nucleotides in tubulin polymerizaton, J. Biol. Chem., 256:11887-11894.

Horwitz, S.B., Parness, J., Schiff, P.B. and Manfredi, J.J., 1982, Taxol: A new probe for studying the structure and function of microtubules, Cold Spring Harbor Symp. Quant. Biol., 46:219-226.

Kumar, N., 1981, Taxol-induced polymerization of purified tubulin, J. Biol. Chem., 256:10435-10441.

Kuo, T., Pathak, S., Ramagli, L., Rodriquez, L. and Hsu, T.C., 1982, Vincristine-resistant Chinese hamster ovary cells, in "Gene Amplification," R.T. Schimke, ed., pp. 53-57, Cold Spring Harbor Laboratory, New York.

Lataste, H., Senilh, V., Wright, M., Guenard, D. and Potier, P., 1984, Relationships between the structures of taxol and baccatine III derivatives and their in vitro action on the disassembly of mammalian brain and Physarum amoebal microtubules, Proc. Natl. Acad. Sci. USA, 81:4090-4094.

Manfredi, J.J. and Horwitz, S.B., 1984, Taxol: An antimitotic agent with a new mechanism of action, Pharmac. Ther., 25:83-125.

Manfredi, J.J., Parness, J. and Horwitz, S.B., 1982, Taxol Binds to cellular microtubules, J. Cell Biol., 94:688-696.

Masurovsky, E.B., Peterson, E.R., Crain, S.M. and Horwitz, S.B., 1981, Microtubule arrays in taxol-treated mouse dorsal root ganglion-spinal cord cultures, Brain Res., 217:392-398.

Masurovsky, E.B., Peterson, E.R., Crain, S.M. and Horwitz, S.B., 1982, Taxol-induced microtubule formations in fibroblasts of fetal mouse dorsal root ganglion-spinal cord cultures, Biol. Cell, 46:213-216.

Masurovsky, E.B., Peterson, E.R., Crain, S.M. and Horwitz, S.B., 1983, Morphologic alterations in dorsal root ganglion neurons and supporting cells of organotypic mouse spinal cord-ganglion cultures exposed to taxol, Neuroscience, 10:491-509.

McLaughlin, J.L., Miller, R.W., Powell, R.G. and Smith, C.R., Jr., 1981, 19-Hydroxybaccatin III, 10-deacetylcephalomannine, and 10-deacetyltaxol: New antitumor taxanes from Taxus wallichiana, J. Nat. Prod., 44:312-319.

Mellado, W., Magri, N.F., Kingston, D.G.I., Garcia-Arenas, R., Orr, G.A. and Horwitz, S.B., 1984, Preparation and biological activity of taxol acetates, Biochem. Biophys. Res. Commun., 124:329-336.

Miller, R.W., Powell, R.G., Smith, C.R., Jr., Arnold, E. and Clardy, J., 1981, Antiluekemic alkaloids from Taxus wallichiana Zucc, J. Organic Chem., 46:1469-1474.

Parness, J. and Horwitz, S.B., 1981, Taxol binds to polymerized tubulin in vitro, J. Cell Biol., 91:479-487.

Parness, J., Asnes, C.F. and Horwitz, S.B., 1983, Taxol binds differentially to flagellar outer doublets and their reassembled microtubules, Cell. Motility, 3:123-130.

Parness, J., Kingston, D.G.I., Powell, R.G., Harracksingh, C. and Horwitz, S.B., 1982, Structure-activity study of cytotoxicity and microtubule assembly in vitro by taxol and related taxanes, Biochem. Biophys. Res. Commun., 105:1082-1089.

Powell, R.G., Miller, R.W. and Smith, C.R., Jr., 1979, Cephalomannine; a new antitumor alkaloid from Cephalotaxus mannii, J. Chem. Soc. Chem. Commun., 102-104.

Riordan, J.R. and Ling, V., 1979, Purification of P-glycoprotein from plasma membrane vesicles of Chinese hamster ovary cell mutants with reduced colchicine permeability, J. Biol. Chem., 254:12701-12705.

Robertson, S.M., Ling, V. and Stanner, C.P., 1984, Coamplification of double minute chromosomes, multiple drug resistance, and cell surface P-glycoprotein in DNA-mediated transformants of mouse cells, Mol. Cell. Biol., 4:500-506.

Roninson, I.B., Albelson, H.T., Housman, D.E., Howell, N. and Varshavsky, A., 1984, Amplification of specific DNA sequences correlates with multi-drug resistance in Chinese hamster cells, Nature, 309:626-628.

Roy, S.N. and Horwitz, S.B., 1985, A phosphoglycoprotein associated with taxol-resistance in J774.2 cells, (submitted for publication).

Schiff, P.B. and Horwitz, S.B., 1980, Taxol stabilizes microtubules in mouse fibroblast cells, Proc. Natl. Acad. Sci. USA, 77:1561-1565.

Schiff, P.B. and Horwitz, S.B., 1981a, Taxol assembles tubulin in the absence of exogenous guanosine 5'-triphosphate or microtubule-associated proteins, Biochemistry, 20:3247-3252.

Schiff, P.B. and Horwitz, S.B., 1981b, Tubulin: a target for chemotherapeutic agents, in: "Molecular Actions and Targets for Cancer Chemotherapeutic Agents," A.C. Sartorelli, J.S. Lazo and J.R. Bertino, eds., Bristol-Myers Cancer Symposium, Vol. 2, pp. 483-507, Academic Press, New York.

Schiff, P.B., Fant, J. and Horwitz, 1979, Promotion of microtubule assembly in vitro by taxol, Nature, 277:665-667.

Soifer, D. and Czosnek, H., 1980, Association of new synthesized tubulin with brain microsomal membranes, Int. Soc. Neurochem., 35:1128-1136.

Thompson, W.C., Wilson, L. and Purich, D.L., 1981, Taxol induces microtubule assembly at low temperature, Cell Motility, 1:445-454.

Tokunaka, S., Friedman, T.M., Toyama, Y., Pacifici, M. and Holtzer, H., 1983, Taxol induces microtubule-rough endoplasmic reticulum complexes and microtubule-bundles in cultured chondroblasts, Differentiation, 24:39-47.

Vallee, R.B., 1982, A taxol dependent procedure for the isolation of microtubules and microtubule-associated proteins (MAPs), J. Cell Biol., 92:435-447.

Wani, M.C., Taylor, H.L., Wall, M.E., Coggon, P. and McPhail, A.T., 1971, Plant antitumor agents. VI. The isolation and structure of taxol, a novel antileukemic and antitumor agent from Taxus brevifolia, J. Am. Chem. Soc., 93:2325-2327.

Warr, J.R., Flanagan, D.J. and Anderson, M., 1982, Mutants of Chinese hamster ovary cells with altered sensitivity to taxol and benzimidazole carbamates, Cell Biol. Intl. Rep., 6:455-460.

Zieve, G. and Solomon, F., 1982, Proteins specifically associated with the microtubules of the mammalian mitotic spindle, Cell, 28:233-242.

REORGANIZATION OF AXONAL CYTOSKELETON FOLLOWING β,β'-IMINODIPROPIONITRILE (IDPN) INTOXICATION

Sozos Ch. Papasozomenos

Department of Pathology and Laboratory Medicine, Medical School, The University of Texas Health Science Center at Houston, Texas, USA

INTRODUCTION

β,β'-Iminodipropionitrile (IDPN), $NH=(CH_2CH_2CN)_2$, is a synthetic compound closely related to the osteolathyrogen β-aminopropionitrile ($NH_2CH_2CH_2CN$). Intoxication of various experimental animals with IDPN produces the excitement, circling and choreathetosis (ECC) or waltzing syndrome (Selye, 1957), a permanent symptom complex indicating irreparable damage to the central nervous system (CNS). Early histopathological studies have described large, amorphous bodies in the anterior horn cells of spinal cord and in other large neurons throughout the nervous system. These amorphous bodies were originally misinterpreted as degenerated neuronal perikarya and were called "ghost cells" (Bachhuber et al., 1955; Ule, 1962). It was shown subsequently by Chou and Hartman (1964; 1965) that the "ghost cells" were actually huge balloons and swollen axons connected to the cell body by an apparently normal initial segment (Fig. 1). They suggested that "axostasis," caused by a "plug" of particulate organelles in the proximal axon, produced the axonal swellings.

More recently, however, Griffin and co-workers (1978) have demonstrated that IDPN severely impairs the transport of neurofilament proteins, and to a lesser degree tubulin and actin, while the rates of fast anterograde and retrograde transport remain normal. Because of the impairment of their transport, neurofilaments accumulate in the proximal axon. These investigators also noted that following acute intoxication with IDPN, proteins of the slow component already en route in the distal sciatic nerve stop moving. It was shown subsequently by Japanese workers (Yokoyama et al., 1980) that in small, unmyelinated axons with a low content of neurofilaments the transport of tubulin and actin was not affected, and only the transport of neurofilament proteins was selectively impaired. It thus became evident that the marked accumulation of neurofilaments in the proximal axon secondarily impedes the movement of the other axonal components moved by slow transport. These investigators have also noted that in spinal motor neurons, accumulation of neurofilaments begins at the junction between the CNS and peripheral nervous system (PNS) parts of the axons, and proceeds towards the perikaryon. These findings indicate that the block in neurofilament transport takes place in the extraspinal portion of motor axons. Thus, accumulation will occur as neurofilament masses although synthesis in the cell body continues unimpaired (Chou and Klein, 1972). From these observations, it is then reasonable to ask what are the effects of IDPN on the distal axon.

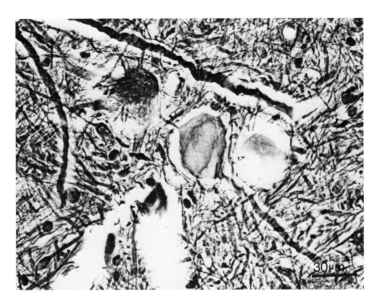

Fig. 1. Bodian's silver stain on rat lumbar spinal cord 3 weeks after IDPN injection. A swollen axon with a proximal balloon is connected to the cell body of a motor neuron by its initial segment.

IDPN-CAUSED SEGREGATION OF MICROTUBULES AND MEMBRANOUS ORGANELLES FROM NEUROFILAMENTS IN AXONS

Electron Microscopy

Following acute systemic intoxication of adult rats with IDPN, (2g/kg injected i.p.) the axonal cytoskeleton became segregated, with displacement of neurofilaments towards the periphery and of microtubules, together with mitochondria and a large portion of smooth endoplasmic reticulum, towards the center of the axon (Papasozomenos et al., 1981; 1982a) (Fig. 2a). While most of the profiles of smooth endoplasmic reticulum were found at the central region, a significant number remained just under the axolemma. Side-arms between neurofilaments and "wispy" material among microtubules were easily recognizable. No evidence of excessive cross-linking or any other structural abnormality of neurofilaments was evident. In control axons, all axoplasmic organelles were spread evenly over the cross section of axons (Fig. 2b).

Immunohistochemical Study of the Spatio-temporal Evolution of the IDPN Axon

Using the peroxidase-antiperoxidase (PAP) technique at the light and electron microscopic levels, and an antiserum against the 68,000 dalton subunit of neurofilaments (provided by Dr. P. Gambetti), an affinity purified polyclonal antibody to α- and β-subunits of tubulin, and a monoclonal antibody against β-tubulin (gifts from Drs. P.K. Bender and L.I. Binder, respectively), we have studied the spatio-temporal evolution of the segregation of microtubules from neurofilaments in axons of the lumbar segment of spinal cord, ventral roots and along the sciatic nerve.

In controls, all three antibodies stained the axons uniformly and intensely (Figs. 3c and 5b). In IDPN affected axons, as was expected from the electron microscopic findings, the neurofilament antiserum stained only the peripheral axoplasm, while the axon center remained unstained (Figs. 3a, 4a and 5a). With both antibodies to tubulin only the centers of the axons

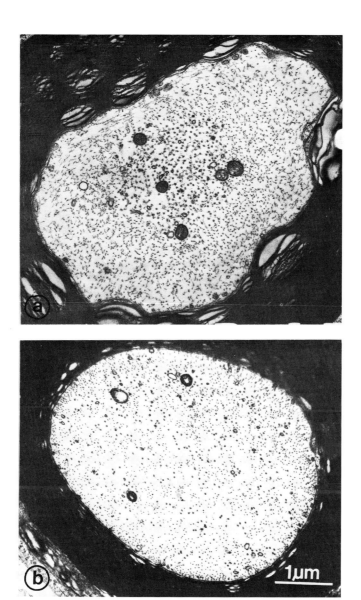

Fig. 2. Electron micrographs of rat L_5 distal ventral root. (a) Two weeks after IDPN intoxication, neurofilaments are displaced toward the periphery of the axon, while microtubules, mitochondrial and a large portion of smooth endoplasmic reticulum are found in the center; there are no structural alterations of neurofilaments. (b) In the control, all types of axoplasmic organelles are present over the cross-section of the axon.

stained, while the peripheral axoplasm was negative (Figs. 3b and 4b). Thus, antibodies to neurofilament proteins and to tubulin produced staining patterns complementary to each other. The segregation of neurofilaments from microtubules, so clearly demonstrated by immunocytochemistry, was shown to occur simultaneously along the entire length of peripheral axons, from the CNS-PNS junction at the ventral root to the plantar branches of the sciatic nerve. In the sciatic nerve and its branches, all axons, including sensory and motor, small and large showed this differential, topographic localization of the neurofilament and tubulin antibodies (Fig. 6).

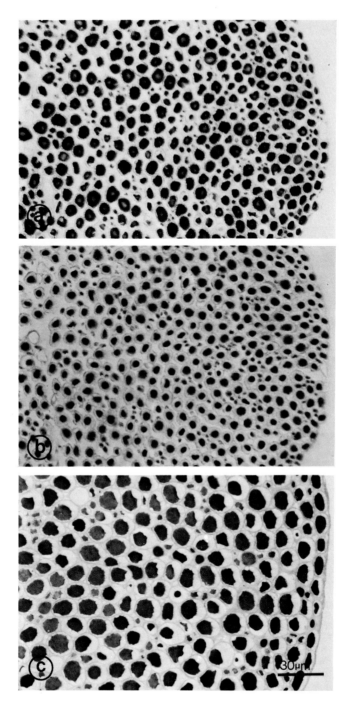

Fig. 3. Cross sections of Epon-embedded rat L_5 distal ventral root, immunostained by the PAP method. (a) Four weeks after IDPN administration, an antiserum to the 68,000 dalton subunit of neurofilaments stained only the periphery of axons, and (b) a polyclonal antibody to tubulin stained only the central region of axons. (a) and (b) are adjacent sections. (c) Control axons uniformly stained with the neurofilament antiserum. Note the reduction in size of IDPN affected axons.

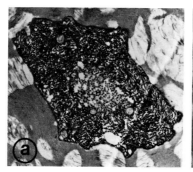 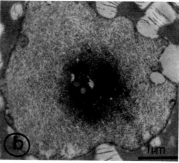

Fig. 4. Electron micrographs of vibratome sections of rat L5 distal ventral root two weeks after IDPN injection, immunostained by the PAP method. (a) Antiserum to the 68,000 dalton subunit of neurofilaments stained only neurofilaments, displaced into the peripheral axoplasm; (b) monoclonal antibody to β-tubulin stained only the central region, occupied by microtubules. Membranous organelles remained unstained.

This axonal change was detectable 4 days after IDPN treatment, and became more apparent during the following 6 weeks. Between the 6th and 16th week, the "doughnut"- or "tube"-like appearance of axons, which was present following immunostaining with the antiserum raised to the 68,000 dalton subunit of neurofilaments, gradually disappeared in a proximo-distal direction, at a rate of 1-2 mm/day (Fig. 7) (Papasozomenos et al., 1981). After the 16th week, IDPN axons immunostained identically to the controls.

It is important to note that the segregation of microtubules from neurofilaments distal to the CNS-PNS junction of spinal motor axons occured contemporaneously with the impairment of transport of neurofilament proteins in the same regions (Griffin et al., 1978). In the intraspinal portion, proximal to the CNS-PNS junction, no microtubule-neurofilament segregation was found. In this portion of the axon neurofilament proteins continued to move and accumulate because their perikaryal synthesis and export into the axon went on uninterrupted. In our model, axonal balloons in the most proximal segment of the axon were first noticed 2 weeks after IDPN treatment; they were still present after one year, although in decreased number. At this time, these axonal balloons were the only detectable histopathologic abnormality still present. They were filled with neurofilaments (Fig. 8).

In the posterior columns of lumbar spinal cord, following immunostaining with the neurofilament antiserum, almost all axons had a ring-like appearance at about 2 weeks after injection of IDPN (Fig. 9). Occasional doughnut-like axons and axonal swellings were also found in the anterior and lateral spinal columns throughout the study. Swollen axons were especially prominent in the spinocerebellar tracts at about 2 months after IDPN treatment. At these sites, axons later resumed their normal appearance.

Bodian's Silver Method

Staining of the peripheral axoplasm only of IDPN affected axons was obtained using the Bodian's silver method (Fig. 10). Starting from this observation, it was shown subsequently that Bodian's silver method stains neurofilament polypeptides (Gambetti et al., 1981).

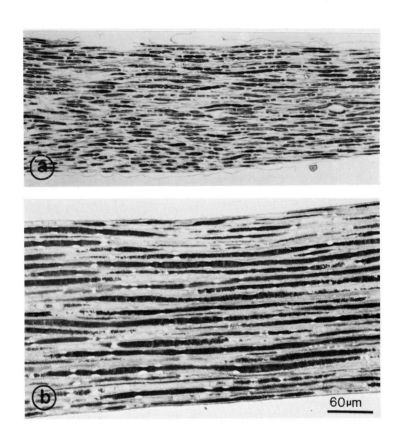

Fig. 5. Longitudinal Epon-embedded sections of rat L5 distal ventral root, immunostained with antiserum to the 68,000 dalton subunit of neurofilaments using the PAP method. (a) Two weeks after IDPN injection, only the peripheral axoplasm along the entire length of the axon is stained; (b) control axons are uniformly stained.

Morphometric Analysis at the Light and Electron Microscope

Numerical morphometric analysis of the distribution of axoplasmic organelles at the electron microscopic level was carried out in four concentric compartments, drawn at 0.37 μm intervals from the axolemma, over the entire cross section of the axon (Papasozomenos et al., 1981).

Control and IDPN affected axons were selected randomly and divided into three groups of comparable sizes. As is shown in Fig. 11 for the medium sized axons, there was a statistically significant increase in the density of neurofilaments towards the periphery and in the densities of microtubules, smooth endoplasmic reticulum and mitochondria towards the center of IDPN affected axons. The neurofilament density in the subaxolemmal compartment was more than twice the density at the center. In contrast, the densities of microtubules, smooth endoplasmic reticulum and mitochondria were about 10, 2.5 and 8 times higher, respectively, at the center than at the periphery.

The total densities of neurofilaments, microtubules, smooth endoplasmic reticulum and mitochondria over the cross section of medium-size axons are shown in Table 1. There was a 10-15 percent increase in the total densities of neurofilaments and microtubules in IDPN affected axons. This increase was not, however, statistically significant, and the neurofilament/microtubule ratio remained unchanged. Only the density of mitochondria in

Table 1. Total Densities of Axonal Organelles (Number/μm^2)[a]

	Neurofilaments[b]	Microtubules[b]	Neurofilament/Microtubule Ratio	Smooth Endoplasmic Reticulum[b]	Mitochondria[c]
Control	114.8 ± 8.7	17.8 ± 2.5	6.55 ± 1.37	3.83 ± 0.51	0.187 ± 0.018
IDPN	128.6 ± 20.3	20.6 ± 5.3	6.50 ± 1.43	3.97 ± 0.60	0.250 ± 0.021
P	NS	NS	NS	NS	< 0.05

[a]The data were obtained from the most distal portion of L5 ventral root from three control and three IDPN-treated rats, two weeks after IDPN intoxication.

[b]The average size of five control and five IDPN axons were: Control, 23.1 ± 0.6 μm^2; IDPN, 20.2 ± 2.5 μm^2. Values are expressed as means ± standard deviations. NS, $P > 0.05$.

[c]The average size of 32 control and 32 IDPN axons were: Control, 21.4 ± 0.4 μm^2; IDPN, 21.2 ± 0.4 μm^2. Values are expressed as means ± standard errors.

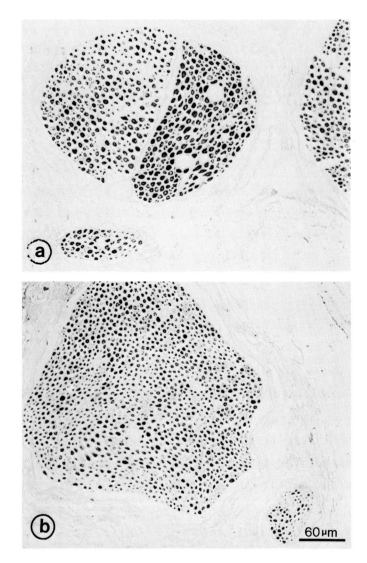

Fig. 6. Cross sections of Epon-embedded of rat: (a) tibial nerve and its branches at the level of the popliteal fossa; and (b) plantar nerve, two weeks after IDPN injection, immunostained as in Fig. 5. Only the periphery of axons is stained.

IDPN affected axons was significantly higher than that of the controls ($p<0.05$). In smaller and larger axons taken for measurement, the total axonal densities, as well as the distribution of axoplasmic organelles, were similar to those found for medium size axons.

Morphometric analyses at the light microscopic level of the size of axons present in the distal portion of L_5 ventral root of a control rat and of a rat 2 weeks after IDPN treatment, showed a reduction in size and an increase in the irregularity in axon shape (Fig. 12). The large ($>14 \mu m^2$) axons were more severely affected (24 percent) than the small ones (10 percent) (Papasozomenos et al., 1981). No differences were found between control and IDPN treated animals either in the total number of axons (control, 2,118 axons/whole root; IDPN, 2,225 axons/whole root) or in the number of axons present in the groups of small and large axons.

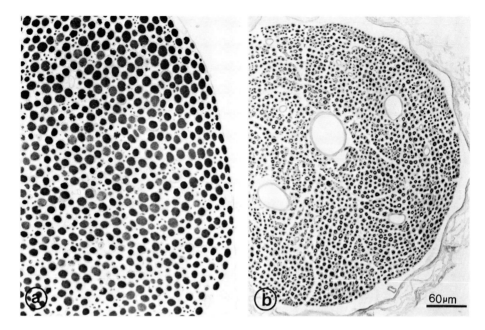

Fig. 7. Cross sections of Epon-embedded of rat: (a) L_5 distal ventral root; and (b) posterior tibial nerve, 12 weeks after IDPN injection, immunostained as in Fig. 5. At the level of posterior tibial nerve neurofilaments are still displaced to the periphery of the axons; at the level of the ventral root, the axons are uniformly stained. Note also the difference in size of the axons at these two levels.

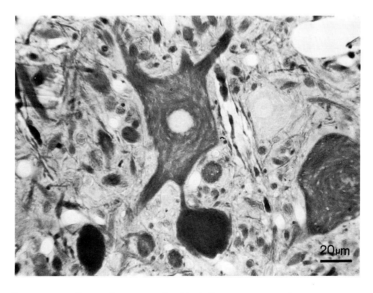

Fig. 8. Cross section of Epon-embedded lumbar spinal cord of an IDPN-treated rat, immunostained as in Fig. 5. An axonal balloon is connected to a motor neuron by an apparently normal initial segment.

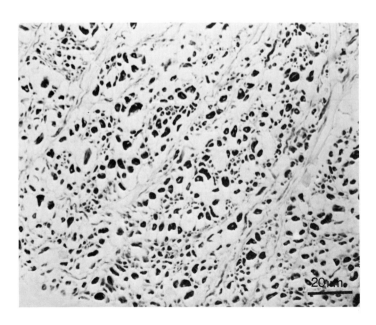

Fig. 9. Cross section of Epon-embedded rat posterior columns of lumbar spinal cord, two weeks after IDPN injection, immunostained as in Fig. 5. Although the axons are distorted, unstained central regions are present.

Immunocytochemical Studies Using Monoclonal Antibodies Against Microtubule-Associated Protein 2 (MAP2)

To investigate the pathogenesis of the microtubule-neurofilament segregation in IDPN affected axons, we used two monoclonal antibodies directed against different epitopes on the high molecular weight microtubule-associated protein 2, MAP2, molecule. MAP2 is believed to mediate interactions between microtubules and other organelles (see Vallee, 1984 for review). Contrary to the current view (Matus et al., 1981), we found that spinal and other motor axons and certain other types of axons contain significant amounts of immunocytochemically detectable MAP2. While in the CNS portion of IDPN affected axons both antibodies to MAP2 localized to microtubules, in the PNS portion one localized to neurofilaments and the other to microtubules (Fig. 13) (Papasozomenos et al., 1985).

WHAT IS THE SITE AND MODE OF ACTION OF IDPN

In our studies, animals were intoxicated acutely by a single intraperitoneal injection, but similar reorganizations of axoplasmic organelles have been produced also after direct injection of IDPN into the endoneurial space of sciatic nerve (Griffin et al., 1983). It is thus evident that IDPN or a metabolite can exert a direct, "local" effect on both PNS and CNS axons. In the systemic intoxication model, the effect of IDPN on the axon is irreversible; the axons resumed a normal organization in a proximo-distal direction, at a rate of 1-2 mm/day, consistent with that of the slow component (SCa) of axonal transport. It appears, therefore, that the altered cytoskeleton must be replaced by axonal transport for the axon to become normal again. On the other hand, because of the limited size of lesions after local endoneurial injection, it is difficult to determine the reversibility or irreversibility of these lesions. It is worth noting the remarkable stability of axonal neurofilaments; they preserved their normal

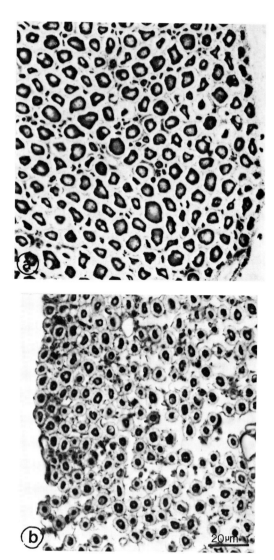

Fig. 10. (a) Epon-embedded and (b) paraffin-embedded cross sections of rat L_5 distal ventral root, three weeks after IDPN injection, stained with Bodian's silver method. Only the peripheral axoplasm, occupied by neurofilaments, is stained. The central region, occupied by microtubules and membranous organelles, remains unstained. The difference in size of axons between the two figures is due to the marked shrinkage of axons following embedding in paraffin.

morphology and immunoreactivity even after derangement of their interactions with other axoplasmic organelles and cessation of their transport for at least four months.

The molecular basis of microtubule-neurofilament segregation is still a matter of speculation. It has been suggested (Anthony et al., 1983) that a direct action of IDPN on neurofilament proteins leads to excessive cross linking between neurofilaments and renders them incapable of being transported. But, in our model, we obtained no evidence of their structural

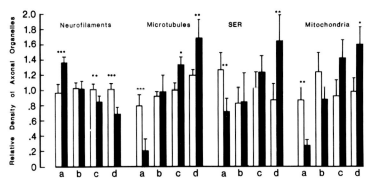

Fig. 11. Distribution of organelles in four concentric axonal compartments designated a, b, c, and d from the axolemma to the center of the axon. Relative densities were obtained by dividing the densities in each compartment by the total axonal density. □, control; ■, IDPN. In IDPN-treated animals, the density of neurofilaments is highest at the periphery and decreases toward the center, while the densities of microtubules, smooth endoplasmic reticulum and mitochondria are lowest at the periphery and increase toward the center of the axon. Note also the higher density of smooth endoplasmic reticulum in compartment a and of mitochondria in compartment b in the controls. Size of axons and experimental design are as in Table 1. *, $P<0.05$; **, $P<0.01$; and ***, $P<0.0001$ for IDPN data as compared with controls for each compartment. (From Papasozomenos et al., 1981).

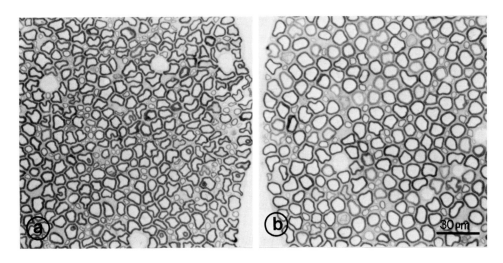

Fig. 12. (a) Cross sections of Epon-embedded L_5 distal ventral root of rat (a) two weeks after IDPN administration, and (b) control, stained with toluidine blue. The IDPN affected axons are atrophic and more irregular in shape.

alteration. It has also been hypothesized that an inhibition of glycolysis, which would result in insufficient energy supply, may be a possible mechanism for neurotoxin-induced distal axonopathies (Spencer et al., 1979). But recent biochemical data have shown that the production of glucose-dependent lactate in rats treated with various neurotoxins was not

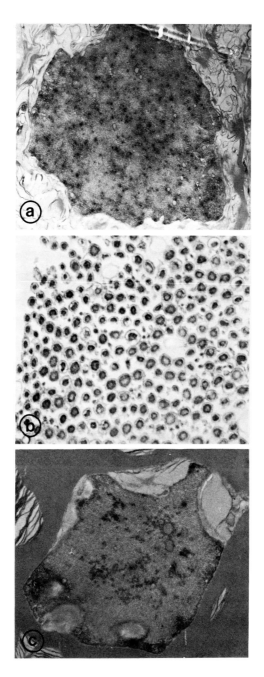

Fig. 13. (a) Electron micrograph of a vibratome section stained with monoclonal antibodies to MAP2 which localizes to microtubules in the intraspinal portion of IDPN affected motor axons; (b) cross section of Epon-embedded L5 distal ventral root two weeks after IDPN injection, immunostained with one monoclonal antibody to MAP2. Only the peripheral axoplasm, occupied by neurofilaments, is stained; (c) in this electron micrograph, of a vibratome section of L5 distal ventral four days after IDPN injection, the other monoclonal antibody to MAP2 localized to microtubules. All sections were immunostained by the PAP method.

different from that of controls (Lopachin et al., 1983) arguing against the above hypothesis. In our study, however, we have found a significant increase in the number of mitochondria in IDPN affected axons in distal ventral roots; Chou and Hartmann (1964) also had noted that mitochondria were increased markedly in number in the proximal swollen axons. In addition, Goldberg (1977) found a significant increase in the uptake of radioactive protein precursors by the brain mitochondrial fraction of IDPN-treated mice. Assuming that there is no energy deficit in IDPN intoxication, the above data suggest that there is an adaptation of energy metabolism in the presence of IDPN, possibly of no pathogenic significance.

A most likely target of IDPN are the microtubule-neurofilament interactions which our data suggest are at least partially mediated by MAP2 in spinal motor axons. The nature of these interactions and the mechanism of damage by IDPN, whether structural or biochemical, remain to be determined. The differential localization of the two MAP2 antibodies with microtubules and neurofilaments in the PNS axons after IDPN treatment is intriguing because these two antibodies bind different epitopes on the MAP2 molecule. It is conceivable that IDPN, presumably by activating a protease, cleaves the MAP2 molecule between these epitopes. It may also indicate that the interactions between microtubules and membranous organelles, on one hand, and microtubules and neurofilaments, on the other, are different although MAP2 may play a role in both. It appears, therefore, that the very nature of these interactions defines their differential vulnerability to IDPN. We suggest that in IDPN affected axons, microtubule-neurofilament interactions are altered, resulting in the segregation of microtubules from neurofilaments and the impairment of neurofilament transport. Interactions between microtubules and membranous organelles, comprising the fast component of axonal transport, by contrast, remain normal.

It is also important to note that in the CNS portion of spinal motor axons both MAP2 antibodies localized with microtubules. In this part of the axon no microtubule-neurofilament segregation was found and the transport of neurofilaments appeared to continue unimpaired. This suggests that microtubules may play a role in the transport of neurofilaments. Although it is possible that differences in the organization of the axonal cytoskeleton and its surroundings in extraspinal and intraspinal portions of spinal motor axons may determine the latter's vulnerability to IDPN, we suggest that the accessibility of IDPN to the axons as well as the distance from the cell body determine the type of lesion. The presence of a blood-brain barrier, thus, may be a rate-limiting factor. Cell culture studies, now in progress, may clarify these points.

Several other neurotoxins, such as aluminium, doxorubicin hexacarbons, acrylamide, carbon disulfide, 3,4-dimethyl-2,5 hexanedione, colchicine, vinca alkaloids, podophylotoxin and maytansinoids produce neurofibrillary accumulations. Although all these neurotoxic agents may share some morphologic manifestations with IDPN, we believe that IDPN is unique in its mode of action.

AXONAL ATROPHY IN IDPN INTOXICATION

In both acute (Papasozomenos et al., 1981) and chronic (Clark et al., 1980) IDPN intoxication, distal axonal atrophy takes place. In the above study of chronic IDPN intoxication, reduction in the delivery of neurofilaments was suggested as the cause of axonal atrophy although no morphometric analysis of neurofilaments was carried out. In our investigation, there was both a reduction in the size and an increase in the irregularity in the shape of IDPN axons; but in the same axons the densities of microtubules and neurofilaments were higher than that of controls. In addition, there was no increase in the number of IDPN affected axons present in the small axon group. While the caliber of normal axons correlates best with the sum of the number of microtubules and neurofilaments and also correlates well with the number of neurofilaments alone (Friede and Samorajski, 1970), it appears that a normal microtubule-neurofilament network may be required for axons to maintain their proper size and shape. It is thus reasonable to suggest that disorganization of the microtubule-neurofilament network in IDPN intoxication could cause a sudden collapse of all axons.

USEFULNESS OF THE IDPN MODEL

The rearrangement of axoplasmic organelles that is found following IDPN intoxication is highly reproducible. As an application of the IDPN model, we (Papasozomenos et al., 1982b; 1983) studied the cross sectional distribution of ^{3}H-labeled proteins migrating with the front of the fast component of axonal transport. While in control axons the radioactivity was uniformly distributed, in IDPN affected axons virtually all the radioactivity was found at the central region occupied by microtubules, smooth endoplasmic reticulum and mitochondria; no significant radioactivity was present in the neurofilament-containing peripheral axoplasm. This suggests that fast axonal transport is preferentially, and perhaps necessarily, associated with microtubules and that neurofilaments play no role in fast axonal transport. They also indicate that although a normal microtubule-neurofilament network may be necessary for the maintenance of the shape and size of axons, this network is not required for fast axonal transport.

THe IDPN affected axon is a simple and reproducible model. It could be very useful in illuminating the organization of axonal cytoskeleton and its role in axonal transport and in modulating the size and shape of axon. A better understanding of these basic processes may help in elucidating the pathogenesis of other human and experimental neurofibrillary degenerations.

ACKNOWLEDGMENTS

Part of this work was accomplished in Dr. P. Gambetti's laboratory at the Institute of Pathology of Case Western Reserve University. I thank Drs. P. Gambetti, L.I. Binder and P.K. Bender for the gifts of antibodies. I also thank Ms. Lynn Mohr for typing the manuscript. This work was supported by NIH grants NS19351 and BRSGRR05431.

REFERENCES

Anthony, C.D., Giangaspero, F. and Graham, D.G., 1983, The spatiotemporal pattern of the axonopathy associated with the neurotoxicity of 3,4-dimethyl-2,5-hexamedinoe in the rat, J. Neuropathol. Exp. Neurol. 42:548-560.
Bachhuber, R.E., Lalich, J.J., Angevine, D.M., Schilling, E.D. and Strong, F.M., 1955, Lathyrus factor activity of BAPN and related compounds, Proc. Soc. Exp. Biol. Med. 89:294-297.
Chou, S.M. and Hartmann, H.A., 1964, Axonal lesions and waltzing syndrome after IDPN administration in rats, Acta Neuropathol. 3:428-450.
Chou, S.M. and Hartmann, H.A., 1965, Electron microscopy of focal neuroaxonal lesions produced by β,β'-iminodipropionitrile (IDPN) in rat, Acta Neuropathol. 4:590-603.
Chou, S.M. and Klein, R.A., 1972, Autoradiographic studies of protein turnover in motor neurons of IDPN-treated rats, Acta Neuropathol. 22:183-189.

Clark, A.W., Griffin, J.W. and Price, D.L., 1980, The axonal pathology in chronic IDPN intoxication, J. Neuropathol. Exp. Neurol. 39:42-55.
Friede, R.L. and Samorajski, T., 1970, Axon caliber related to neurofilaments and microtubules in sciatic nerve fibers of rats and mice, Anat. Rec. 167:379-388.
Gambetti, P., Autilio-Gambetti, L. and Papasozomenos, S.Ch., 1981, Bodian's silver method stains neurofilament polypeptides, Science 213:1521-1522.
Goldberg, M.A., 1977, Protein turnover and axonal flow in β,β'-iminodipropionitrile induced axonal dystrophy, J. Neurochem. 28:259-261.
Griffin, J.W., Hoffman, P.N., Clark, A.W., Carroll, P.T. and Price, D.L., 1978, Slow axonal transport of neurofilament proteins: Impairment by β,β'-iminodipropionitrile administration, Science 202:633-635.
Griffin, J.W., Fahnestock, K.E., Price, D.L. and Hoffman, P.N., 1983, Microtubule-neurofilament segregation produced by β,β'-iminodipropionitrile: Evidence for the association of fast axonal transport with microtubules, J. Neurosci. 3:557-566.
Lopachin, R.M., Moore, R.M., Menahon, L.A. and Peterson, R.E., 1983, Glucose-dependent lactate production by homogenates of neuronal tissues prepared from rats treated with 2,4-dithiobiuret, acrylamide, p-bromophenylacetylurea and 2,5-hexanedione, Second International Conference on Neurotoxicology of Selected Chemicals, Chicago, p. 8.
Matus, A., Bernhardt, R. and Hugh-Jones, T., 1981, High molecular weight microtubule-associated proteins are preferentially associated with dendritic microtubules in brain, Proc. Natl. Acad. Sci., 78:3010-3014.
Papasozomenos, S.Ch., Autilio-Gambetti, L. and Gambetti, P., 1981, Reorganization of axoplasmic organelles following β,β'-iminodipropionitrile administration, J. Cell Biol., 91:866-871.
Papasozomenos, S.Ch., Autilio-Gambetti, L. and Gambetti, P., 1982, The IDPN axon: Rearrangement of axonal cytoskeleton and organelles following β,β'-iminodipropionitrile (IDPN) intoxication, in: "Axoplasmic Transport," D.G. Weiss, ed., Springer-Verlag, pp. 241-250, New York.
Papasozomenos, S.Ch., Yoon, M., Crane, R., Autilio-Gambetti, L. and Gambetti, P., 1982b, Redistribution of proteins of fast axonal transport following administration of β,β'-iminodiproprionitrile: A quantitative autoradiographic study, J. Cell Biol., 95:672-675.
Papasozomenos, S.Ch., Autilio-Gambetti, L. and Gambetti, P., 1983, Distribution of proteins migrating with fast axonal transport. Their relationship to smooth endoplasmic reticulum, Brain Res., 278:232-235.
Papasozomenos, S.Ch., Binder, L.I., Bender, P.K. and Payne, M.R., 1985, Microtubule-associated protein 2 within axons of spinal motor neurons: Associations with microtubules and neurofilaments in normal and β,β'-iminodiproprionitrile-treated axons, J. Cell Biol., 100:74-85.
Selye, H., 1957, Lathyrism, Rev. Canad. Biol., 16:1-82.
Spencer, S.P., Sabri, M.I., Schaumburg, H.H. and Moore, C.L., 1979, Does a defect of energy metabolism in the nerve fiber underlie axonal degeneration in polyneuropathies , Ann. Neurol., 5:501-507.
Ule, G., 1962, Zur ultrastruktur der ghost-cells beim experimentellen neurolathyrismus der ratte, Z. Zellforsch., 56:130-142.
Vallee, R.B., 1984, MAP2 (Microtubule-associated protein 2), in: "Cell and Muscle Motility, Vol. 5," J.W. Shay, ed., pp. 289-311, Plenum, New York.
Yokoyama, K., Tsukita, S., Ishikawa, H. and Kurokawa, M., 1980, Early changes in the neuronal cytoskeleton caused by β,β'-iminodipropionitrile: Selective impairment of neurofilament polypeptides, Biomed. Res., 1:537-547.

MOLECULAR AND CELLULAR ASPECTS OF THE INTERACTION OF BENZIMIDAZOLE CARBAMATE PESTICIDES WITH MICROTUBULES

Edward H. Byard* and Keith Gull

Biological Laboratory
University of Kent
Canterbury, Kent, England

INTRODUCTION

For about twenty years, benzimidazole drugs have enjoyed wide commercial applications as systemic fungicides, as broad spectrum anthelmintics against cestode and nematode infections, and as antitumoral agents. The range of benzimidazole derivatives that became available was limited only by the resources of the drug firms, and the skill of the organic chemists. The effectiveness of these drugs is well established: they have performed well in field and clinical trials, and are commercially successful. However, it is only recently, and certainly within the last ten years that the mechanism of their action has been studied in any systematic way. Basic science is now catching up to commercial pragmatism!!

Many laboratories are now interested in the benzimidazole drugs, not only because they are clearly of biological interest in pest control and cancer therapy, but because the drugs appear to have a singular action--they bind to tubulin, the major polypeptide component of the microtubules in the eukaryotic cytoskeleton. Furthermore, the binding is selective--some organisms are sensitive to the drugs, others are not. The selectivity is at once provocative to both the applied scientist and the cell biologist. In the first instance, the potential for a drug that might, for example, kill a parasite without harming the host is appealing; the cell biologist sees the question of selectivity as one which might have an interesting answer in the structure and biochemistry of the target tubulins in the affected organism. In addition, given that a group of drugs has a common effect upon a particular gene product, it is possible to obtain mutants which are resistant to the drugs, and thus acquire a useful tool in understanding the genetics of resistance and sensitivity.

In this paper, we will briefly review the biological effects of the benzimidazole drugs in a variety of organisms. We will then relate these effects to the more recent data, from our laboratory and others, pertaining to the selective binding of some benzimidazoles to purified tubulins from fungi, nematodes and mammalian cells. Finally, we will describe some recent studies on the genetics of drug resistance.

---
*Permanent address: Department of Biology, University of Winnipeg, Winnipeg, Canada, R3B 2E9.

BENZIMIDAZOLE CHEMISTRY

The range of benzimidazole compounds that has been produced are derivatives of the benzimidazole group which is normally substituted at one of three positions (Fig. 1). Benzimidazole-2-yl carbamates form the largest group of derivatives that have been in common use. The simplest carbamate is methyl benzimidazole 2-yl carbamate (MBC) which is a hydrolysis product of, and the active component of benomyl (Benlate), a widely used fungicide. Various 5-substitutions of the MBC moiety produce the anthelmintics parbendazole, mebendazole, oxfendazole and fenbendazole, and the antitumoral drug nocodazole (sometimes referred to as oncodazole). Other substitutions of the benzimidazole structure are possible, and these produce compounds that have been exploited with some success; in this latter group are the fungicides thiabendazole and cambendazole (Fig. 1).

BIOLOGICAL ACTION OF BENZIMIDAZOLE DERIVATIVES

Although clearly similar in structure, the benzimidazole drugs have a wide range of biological action, depending, in many cases on the concentration of the drug used, and of course, the target organism. Since the drugs were originally designed as fungicides and anthelmintics, the body of information on their biological action comes from studies done on fungi and nematodes or cestodes, though some information has been obtained from other systems. In general, the action of the drugs appears to be either to inhibit growth via a block of nuclear division, or to prodce a disruption of the cytoskeletal network of particular cell types within an organism; either effect may lead to death, or at least to a perturbation in an essential physiological process or developmental stage.

Toxicity in Fungi

In fungi, the drugs act to produce mitotic instability at sub-lethal concentrations, or complete growth inhibition at higher concentrations. Hastie (1970) showed that diploid strains of Aspergillus nidulans treated

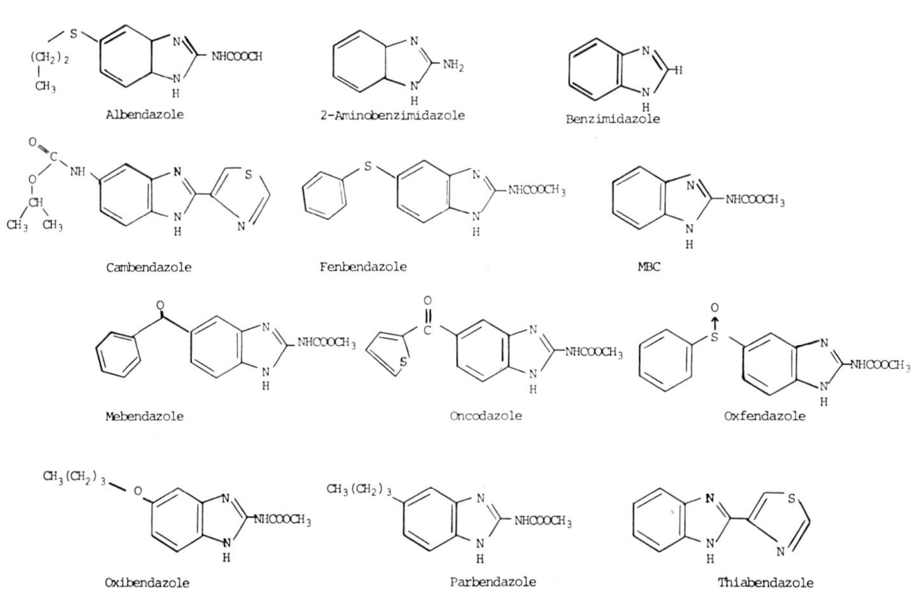

Fig. 1. Chemical structures of some benzimidazole compounds.

with low concentrations of benomyl became mitotically unstable; haploids and aneuploids were often produced. Higher levels of MBC inhibited mitosis, DNA synthesis, and growth (Davidse, 1973). Similarly, in the cellular slime molds Dictyostelium discoideum and Polysphondylium pallidum, diploid amoebae broke down to haploids when treated with low concentrations of several different benzimidazoles (Williams and Barrrand, 1978; Walker and Williams, 1980), whereas higher drug levels inhibited mitosis (Capucinelli et al., 1979; Williams, 1980). The true slime mold, Physarum polycephalum has a uninucleate myxamoeba and a multinucleate plasmodium as distinct phases of the normal life cycle. Treatment of myxamoebal cultures with 2 μm MBC resulted in growth inhibition (Wright et al., 1976; Mir and Wright, 1977; Quinlan et al., 1981). Treatment of the plasmodium at this concentration of drug had no effect; higher concentrations, however, inhibited growth, and more particularly, produced a vast increase in nuclear size. These nuclei were most likely polyploid (Wright et al., 1976) arising from nuclear replication in the absence of nuclear division. Benomyl and MBC also generated diploids in the yeast Saccharomyces cerevisiae (Wood, 1982). In addition, in both S. cerevisiae or the fission yeast Schizosaccharomyces pombe, the drugs arrested progression through the cell cycle by affecting cells at, or shortly before mitosis (Quinlan et al., 1980; Fantes, 1982; Walker, 1982; Wood and Hartwell, 1982).

From these results, it seems clear that the observed effects of the benzimidazole drugs can be satisfactorily explained by a primary block on the formation of the mitotic spindle. In other words, the microtubule assembly necessary to form a proper spindle was disrupted, resulting in the improper movement of the chromosomes (leading to diploidization of haploids, polyploids, etc.) or a complete block in mitosis (leading to growth inhibition). Microtubules, of course, exist outside the nucleus as part of the cytoplasmic cytoskeleton. These microtubules are in a dynamic equilibrium with the soluble tubulin pool; thus, if the benzimidazoles do, in fact, prevent microtubule assembly, drug treatment should disrupt the assembly of cytoplasmic microtubules. In Physarum, the myxamoebae will develop a pair of flagella in moist conditions. This is a reversible process which requires microtubule assembly. Flagellar development was inhibited by both MBC and nocodazole (Mir and Wright, 1977). In Fusarium acuminatum, treatment of hyphae with MBC resulted in a loss of microtubules from the cytoplasm, particularly at the elongating hyphal tips (Howard and Aist, 1977; 1980). Thus it seems clear that both cytoplasmic and nuclear microtubules are affected by benzimidazole treatment.

Toxicity in Nematodes and Cestodes

It is more difficult to assess the effects of a benzimidazoles as anthelmintics partly because a detailed cellular analysis of the drug effects has not been done and partly because the target organisms are often successful parasites with complex life cycles. Drug treatment can kill vulnerable larval stages or eggs, for example, but the same drug might have no effect on adults (Bannerjee and Prakash, 1972; Wagner and Chavarria, 1974; Fernando and Denham, 1976). Cysticercoid development in two species of tapeworms (Hymenolepis nana and H. diminuta) was effectively inhibited by treating the host beetles with a variety of benzimidazole drugs, but a third species (H. microstoma) was not affected at all (Novak et al., 1982).

It may be argued that some of the variability in effect of drug treatment might be due to metabolism of the drug by the host. This problem can be alleviated by studying the effects of the drug on a free-living form. One such worm is the soil nematode, Caenorhabditis elegans, which can be grown axenically, or in monoxenic culture with a bacterial food source (Brenner, 1974). Platzer et al. (1977) reported that a wide range of benzimidazoles, and particularly the benzimidazole-2-yl carbamates substituted in the 5-position (Fig. 1), inhibited growth in C. elegans. In a more detailed study of the effects of mebendazole on C. elegans (Spence et al., 1982), it was found that the drug inhibited growth, drastically reduced egg laying rates, and caused paralysis in the later larval stages and adults.

Some studies have attempted to address the cellular basis for benzimidazole action in the helminths. Rapson et al. (1981) found that in vivo treatment of host rats with oxfendazole or mebendazole caused a marked accumulation of acetylcholinesterase in the parasitic nematode, Nippostrongylus brasiliensis, due to an inhibition, by the benzimidazole carbamates, of the normal secretion of acetylcholinesterase to the exterior of the worm. Acetylcholinesterase secretion is necessary for the successful attachment of N. brasiliensis to the host intestine, thus benzimidazole treatment leads, eventually, to rejection of the parasite by the host. Atkinson et al. (1980) reported that mebendazole treatment impaired the secretion of digestive enzymes in the intestinal cells of Ascaridia galli and that this effect was due to a loss of cytoplasmic microtubules from the gut epithelium. Other studies have confirmed this observation in other nematodes and have suggested, in addition, that drug treatment would lead eventually to intestinal collapse as a result of the loss of microtubules (Borgers et al., 1975; Comley, 1980).

Recently, it has become clear that in appraising the anthelmintic activity of the benzimidazole drugs, one must take into account an unusual property of nematode microtubules. Microtubules from mammalian or fungal cells are made up of 13 protofilaments of tubulin subunits, whereas ultrastructural studies have shown that, in nematodes, there were no microtubules with 13 protofilaments at all. Rather, the majority of cells had microtubules made up of 11 or 12 protofilaments, and a small group of sensory neurons (the so-called microtubule cells) had microtubules made of either 14 protofilaments (Davis and Gull, 1983) or 15 protofilaments (Chalfie and Thomson, 1982). Furthermore, the protofilament number in a microtubule altered its sensitivity to benzimidazole treatment. In Caenorhabditis elegans, benomyl or nocodazole treatment lead to a selective loss of the 11-protofilament microtubules (Chalfie and Thomson, 1982), whereas there was no observable effect on the number or occurrence of the 15-protofilament subset of microtubules. In the parasitic nematode, Trichostrongylus colubriformis, nerve, intestinal, and pharyngeal cells have microtubules with either 11 or 12 protofilaments (Davis and Gull, 1983). More recently, in our laboratory, it was found that treatment of these worms with albendazole resulted in the selective loss of the 11-protofilament subset of microtubules (C. Davis, unpublished observations).

Toxicity in Mammalian Cells

To workers outside the realm of parasitology and mycology, and particularly those working with mammalian systems, the benzimidazoles were appealing since they offered an alternative to the use of the plant alkaloid colchicine, a well known microtubule inhibitor in higher eukaryotes (Borisy and Taylor, 1967). It remained only to show that the benzimidazole drugs would disrupt microtubules in mammalian cell lines in vivo or in vitro. Styles and Garner (1974) showed that, in several cultured mammalian cell lines, MBC at low concentrations (down to 10^{-6}M) caused metaphase arrest, and the occurrence of a high proportion of abnormal and pyknotic nuclei. Seiler (1976) tested the effect of several benzimidazole compounds on mouse bone marrow cells and found that MBC, in particular, produced 3-4 times as many cells in metaphase arrest as control cells. These two studies point to a direct interaction between the drugs and the microtubules of the mitotic spindle.

Other studies on mammalian cells have looked more specifically at the cytoplasmic microtubules in mammalian cells after drug treatment.

DeBrabander et al. (1977) showed that nocodazole treatment resulted in the disappearance of cytoplasmic and mitotic microtubules from a mouse embryonal cell line. Similarly, Havercroft et al. (1981), in our laboratory, showed that parbendazole effectively depolymerized the cytoplasmic microtubules of Vero cells, leaving only one or two isolated microtubules (Fig. 2). Birkett et al. (1981) treated primary hepatocyte cultures with nocodazole, MBC, and parbendazole as well as colchicine; they used the rate of secretion of newly synthesized lipids and proteins into the culture medium as an assay for the integrity of the cytoplasmic microtubules. Colchicine and nocodazole caused a similar decrease in the secretion of both triacylglycerol and albumin, whilst parabendazole was slightly less effective. An interesting result of this experiment was that MBC had no effect--a point which emphasizes the generic selectivity of benzimidazole action (MBC is highly toxic in fungi but not in mammalian cells) and indicates that perhaps 5-substitution of the benzimidazole structure is important in their relative potency.

Benzimidazoles have also been tested as possible agents to prevent the migration of mammalian tumor cells into healthy tissue. Mareel et al. (1980) evaluated the effects of several drugs, among them two benzimidazole carbamates (nocodazole and the compound methyl [5-(2-(4-fluorophenyl) 1,3-dioxolan-2-yl) -1H-benzimidazol-2-yl] carbamate), in the disruption of the migration of malignant mouse fibroblastic cells into chick heart cells in three-dimensional cultures. They found that the drugs were effective in preventing this invasion, and that this effect was due to an abolishment of the cytoplasmic microtubule complex, thus hampering the directional migration of the malignant cells. The anti-invasive effect seems to be distinct from a mitostatic action.

EFFECTS OF THE BENZIMIDAZOLES IN VITRO

The benzimidazoles seem to act specifically on microtubules in a variety of organisms and cells. It is clearly of interest to further study the direct interaction of the drugs with microtubules, preferably under

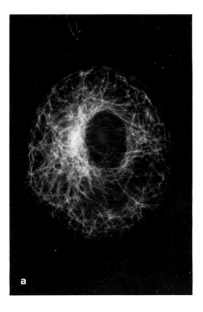
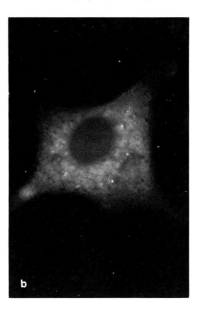

Fig. 2. The effects of parbendazole on mammalian cells in culture. The microtubule network of Vero cells was observed by indirect immunofluorescence using a fluorescein-conjugated antibody to tubulin. (a) Control cell; (b) Cell treated with 2 μM parbendazole for 26 hrs.

conditions whereby assembly of tubulin subunits into microtubules can be controlled in vitro. The initial studies of this type were done on purified mammalian brain tubulin simply because the method for the isolation and in vitro polymerization of this tubulin was the only method available (Shelanski et al., 1973).

Hoebeke et al. (1976) tested the effects of nocodazole on the assembly in vitro of rat brain tubulin into microtubules. The drug did not cause the disassembly of preformed microtubules, but did inhibit assembly, with an IC_{50} of 10^{-6} M; in fact, this drug was even more effective in the assay than colchicine. When radioactive nocodazole was mixed with tubulin, the radioactivity and tubulin co-eluted from a Sephadex G50 column. Furthermore, nocodazole competitively inhibited the binding of colchicine, confirming that nocodazole was a true microtubule inhibitor (Hoebeke et al., 1976). The inhibition of assembly of mammalian brain tubulin into microtubules in vitro has been similarly demonstrated for a number of benzimidazoles. Nocodazole, oxibendazole, parbendazole, mebendazole, and fenbendazole were more effective than colchicine in the inhibition of bovine brain microtubule assembly; benomyl, cambendazole, MBC, and thiabendazole were less effective (Friedman and Platzer, 1978; Ireland et al., 1979).

Until Roobol et al. (1980) successfully achieved microtubule assembly in vitro from Physarum polycephalum myxamoebae, the techniques for in vitro microtubule assembly from lower eukaryotes were not available, largely because the concentration of tubulin in non-neural tissue is very low. Nevertheless, Davidse and Flach (1977) were able to show that MBC formed a complex in vitro with a tubulin-like protein in mycelial extracts of Aspergillus nidulans, at concentrations which inhibited mycelial growth. This complex eluted as a single peak from a Sephadex column, and had a molecular weight (~100,000) characteristic of the $\alpha\beta$-tubulin heterodimer. As found by Hoebeke et al. (1976) for brain tubulin, colchicine competitively inhibited MBC binding (as did nocodazole). Notably, strains of Aspergillus that were resistant to MBC had a lower affinity for MBC in the binding assay, whereas supersensitive strains had higher affinity for the drug.

In our laboratory, Quinlan et al. (1981), using the purified preparations of Physarum polycephalum myxamoebal tubulin prepared according to the method of Roobol et al. (1980), showed that the benzimidazole carbamates MBC, parabendazole, and nocodazole inhibited the in vitro assembly of microtubules. Furthermore, the effective concentrations were exactly those found to produce inhibition of myxamoebal growth (Fig. 3). Kilmartin (1981) successfully isolated tubulin from two species of yeast, Saccharomyces cerevisiae and S. uvarum and demonstrated that both MBC and nocodazole were effective inhibitors of microtubule assembly in vitro. Both Quinlan et al. (1981) and Kilmartin (1981) showed that the slime mold and yeast tubulins, respectively, were virtually insensitive to colchicine. The importance of these two studies, taken together, is that they clearly demonstrate the selective action of the antimicrotubule agents. Colchicine is a potent inhibitor of mammalian microtubule assembly, whilst MBC is ineffective. In contrast, MBC inhibits yeast and slime mold microtubule assembly, but colchicine is a poor inhibitor.

There are few reports of benzimidazole binding to helminth tubulin, and until recently, none at all on the effects of the drug on in vitro microtubule assembly of helminth tubulin. As with the early fungal work, most of the studies in the literature concern binding of the drug to worm extracts in which tubulin has been identified by various criteria, including gel filtration or sodium dodecyl sulphate (SDS) polyacrylamide gel electrophoresis (Friedman and Platzer, 1980; Kohler and Bachmann, 1981; Watts, 1980, 1981; Ireland et al., 1982). These studies generally confirm the observation, already made, that the benzimidazole will bind to putative worm tubulins, and may compete with colchicine in this action.

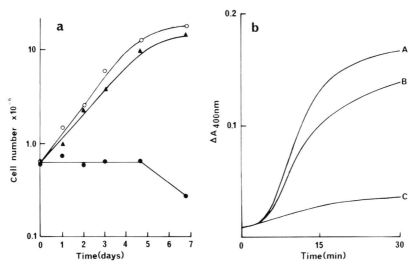

Fig. 3. The effects of nocodazole on: (a) Physarum amoebal growth; and (b) the assembly of amoebal tubulin in vitro. In (a), control cultures (o) were grown in a liquid, semi-defined medium. Drug was added to give final concentrations of 0.4 μM (▲) or 2 μM (●). In (b) dilute solutions of amoebal microtubule protein (1.45 mg/ml) were warmed from 0°C to 30°C, and the assembly of microtubules was estimated by monitoring the increase in absorbance at 400 nm. A - control, with no drug; B - 0.4 μM nocodazole added; C - 2.0 μM nocodazole added. (Modified from Quinlan et al., 1981.)

In embryonic extracts of Ascaris suum, colchicine binding was inhibited by benzimidazole carbamates (Friedman and Platzer, 1980). Furthermore, it appears that the nematode extracts had a much higher affinity for mebendazole and fenbendazole than did purified bovine brain tubulin. However, Kohler and Bachmann (1981), using extracts of adult nematodes showed very little difference in mebendazole binding between worm extracts and mammalian brain tubulin. There are insufficient data available to discern, as yet, whether this discrepancy was a result of the source of the tubulin (embryonic tubulin may have a real and significantly different affinity for the drugs; for example see Spence et al., 1982), or whether there was merely a trivial difference in the method of the experiments and the preparation of the extracts. This latter problem is effectively removed by preparing purified tubulin extracts from worms and testing drug action on microtubule assembly in vitro.

In our laboratory, Dawson et al. (1983) successfully isolated tubulins from the parasitic nematode, Ascaridia galli, and developed a method for the assessment of the effects of various benzimidazoles and colchicine on microtubule assembly in vitro (Dawson et al., 1984). Generally, all drugs tested were more effective in preventing worm microtubule assembly than they were in inhibiting mammalian brain microtubule assembly. Oxfendazole and thiabendazole were particularly ineffective in preventing mammalian brain microtubule assembly, although they were both good inhibitors of worm tubulin polymerization (Table 1). It did not appear however, that any benzimidazole was significantly more or less inhibitory than colchicine in worm tubulin polymerization assays.

From the various studies reviewed above, it is clear that there is selectivity in the toxicity of the benzimidazole drugs in vivo and in vitro.

Table 1. Comparison of the effectiveness of anthelmintic benzimidazoles and colchicine on the in vitro assembly of microtubules in purified tubulin preparations from mammalian brain and the nematode, Ascaridia galli. [Modified from Dawson et al., 1984.]

Drug	IC_{50}* Mammal	A. galli
Fenbendazole	3.5	4.5
Parbendazole	6.5	6
Mebendazole	7	5
Oxfendazole	200+	6
Thiabendazole	950+	8
Colchicine	9	4

*Concentration of drug (X 10^{-6} M) causing 50 percent inhibition of microtubule assembly.
+Limit of solubility of the drug under conditions used for assembly.

Generally, the fungi and the tubulins isolated from them were sensitive to the benzimidazoles, but not to colchicine (Kilmartin, 1981; Quinlan et al., 1981), whereas mammalian cells and helminths, and the appropriate in vitro microtubule preparations from each, were sensitive to colchicine as well as to many, but not all, benzimidazoles (Hoebeke et al., 1976; Friedman and Platzer, 1978; Ireland et al., 1979; Havercroft et al., 1981; Chalfie and Thomson, 1982; Dawson et al., 1984). This selective drug action could be explained by heterogeneity in the tubulins of the target organism. Although there is some evidence, discussed earlier, of heterogeneity in the overall protofilament structure of microtubules (Chalfie and Thomson, 1982; Davis and Gull, 1983) as seen with the electron microscope, it is more likely that heterogeneity at the molecular level is responsible for alterations in drug binding. In fact, now that methods have been developed to allow purification of tubulins from the lower eukaryotes, it is possible to characterize the tubulins biochemically. The work of Kilmartin (1981) on purified yeast tubulins, and our work on Physarum (Clayton et al., 1980; Roobol et al., 1984) and Ascaridia (Dawson et al., 1983) tubulins has revealed molecular differences in tubulins from different organisms.

It has been reported that, in each of these cases, tubulins from the lower eukaryotes have a different electrophoretic mobility from that of mammalian brain tubulins. In yeast, Kilmartin (1981) showed that the β-tubulin subunit migrated slightly ahead of the mammalian brain β-tubulin on SDS polyacrylamide gels. Tubulin from the myxamoebae of the slime mold, Physarum polycephalum has been shown, using SDS polyacrylamide gel electrophoresis and limited peptide mapping, to have significant differences from mammalian tubulin (Clayton et al., 1980; Burland et al., 1983). The Physarum β-tubulin closely resembled mammalian brain β-tubulin, but amoebal α-tubulin was different than the α-tubulin subunit of brain. In the plasmodium of Physarum, four electrophoretically separable tubulin species were seen on SDS gels. These have been designated $α_1$ and $β_1$, representing the same species as found in the amoeba, and $α_2$ and $β_2$, which were only expressed in the plasmodium (Burland et al, 1983; Roobol et al., 1984). In the nematode, Ascaridia galli, Dawson et al. (1983) showed, using peptide mapping, that the α-tubulin subunit had some differences from the mammalian brain standard, whereas the β-tubulin was quite similar to mammalian brain β-tubulin. These differences between mammalian brain tubulin and helminth tubulin also give quite different electrophoretic behaviour on SDS polyacrylamide gels (Fig. 4).

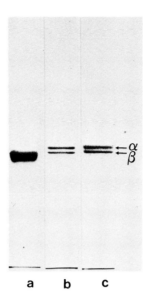

Fig. 4. A comparison of SDS polyacrylamide gels of purified tubulin from: (a) mammalian brain; (b) the parasitic nematodes Ascaridia galli, and (c) Ascaris suum. The α- and β-tubulin subunits are clearly separated in both (b) and (c), but not in (a).

Since the documentation of differences in tubulin polypeptides between evolutionarily diverse organisms, other examples of tubulin heterogeneity have been established. It now seems clear that many organisms--both eukaryotic microbes and metazoans--have multiple genes for tubulin and that these genes can be expressed in a cell type dependent manner (Raff, 1984). It follows that variations in the benzimidazole binding site may, in future, become apparent within these multiple tubulins expressed in the different tissues or developmental stages of target organisms.

BENZIMIDAZOLE DRUGS AND THE GENETICS OF RESISTANCE

The wide use of the benzimidazole drugs against fungi and helminths coupled with the specificity of their action raises the possibility that resistant strains may be spontaneously produced. This resistance is obviously a problem in the application of the drugs since the existence of resistant strains precludes effective drug therapy. Given the known action of benzimidazoles on microtubules, it is reasonable that at least some of the mutations conferring benzimidazole resistance should be in tubulin genes. This is in fact the case, and drug-induced mutants have proven invaluable in the analysis of tubulin structure and function. Again, as with the in vitro studies on benzimidazole-tubulin interactions, most of the work on the genetics of drug resistance has been done in fungi, and especially on Aspergillus nidulans (Sheir-Ness et al., 1978; Morris, 1980), Physarum polycephalum (Burland et al., 1984) and Schizosaccharomyces pombe (Yamamamoto, 1980; Roy and Fantes, 1982; Umesono et al., 1983).

Sheir-Ness et al. (1978) identified tubulins from wild-type Aspergillus using two-dimensional gel electrophoresis. In 26 benomyl resistant mutants, 18 showed a β-tubulin subunit that had an altered electrophoretic mobility, suggesting that the mutation was in a β-tubulin structural gene. The mutations were mapped to a locus called benA. The fact that an alteration in a β-tubulin structural gene confers benomyl resistance is strong support for the view that the benzimidazoles act directly on tubulin. Morris et al. (1979) further showed that some mutants which suppress MBC resistance map to another locus, tubA, a structural gene for α-tubulin. Thus, in Aspergillus, α-tubulins may also have a role in benzimidazole sensitivity.

In Physarum, mutations conferring resistance of myxamoebae to MBC have been mapped to four, unlinked loci called benA, benB, benC, and benD (Burland et al., 1984). BenD has been confirmed as a structural gene for β-tubulin in that one mutation in the benD locus, benD210, resulted in the production of a β_1 tubulin with an altered electrophoretic mobility in two-dimensional gels (Fig. 5).

In the myxamoebae, the wild type β_1 tubulin was still present on the gels, so the simplest explanation of this result was that there are at least two gene products in the wild-type β_1 tubulin "spot," one of which has been mutated. In the plasmodium, which was also resistant, only the mutant β_1 tubulin was seen on two-dimensional gels. The altered mobility of polypeptides on two-dimensional gels was clearly useful in defining the tubulin gene products and the heterogeneity of their expression.

Benzimidazole resistant mutants have also been raised in the free-living nematode Caenorhabditis elegans (Woods et al., unpublished data), but the analysis of these has not progressed to the point where they have been related directly to structural genes for tubulins. In mammalian systems, the interest in genetics of anti-tubulin resistance has centered around colchicine sensitivity. Here, it has been shown that Chinese hamster ovary (CHO) cells that are resistant to colchicine inhibition can be isolated (Cabral et al., 1980). An additional β-tubulin was present in these mutant cells and in vitro translation of RNA from them suggests that the mutant tubulin was not simply the result of a post-translational modification, and thus, may be a mutation in a β-tubulin structural gene (Cabral et al., 1980). It is intriguing that colchicine resistant mutants also seem to produce β-tubulin modifications.

The analysis of the genetics of drug resistance is a rapidly developing field, but it is too early to speculate just exactly where the binding site for the benzimidazole drugs lies. It is tempting to suggest that, since β-tubulin modifications are produced by benzimidazole resistant mutants, the binding site for the drugs is on the β-tubulin subunit. However, since many

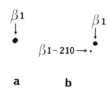

    a   b

Fig. 5. Identification of an electrophoretically novel tubulin in a Physarum mutant resistant to benzimidazole carbamates. Cell lysates of myxamoebae of wild type and the mutant resistant to benzimidazole carbamate, BenD210, were separated by two-dimensional electrophoresis; the proteins were transferred to nitrocellulose paper by Western blotting. These blots were probed with a monoclonal antibody that specifically recognizes β-tubulin. One β_1-tubulin was detected in wild type myxamoebae (a), whilst the BenD210 mutant (b) possessed an electrophoretically novel β-tubulin (β_1-210) in addition to the β_1 species.

molecules interact with β-tubulin, drug action may be a complex interaction involving α-tubulin, associated proteins, or other, as yet unidentified cell products (Burland and Gull, 1984). Nevertheless, it is clear that, in addition to well-characterized biological action, the benzimidazole drugs have an expanding potential for providing insight into the structure and function of microtubules.

ACKNOWLEDGMENTS

Work in our laboratory has been supported by grants from the Science and Engineering Research Council, the Agricultural and Food Research Council, the Wellcome Trust and the Royal Society. We would like to thank Dr. P.J. Dawson and Dr. J. Havercroft for providing Figures 2 and 4.

REFERENCES

Atkinson, C., Newsam, R.J. and Gull, K., 1980, Influence of the anti-microtubule agent, mebendazole, on the secretory activity of intestinal cells of Ascaridia galli, Protoplasma, 105:69-76.

Bannerjee, D. and Prakash, O., 1972, In vitro action of a new anthelmintic, mebendazole (R-17635), on the development of hookworms, Indian J. Med., 60:363-366.

Birkett, C.R., Coulson, C., Pogson, C.I. and Gull, K., 1981, Inhibition of secretion of proteins and triacylglyercol from isolated rat hepatocytes mediated by benzimidazole carbamate antimicrotubule agents, Biochem. Pharmacol., 30:1629-1633.

Borgers, M., deNollin, S, deBrabander, M. and Thienpoint, D., 1975, Influence of the anthelmintic, mebendazole, on microtubules and intracellular organelle movement in nematode intestinal cells, Am. J. Vet. Res., 36:1153-1161.

Borisy, G.G. and Taylor, E.W., 1967, The mechanism of action of colchicine. Binding of colchicine - ^{3}H to cellular protein, J. Cell Biol. 34:525-533.

Brenner, S., 1974, The genetics of Caenorhabditis elegans, Genetics, 77:71-94.

Burland, T.G. and Gull, K., 1984, Molecular and cellular aspects of the interaction of benzimidazole fungicides with tubulin and microtubules, in: "Mode of Action of Antifungal Agents," A.P.J. Trinci and J.P. Ryley (eds.), British Mycological Symposium, pp. 299-320, Cambridge University Press, Cambridge.

Burland, T.G., Gull, K., Schedl, T., Boston, R.S. and Dove, W.F., 1983, Cell type dependent expression of tubulins in Physarum, J. Cell Biol., 97:1852-1859.

Burland, T.G., Schedl, T., Gull, K. and Dove, W.E., 1984, Genetic analysis of resistance to benzimidazoles in Physarum: differential expression of β-tubulin genes, Genetics, 108:123-141.

Cabral, F., Sobel, M.E. and Gottesman, M.M., 1980, CHO mutants resistant to colchicine, colcemid, or griseofulvin have an altered β-tubulin, Cell, 20:29-36.

Cappucinelli, P., Fighetti, M. and Rubino, S., 1979, A mitotic inhibitor for chromosomal studies in slime molds, FEMS Microbiol Letters, 5:25-27.

Chalfie, M. and Thomson, J.N., 1982, Structural and functional diversity in the neuronal microtubules of Caenorhabditis elegans, J. Cell Biol., 93:15-23.

Clayton, L., Quinlan, R.A., Roobol, A., Pogson, C.I. and Gull, K., 1980, A comparison of tubulins from mammalian brain and Physarum polycephalum using SDS-polyacrylamide gel electrophoresis and peptide mapping, FEBS Lett., 115:301-305.

Comley, J.W., 1980, Ultrastructure of the intestinal cells of Aspicularis tetraptera after in vivo treatment of mice with mebendazole and thiabendazole, Int. J. Parasitol., 10:145-150.

Davidse, L.C., 1973, Antimitotic activity of methylbenzimidazol-2-yl carbamate (MBC) in Aspergillus nidulans, Pesticide Biochem. Physiol., 3:317-325.

Davidse, L.C. and Flach, W., 1977, Differential binding of benzimidazol-2-yl carbamate to fungal tubulin as a mechanism of resistance to this antimitotic agent in mutant strains of Aspergillus nidulans, J. Cell Biol. 72:174-193.

Davis, C. and Gull, K., 1983, Protofilament number in microtubules in cells of two parasitic nematodes, J. Parasitol., 69:1094-1099.

Dawson, P.J., Gutteridge, W.E. and Gull, K., 1983, Purification and characterisation of tubulin from the parasitic nematode, Ascaridia galli, Molec. Biochem. Parasitol., 7:267-277.

Dawson, P.J., Gutteridge, W.E. and Gull, K., 1984, A comparison of the interaction of anthelmintic benzimidazoles with tubulin isolated from mammalian tissue and the parasitic nematode Ascaridia galli, Biochem. Pharmacol., 33:1069-1074.

DeBrabander, M., DeMey, J., Joniau, M. and Gevens, G., 1977, Ultrastructural immunocytochemical distribution of tubulin in cultured cells treated with microtubule inhibitors, Cell Biol. Int. Reports, 1:177-183.

Fantes, P.A., 1982, Dependency relations between events in mitotis in Schizosaccharomyces pombe, J. Cell Sci., 55:383-402.

Fernando, S.S.E. and Denham, D.A., 1976, The effects of mebendazole and fenbendazole on Trichinella spiralis in mice, J. Parasitol., 62:874-876.

Friedman, P.A. and Platzer, E.G., 1978, Interaction of anthelmintic benzimidazoles and benzimidazole derivatives with bovine brain tubulin, Biochim. Biophys. Acta, 544:605-614.

Friedman, P.A. and Platzer, E.G., 1980, Interaction of anthelmintic benzimidazoles with Ascaris suum tubulin, Biochim. Biophys. Acta, 630:271-278.

Hastie, A.C., 1970, Benlate-induced instability of Aspergillus diploids, Nature, 266:771-772.

Havercroft, J.C., Quinlan, R.A. and Gull, K., 1981, Binding of parbendazole to tubulin and its influence on microtubules in tissue-culture cells as revealed by immunofluorescence microscopy, J. Cell Sci., 49:195-204.

Hoebeke, J., Van Nijen, G. and DeBrabander, M., 1976, Interaction of oncadazole (R17934), a new anti-tumoral drug, with rat brain tubulin, Biochem. Biophys. Res. Commun., 69:319-324.

Howard, R.J. and Aist, J.R., 1977, Effects of MBC on hyphal tip organisation growth and mitosis of Fusarium acuminatum, and their antagonism by D_2O, Protoplasma, 92:195-210.

Howard, R.J. and Aist, J.R., 1980, Cytoplasmic microtubules and fungal morphogenesis: ultrastructural effects of methyl benzimidazole-2-yl carbamate determined by freeze-substitution of hyphal tip cells, J. Cell Biol. 87:55-64.

Ireland, C.M., Gull, K., Gutteridge, W.E. and Pogson, C.I., 1979, The interaction of benzimidazole carbamates with mammalian microtubule protein, Biochem. Pharmacol., 28:2680-2682.

Ireland, C.M., Clayton, L., Gutteridge, W.E., Pogson, C.I. and Gull, K., 1982, Identification and drug binding capabilities of tubulin in the nematode Ascaridia galli, Molec. Biochem. Parasitol., 6:45-53.

Kilmartin, J.V., 1981, Purification of yeast tubulin by self-assembly in vitro, Biochemistry 20:3629-3633.

Kohler, P. and Bachmann, R., 1981, The intestinal tubulin as possible target for the chemotherapeutic action of mebendazole in parasitic nematodes, Molec. Biochem. Parasitol., 4:325-336.

Mareel, M., Storme, G., Debruyne, G. and Van Cauwenberge, R., 1980, Anti-invasive effect of microtubule inhibitors in vitro, in: "Microtubules and Microtubule Inhibitors 1980," M. DeBrabander and J. DeMey (eds.), pp. 534-544, Elsevier North-Holland, Oxford.

Mir, L. and Wright, M., 1977, Action of microtubular drugs on Physarum polycephalum, Microbios. Letters, 5:39-44.

Morris, N.R., 1980, The genetics of microtubule polymerization and the mechanism of action of antimicrotubule drugs in Aspergillus nidulans, in: "Microtubules and Microtubule Inhibitors 1980," M. DeBrabander and J. DeMey (eds.), pp. 511-521, Elsevier North-Holland, Oxford.

Morris, N.R., Lai, M.H. and Oakley, C.E., 1979, Identification of a gene for α-tubulin in Aspergillus nidulans, Cell, 16:437-442.

Novak, M., Hardy, M., Evans, W.S., Blackburn, B.J. and Ankrom, D., 1982, A comparison of the effects of ten 5-substituted benzimidazolyl carbamates on larval development of three hymnolepidid tapeworms, J. Parasitol., 68:1165-1167.

Platzer, E.G., Eby, J.E. and Friedman, P.A., 1977, Growth inhibition of Caenorhabditis elegans with benzimidazoles, J. Nematol., 9:280-290.

Quinlan, R.A., Pogson, C.I. and Gull, K., 1980, The influence of the mitotic inhibitor methyl benzimidazole-2-yl carbamate (MBC) on nuclear division and the cell cycle in Saccharomyces cerevisiae, J. Cell Sci., 46:341-352.

Quinlan, R.A., Roobol, A., Pogson, C.I. and Gull, K., 1981, A correlation between in vivo and in vitro effects of the microtubule inhibitors colchicine, parbendazole, and nocodazole on myxamoebae of Physarum polycephalum, J. Gen. Microbiol., 122:1-6.

Raff, E.C., 1984, The genetics of microtubule systems, J. Cell Biol., 99:1-10.

Rapson, E.B., Lee, D.L. and Watts, S.D.M., 1981, Changes in the acetyl-cholinesterase activity of the nematode Nippostrongylus brasiliensis following treatment with benzimidazoles in vivo, Molec. Biochem. Parasitol., 4:9-15.

Roobol, A., Pogson, C.I. and Gull, K., 1980, In vitro assembly of microtubule proteins from myxamoebae of Physarum polycephalum, Exp. Cell Res., 130:203-215.

Roobol, A., Wilcox, M., Paul, E.C.A. and Gull, K., 1984, Identification of tubulin isoforms in the plasmodium of Physarum polycephalum by in vitro microtubule assembly, Eur. J. Cell Biol., 33:24-28.

Roy, D. and Fantes, P.A., 1982, Benomyl resistant mutants of Schizo-saccharomyces pombe cold-sensitive for mitosis, Curr. Gen., 6:195-201.

Seiler, J.P., 1976, The mutagenicity of benzimidazole and benzimidazole derivatives. VI. Cytogenetic effects of benzimidazole derivatives in the bone marrow of the mouse and the Chinese hamster, Mutat. Res., 40:339-348.

Sheir-Ness, G., Lai, M.H. and Morris, N.R., 1978, Identification of a gene for β-tubulin in Aspergillus nidulans, Cell 15:639-647.

Shelanski, M.L., Gaskin, F. and Cantor, C.R., 1973, Microtubule assembly in the absence of added nucleotides, Proc. Natl. Acad. Sci. USA, 70:765-768.

Spence, A.M., Malone, K.M.B., Novak, M. and Woods, R.A., 1982, The effects of mebendazole on the growth and development of Caenorhabditis elegans, Can. J. Zool., 60:2616-2630.

Styles, J.A. and Garner, R., 1974, Benzimidazole carbamate methylester-evaluation of effects in vivo and in vitro, Mutation Res., 26:177-187.

Umesono, K., Toda, T., Hayashi, S. and Yanagida, M., 1983, Two cell division cycle genes NDA2 and NDA3 of the fission yeast Schizosaccharomyces pombe control microtubular organization and sensitivity to anti-mitotic benzimidazole compounds, J. Mol. Biol., 168:271-284.

Wagner, E.D. and Chavarria, A.P., 1974, In vivo effects of a new anthelmintic, mebendazole (R-17635), on the eggs of Trichuris trichiura and hookworm, Ann. J. Trop. Med. Hyg., 23:151-159.

Walker, G.M., 1982, Cell cycle specificity of certain antimicrotubular drugs in Schizosaccharomyces pombe, J. Gen. Microbiol., 128:61-71.

Walker, D.L. and Williams, K.L., 1980, Mitotic arrest and chromosome doubling using thiabendazole, cambendazole, nocodazole and benlate in the slime molds Dichtyostelium discoideum, J. Gen. Microbiol., 116:397-407.

Watts, S.D.M., 1980, The preparation of a fraction with tubulin-like properties from the rat tapeworm Hymenolepis diminuta, Biochem. Soc. Trans., 8:71-72.

Watts, S.D.M., 1981, Colchicine binding in the rat tapeworm, Hymenolepis diminuta, Biochim. Biophys. Acta, 667:59-69.

Williams, K.L., 1980, Examination of the chromosomes of Polyspondylium pallidum following metaphase arrest by benzimidazole derivatives and colchicine, J. Gen. Microbiol., 116:409-415.

Williams, K.L. and Barrand, P., 1978, Parasexual genetics in the slime mold Dictyostelium discoideum: haploidization of diploid strains using benlate, FEMS Microbiol. Lett., 4:155-159.

Wood, J.S., 1982, Genetic effects of methyl benzimidazole-2-yl carbamate on Saccharomyces cerevisiae, Mol. Cell. Biol., 2:1064-1079.

Wood, J.S. and Hartwell, L.H., 1982, A dependent pathway of gene functions leading to chromosome segregation in Saccharomyces cerevisiae, J. Cell Biol., 94:718-726.

Wright, M., Moisand, A., Tollon, Y. and Oustrin, M.L., 1976, Mise en evidence de l'action due methylbenzimidazole-2-yl carbamate (MBC) et du methyl [5(-2-thienyl carbonyl)1H benzimidazole-2-yl carbamate] (R17934) sur le noyau de Physarum polycephalum (Myxomycetes), Comptes Rendus Acad. Sci. Paris, 283:1361-1364.

Yamamoto, M., 1980, Genetic analysis of resistant mutants to antimitotic benzimidazole compounds in Schizosaccharomyces pombe, Molec. Gen. Genetics, 180:2131-2134.

DISRUPTION OF MICROTUBULES BY METHYLMERCURY

Polly R. Sager[a] and Tore L.M. Syversen[b]

[a]Department of Physiology
University of Connecticut Health Center
Farmington, Connecticut, USA

[b]Department of Pharmacology and Toxicology
University of Trondheim, School of Medicine
Trondheim, Norway

INTRODUCTION

Methylmercury is the only known environmental chemical teratogen. Accidental poisonings caused by contaminated fish and grain in Japan and Iraq resulted in the exposure of a number of children to methylmercury in utero. From these populations, it was clear that exposure during the prenatal or early postnatal period lead to severe and permanent damage to the central nervous system (Amin-Zaki et al., 1974, 1979; Harada, 1977). While it is well documented in both humans and experimental animals that the developing nervous system is more affected by methylmercury than the adult brain (reviewed by Clarkson et al., 1981), little is known about the mechanisms of damage. Furthermore, it is not clear whether different mechanisms underlie damage to developing and mature tissues or how these mechanisms are related.

We have suggested that one cellular insult caused by methylmercury in developing brain involves the disruption of microtubules. This hypothesis was suggested by observations made in two general areas. First, the effects of methylmercury on developing organisms appeared to involve some basic aspect of development. The evidence includes: reduced weights in animals exposed to methylmercury in utero (Spyker and Smithberg, 1972; Mottet, 1974; Su and Okita, 1976; Chen et al., 1979); and the reduction in brain size and loss of organization in the brains of two children also exposed in utero (Choi et al., 1978). These studies suggest that cell proliferation and perhaps neuronal migration are affected by methylmercury. Second, a number of functions related to cytoskeleton are altered by methylmercury. Ramel (1967) and Miura et al. (1978) reported that mitosis in cultured cells was disrupted by methylmercury. Abe et al. (1975) also reported that axonal transport was inhibited, which the authors related to the capacity of methylmercury to depolymerize in vitro assembled microtubules. Additionally, tubulin is an abundant protein in brain, comprising up to 10 percent of the total protein during development (Bamburg et al., 1973; Fellous et al., 1975; Lennon et al., 1980). Thus, we investigated the interaction of methylmercury with microtubules in several model systems using: in vitro polymerization of microtubules; cultured cells to examine cytoplasmic microtubules; and developing brain to monitor mitotic activity.

INTERACTION OF METHYLMERCURY WITH MICROTUBULES IN VITRO

A number of drugs, physical and chemical agents, including a number of heavy metals (Imura et al., 1980; Miura et al., 1984), have been shown to inhibit the assembly of purified microtubule protein. The studies described here were undertaken to characterize the inhibition of microtubule polymerization by methylmercury. Microtubule proteins were derived from several sources. Extracts and purified microtubule protein from newborn mouse brain were used initially; for later studies, protein was purified from porcine brain. Other groups have also used adult rat and bovine brain microtubule proteins.

In vitro assembly of microtubules is indeed inhibited by methylmercury in a concentration dependent fashion (Imura et al., 1980; Sager et al., 1983; Miura et al., 1984). As shown in Fig. 1, both the extent and initial rate of polymerization are affected in high speed extract of 7-day-old mouse brain (Sager et al., 1983). Assembly of microtubule proteins purified by successive cycles of assembly-disassembly from newborn mouse brain and porcine brain were similarly affected (see Fig. 2). Little inhibition was seen at 10-12 μM methylmercury and 75 percent inhibition at 50-60 μM methylmercury. The consistent effects on both the rate and extent of polymerization suggest that both the nucleation and elongation steps of polymerization are inhibited. Figure 2 shows the degree polymerization versus the concentration of methylmercury for several sources of microtubule proteins. In all cases, inhibition occurs over a relatively narrow range of concentrations, about one order of magnitude.

At higher concentrations of methylmercury, small increases in absorbance were observed. To show that this was not due to microtubule assembly but rather to nonmicrotubule aggregates, solutions of three-cycle purified microtubule protein, assembled at 37°C with or without methylmercury, were cooled and the disassembly of microtubules was monitored. For example, Fig. 3 shows the increase in absorbance as microtubules assemble at 37°C without methylmercury; when the temperature was shifted to 10°C, the absorbance decreased dramatically as the microtubules disassembled. However, in the presence of 200 μM methylmercury, the increase in absorbance with warming did not reverse at 10°C. In other experiments, microtubule protein was incubated at 37°C for 30 minutes. The resulting pellets, after centrifugation, were resuspended and chilled for 30 minutes and again spun. Protein concentrations in the resulting supernatants and pellets represent protein that had been present as microtubules or as aggregates respectively. As shown in Table 1, a substantial portion of the initially pelletable protein (H_3P) was not organized into temperature-sensitive microtubules in the presence of methylmercury.

Electron micrographs of solutions incubated with methylmercury (Fig. 4) confirm that at low concentrations of methylmercury (<15 μM), microtubules are formed; at intermediate concentrations (15 μM to 100 μM), assembly of microtubules is depressed. However, at higher concentrations of methylmercury (>100 μM) nonmicrotubule aggregates and filaments form that may confound interpretation of turbidity measurements. Thus the increase in absorbance in the presence of high methylmercury must have been due to aggregates rather than temperature insensitive microtubules. The significance of the nonmicrotubule aggregates formed in the presence of higher methylmercury concentrations is not clear. However, other metals such as zinc and cobalt are known to induce aberrant microtubule structures (Gaskin and Kress, 1977; Gaskin 1981); both organic lead (Roderer and Doenges, 1983) and organic tin (Tan et al., 1978) inhibit in vitro assembly and cause clusters at higher concentrations. Thus aggregate formation may represent a general action of metals on in vitro assembly of microtubules, separate from the inhibition of polymerization.

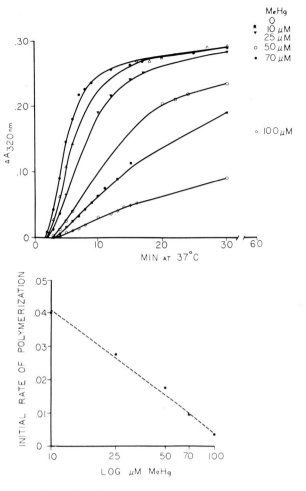

Fig. 1. In vitro polymerization of microtubules in 7-day-old mouse brain extract.
A. Change in absorbance at 320 nm vs. time at 37°C. Aliquots of extract were incubated with methylmercury or buffer before polymerization was initiated by warming. The degree of polymerization was measured by the changes in absorbance at 320 nm (adapted from Sager et al., 1982).
B. Initial rate of polymerization versus the log concentration of methylmercury. The initial rates of polymerization were measured directly from the linear increase in absorbance at 320 nm.

The mechanism underlying the inhibition of polymerization is most likely an interaction of methylmercury with tubulin sulfhydryl groups. Methylmercury is known to have a high affinity for protein sulfhydryls and may act much like other sulfhydryl binding drugs. For example, PCMPS, NEM and DTNB* have been shown to inhibit assembly. Furthermore, the effect of PCMPS can be reversed by addition of DTT* (Kuriyama and Sakai, 1974) or reduced glutathione (Nichida and Kobayashi, 1977, Webb, 1966). It has been known for many years that binding of sulfhydryl reagents to microtubule proteins blocked assembly (Kuriyama and Sakai, 1974; Kuriyama, 1976; Nishida

---

*PCMPS, p-chloromercuribenzenesulfonate; NEM, N-ethyl-maleimide; DTNB, 5,5'-dithio-bis(2-nitrobenzoic acid); DTT, dithiothreitol.

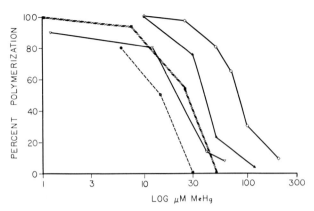

Fig. 2. Degree of polymerization of microtubules vs. log concentration of methylmercury. Degree of polymerization was calculated as the percentage of control polymerization at the plateau.
◇—◇ Extract; 7-day-old mouse brain (Sager et al., 1983)
▲—▲ 1 mg/ml 2 cycle purified; 7-day-old mouse brain (Sager)
▽—▽ 3.2 mg/ml 2 cycle purified; porcine brain (Sager)
▨—▨ 2 mg/ml 2 cycle purified, porcine brain (Imura et al., 1980)
●--● 1.8 mg/ml 3 cycle purified; bovine brain (Vogel et al., 1982)

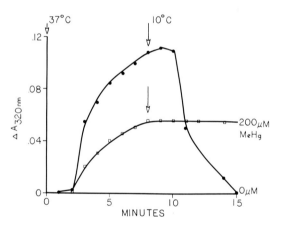

Fig. 3. Temperature reversal of turbidity at high concentrations of methylmercury. Change in absorbance at 320 nm vs. time. Aliquots of microtubule protein were incubated with or without 200 μM methylmercury at 4°C for 30 minutes before polymerization at 37°C. After 7 minutes, the temperature was shifted to 10°C. The turbidity of the control sample increased on warming and decreased when cooled, as expected for microtubules. While the turbidity of the methylmercury sample also increased at 37°C, this was not reversed when the sample was cooled.

and Kobayashi, 1977). Indeed, Abe and coworkers (1975) and Vogel et al. (1982) reported a reduction in the number of free sulfhydryl groups on tubulin in the presence of methylmercury. It appears that equimolar concentrations of methylmercury are necessary to inhibit tubulin assembly into microtubules (Vogel et al., 1982; Sager et al. 1983). Furthermore, the action of methylmercury appears to be distinct from that of microtubule poisons such as colchicine which block assembly at substoichiometric concentrations (see Wilson, this volume).

Table 1. Microtubules Versus Nonmicrotubule Aggregate Formation in the Presence of Methylmercury*

Methylmercury Concentration	Total Protein (H_3P)mg	Cold Stable Aggregates (C_3P)mg	% Total H_3P as Aggregate
25 μM	0.24	0.10	42%
56 μM	0.25	0.13	52%
112 μM	0.29	0.19	66%
250 μM	0.23	0.19	83%

*Microtubule protein (3.2 mg/ml, C_2S) was warmed 30 min in the presence of methylmercury (25 μM to 250 μM). The resulting pellet (H_3P) was resuspended and held at 4°C for 30 min. After centrifugation, the supernatant, containing cold-labile microtubule protein (C_3S), and pellet, containing aggregates (C_3P), were separated and protein concentrations determined by the methods of Lowry et al. (1951).

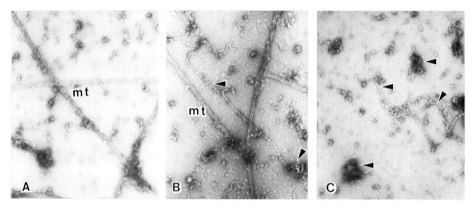

Fig. 4. Electron micrographs of microtubule protein solutions polymerized in the presence of methylmercury. Negatively stained samples were prepared from purified porcine brain microtubule protein; aliquots were incubated with methylmercury or buffer before polymerization. The presence of microtubules (mt) was verified in solutions incubated with 1.2 μM(A) and 40 μM(B) methylmercury; a few filamentous structures (◄) were also seen. The sample incubated with 116 μM methylmercury (C) contained no microtubules although an abundance of filaments and aggregates (◄) were observed. (x30,000 original magnification).

That methylmercury acts via binding to tubulin sulfhydryl groups is also supported by our work with the metal-complexing agent dimercaptosuccinic acid (DMSA) which contains two free sulfhydryl groups. DMSA alone caused partial inhibition of assembly, perhaps by formation of disulfide bonds with tubulin (Fig. 5). The addition of either 20 μM or 40 μM methylmercury severely inhibited polymerization of microtubules. However, assembly of microtubules occurred with the simultaneous addition of 25 μM DMSA. The increases in absorbance noted in the turbidity assay were indeed due to formation of microtubules, rather than aggregates, as confirmed by electron microscopy of the samples (Sager et al., 1983). The assembly of microtubules in the presence of methylmercury and DMSA (Fig. 5) suggests that DMSA reduced the direct binding of methylmercury to microtubule proteins.

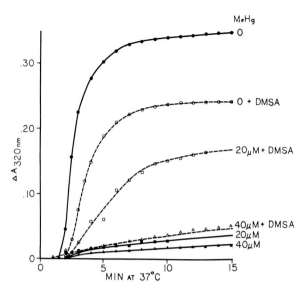

Fig. 5. Effect of methylmercury and DMSA on in vitro polymerization of microtubule proteins. Change in absorbance at 320 nm vs. time at 37°C. Aliquots of microtubule protein, purified from porcine brain, were incubated with methylmercury, DMSA, or methylmercury and DMSA at 4°C for 15 minutes before polymerization at 37°C. Polymerization was measured by the change in absorbance at 320 nm.

Thus, it is evident that methylmercury directly interacts with microtubule proteins to suppress assembly or to disassemble formed microtubules. This action of methylmercury is most likely mediated by binding to tubulin sulfhydryl groups. In extracts or cells, other mechanisms regulating tubulin sulfhydryl oxidation may be involved. For example, methylmercury also binds with a high affinity to both glutathione and cysteine which are abundant in cells and soluble extracts. These two sulfhydryl-containing compounds may function in the regulation of tubulin sulfhydryl groups. Therefore, as an alternative mechanism, methylmercury might act indirectly on tubulin by binding to and shifting the oxidation balance of cysteine and glutathione sulfhydryls.

DISRUPTION OF MICROTUBULES IN CULTURED CELLS

Inhibition of culture growth after treatment of cultured cells with methylmercury has been well documented. For a number of cell lines such as neuroblastoma cells (Prasad et al., 1979) and glioma cells (Miura et al., 1978; Prasad et al., 1979) it has been established that the growth inhibition is due to an antimitotic action rather than direct killing of cells; methylmercury has been reported also to arrest mitosis, similarly to colchicine, in onion root tip (Ramel, 1967). Indeed, the use of organic mercury compounds as fungicides is based on their capacity to block mitosis.

Since mitotic arrest may be caused by a number of mechanisms, some of which are not related directly to microtubules, a more direct measure of microtubule integrity in cultured cells was needed. We chose to visualize cytoplasmic microtubules by indirect immunofluorescence using an antibody raised against tubulin (Van DeWater et al., 1982; Sager et al., 1983). The general questions addressed were: Will exposure of cultured cells to methylmercury cause disruption of cytoplasmic microtubules? What factors modulate the disassembly? Are there differences in sensitivity among cell lines?

In answer to the first question, we established that the disruption of microtubules was both time and concentration dependent (Sager et al., 1983). For example, Table 2 summarizes data for cultured human fibroblasts incubated with methylmercury in the medium. The kinetics of the disruption suggest that a threshold may exist for cellular levels of mercury, above which microtubules begin to disassemble. To test this, cells were incubated with [^{203}Hg] methylmercury and cell-associated mercury was determined for exposure conditions similar to those used for the immunofluorescence studies. In some cases, sulfhydryl containing compounds such as DMSA, glutathione or cysteine were added to the medium along with the methylmercury. The uptake data, summarized in Table 3, indicate that DMSA or cysteine reduced the mercury levels below the apparent threshold concentration associated with loss of microtubule structure; glutathione at equivalent sulfhydryl concentrations only partially depressed the uptake methylmercury. Immunofluorescence staining confirmed the greater protective action of DMSA and cysteine compared to glutathione (Fig. 6).

In addition, DMSA, cysteine, and to a much lesser extent glutathione, reversed the dissociation of microtubules caused by methylmercury, as shown in Fig. 7 (Sager et al., 1983). Cells were preincubated with 10 µM methylmercury; DMSA, cysteine or glutathione was added subsequently and incubation continued in the presence of methylmercury. Within one hour microtubules had reformed in the cells to which DMSA or cysteine had been added (see Fig. 7B-D); only a few microtubules could be seen in some of the cells allowed to recover in the presence of glutathione (see Fig. 7E). As shown in Table 4, this recovery was associated with dramatic reduction (about 90 percent) in cellular mercury concentrations. Again glutathione caused a lesser reduction in cellular mercury and as predicted, microtubules did not reassemble in these cells. Thus, the disruption of cytoplasmic microtubules appeared to be dependent on cellular concentrations of mercury. These levels can be modulated by the addition of sulfhydryl-containing compounds to the medium.

The data presented above document that disruption of cytoplasmic microtubules provides a sensitive indicator of cell damage. Furthermore, when Imura et al. (1980) and Miura et al. (1984) compared the effects of several metals they found that many metal ions such as Hg^{+2}, Cu^{+2}, Ca^{+2}, and Cr^{+3} inhibited in vitro polymerization of microtubules. Only methylmercury selectively disrupted microtubules at concentrations that were growth inhibitory to cultured glioma cells.

Table 2. Kinetics of Microtubule Disruption As a Function of Methylmercury Concentration*

Methylmercury Concentration	Time to Partial Disruption of Microtubules	Time to Total Loss of Microtubules
0.5 µM	22 hrs	-
0.1 µM	15 hrs	-
5.0 µM	1 hr	4 hrs
10 µM	15 min	1 hr
15 µM	15 min	30 min

*Cells grown in Ham's F10 medium supplemented with 15 percent fetal bovine serum were incubated with methylmercury added to the medium. After fixation, cells were stained with a tubulin antibody. Microtubules were scored as present, partially disrupted or absent from the cells (adapted from Sager et al., 1983).

Table 3. Microtubule Structure as a Function of Cell-Associated Mercury*

Microtubules**	Cellular Mercury µM Hg/mg protein	Treatment (1 hr)
+	0.08	10 µM MeHg + 1 mM DMSA
+	0.17	10 µM MeHg + 2 mM cyteine
+	0.22	15 µM MeHg + 1 mM DMSA
±	0.51	10 µM MeHg + 2 mM glutathione
±	0.77	5 µM MeHg
−	1.60	10 µM MeHg
−	4.00	15 µM MeHg

*Cells were incubated with [^{203}Hg] methylmercury, ± DMSA, cysteine, or glutathione, after which the cells were rinsed, dissolved in NaOH; aliquots were used to determine radioactivity and protein concentrations. Mercury concentrations were reported as µg Hg/mg cellular protein, the mean of at least 4 separate cultures (adapted from Sager et al., 1983).
**Microtubules were scored as present (+), partially disrupted (±) or absent (−).

A central feature of methylmercury neurotoxicity is the specificity it shows; neurons appear to be more severely affected than glial or nonneural cells (Hunter and Russell, 1954). We approached this problem by using different cell lines and comparing disruption of microtubules with uptake of methylmercury. While previous studies have compared the effects of methylmercury treatment in several cell lines (Miura et al., 1978; Prasad et al., 1979), difference in culture conditions and assay times confound the interpretation of those results. Therefore we wished to determine whether with comparable growth and assay conditions certain cell types accumulate more mercury, or respond to lower cellular concentrations of mercury. The cell lines chosen for this work were mouse neuroblastoma cells (Nb2a-AB-1), rat glioma cells (ATCC CCL 107) and human fibroblasts (Hb-71-6) grown as described (Sager and Syversen, 1984). When cells were incubated for 6 hours with 1 µM methylmercury, only the neuroblastoma cells showed evidence of microtubule disassembly; incubation with 5 µM methylmercury caused dissociation of microtubules in all 3 cell types. As shown in Table 5, parallel uptake studies using radiolabelled methylmercury demonstrated that over a range of concentrations of methylmercury in the medium (1×10^{-8}M to 5×10^{-6}M), neuroblastoma cells had the highest cellular concentration of mercury, 3 to 6 times those in glioma cells or fibroblasts (Sager and Syversen, 1984).

Various metal-binding agents, such as DMSA, have been used in experimental animals to reduce brain concentrations of methylmercury and to prevent its neurotoxic effects (Magos, 1976; Magos et al., 1978; Hughes and Sparber, 1978); DMSA also has been reported to reverse methylmercury effects in cultured human astrocytes (Choi and Lapham, 1981). We further tested the selectivity of methylmercury by examining the reversal of effects by DMSA; perhaps the chelating agent would demonstrate greater efficacy in certain cell types. As previously shown for fibroblasts, subsequent addition of DMSA to cells preincubated with methylmercury resulted in an efflux of mercury from the cells (see Table 6). While mercury levels in all cell lines were reduced after the addition of fresh medium, as expected, the addition of DMSA further decreased the cell mercury concentrations. It is also of interest that DMSA had the greatest effect on neuroblastoma cells.

Table 4. Reduction in Cellular Mercury After Addition of DMSA, Cysteine or Glutathione*

Pretreatment MeHg (1 hr)	Cellular Mercury μg Hg/mg protein**	Recovery (1 hr)	Cellular Mercury μg Hg/mg protein**	% Reduction
10 μM	1.6(0.8)			
10 μM		+ 1 mM DMSA	0.10(0.02)	94%
10 μM		+ 2 mM Cysteine	0.17(0.03)	89%
10 μM		+ 2 mM Glutathione	0.81(0.37)	49%

*Cells were preincubated for 60 min with 10 μM [203Hg] methylmercury. DMSA, cysteine, or glutathione was then added and the incubation continued for an additional 60 min. Cell-associated mercury was determined as in Table 3.
**Mean (standard deviation) of 4-10 separate cultures.

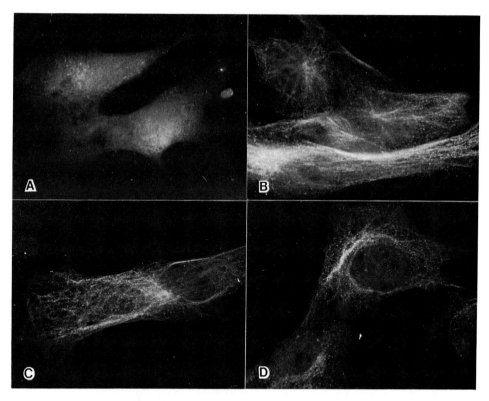

Fig. 6. Immunofluorescence staining of microtubules in fibroblasts exposed to methylmercury and DMSA, cysteine, or glutathione.
A. 10 μM MeHg for 1 hr.
B. 10 μM MeHg and 1 mM DMSA for 1 hr.
C. 10 μM MeHg and 2 mM cysteine for 1 hr.
D. 10 μM MeHg and 2 mM glutathione for 1 hr.

Table 5. Cell-Associated Mercury After 6 Hour Incubation with Methylmercury*

Methylmercury Concentration	Cellular Mercury (μg Hg/mg protein)		
	Neuroblastomas	Gliomas	Fibroblasts
1×10^{-8} M	0.007(0.001)**	0.002(0.0002)	0.001(0.0001)
5×10^{-8} M	0.037(0.011)	0.005(0.0003)	0.006(0.0006)
1×10^{-7} M	0.051(0.037)	0.010(0.0003)	0.011(0.002)
5×10^{-7} M	0.19 (0.01)	0.051(0.003)	0.055(0.005)
1×10^{-6} M	0.42 (0.12)+	0.12 (0.003)	0.13 (0.02)
5×10^{-6} M	2.5 (0.9)+	0.65 (0.04)+	0.56 (0.10)+

*Cells grown in medium containing 10 percent fetal bovine serum were incubated with [^{203}Hg] methylmercury for 6 hours. Mercury/protein concentrations were determined as for Table 3.
**Mean (standard deviation) of 4 cultures.
+Microtubules were disrupted in these cells as determined by immunofluorescence staining (adapted from Sager and Syversen, 1984).

Table 6. Reduction in Cellular Mercury After Addition of DMSA*

Cells	Pretreatment Methylmercury for 6 Hours	Cellular Mercury μg Hg/mg Protein**	2 Hour Recovery ±DMSA	Cellular Mercury μg Hg/mg Protein**	% Reduction
Neuroblastoma	1 μM	0.43(0.12)[+]	0 μM 100 μM	0.14(0.01)[+] 0.04(0.001)	67% 91%
	5 μM	2.5 (0.9)[+]	0 μM 100 μM	1.14(0.06)[+] 0.28(0.04)	54% 89%
Glioma	5 μM	0.65(0.04)[+]	0 μM 100 μM	0.20(0.02)[+] 0.12(0.01)	70% 82%
Fibroblasts	5 μM	0.56(0.10)[+]	0 μM 100 μM	0.27(0.02)[+] 0.17(0.01)	52% 70%

*Cells were preincubated with [203Hg] methylmercury followed by 2 hours incubation in fresh medium ± DMSA. Cellular mercury was determined as in Table 3.
**Mean (standard deviation) of 4 cultures.
[+]Microtubules were partially or totally disrupted in these cells as determined by immunofluorescence staining.

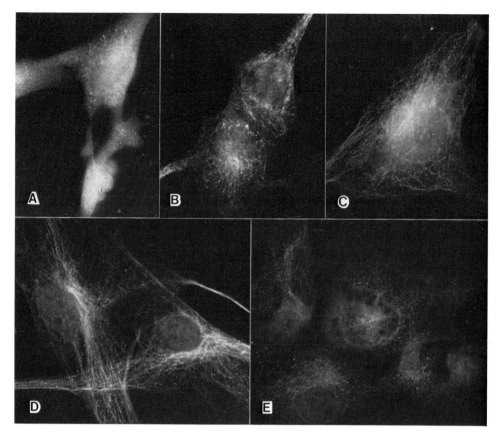

Fig. 7. Immunofluorescence staining of microtubules in fibroblasts exposed to methylmercury. Reversal of disruption with subsequent addition of DMSA or cysteine.
A. 10 μM MeHg for 1 hr.
B. 10 μM MeHg preincubation for 1 hr; 1 mM DMSA added for 15 min.
C. 10 μM MeHg princubation for 1 hr; 1 mM DMSA added for 1 hr recovery.
D. 10 μM MeHg preincubation for 1 hr; 2 mM cysteine added for 1 hr recovery.
E. 10 μM MeHg preincubation for 1 hr; 2 mM glutathione added for 1 hr recovery.

In all cases where methylmercury pretreatment had caused dissociation of mirotubules, the subsequent addition of DMSA resulted in reassembly of at least some cellular microtubules (Fig. 8A,B,D); little recovery was observed without DMSA (Fig. 8C).

The neuroblastoma, glioma and fibroblast cell lines offer a model system for studying mechanisms of damage and selectivity. In our studies, neuroblastoma cells, which are of neuronal origin, appear more sensitive than glioma cells or fibroblasts. This is in contrast to the work of Miura et al. (1978) and Prasad et al. (1979) who found growth of glioma cells to be more affected by methylmercury than other cell lines including neuroblastoma cells; however in these studies differences in culture and assay conditions make comparisons difficult.

The disruption and recovery of microtubule structure is apparently related to cell concentrations of mercury. For short exposures of

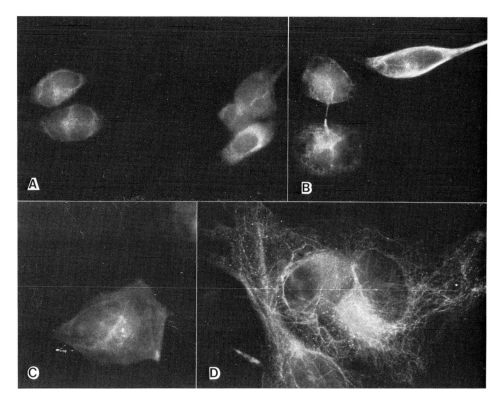

Fig. 8. Immunofluorescence staining of microtubules in neuroblastoma cells, glioma cells and fibroblasts exposed to methylmercury for 6 hours.
  A. Neuroblastoma cells; 1 μM methylmercury 6 hrs, 100 μM DMSA added for 9 hr recovery.
  B. Neuroblastoma cells; 5 μM methylmercury, 100 μM DMSA added for 9 hr recovery.
  C. Fibroblasts; 5 μM methylmercury, 0 μM DMSA added for 9 hr.
  D. Fibroblasts; 5 μM methylmercury, 100 mM DMSA added for 9 hr.

fibroblasts to methylmercury (2 hours or less) the threshold for partial disruption is between 0.3 and 0.5 μg Hg/mg protein with total loss of microtubule structure occurring above 1.2 μg Hg/mg protein (Table 3 and Sager et al., 1983). However after 6 hours of incubation, disassembly was observed in fibroblasts, glioma and neuroblastoma cells with mercury concentrations between 0.4 and 0.7 μg Hg/mg protein (Table 6 and Sager and Syversen, 1984). The basis for the reduction in the apparent threshold with exposure time is not known but may reflect redistribution of mercury with time; the assay does not distinguish between membrane bound and internalized mercury. Indeed with 15 percent serum in the medium, only about 2 percent of the added methylmercury was associated with the cells after 1 hour (Sager et al., 1983), indicating that a substantial portion of the mercury was bound to the serum protein and available for redistribution to the cells.

The action of cysteine and DMSA to prevent or reverse the dissociation of microtubules by methylmercury is almost certainly related to competitive binding of the metal. The basis for the lesser efficiency of glutathione in protecting against or reversing the disassembly of microtubules is not clear, especially since cellular concentrations of glutathione range from 0.1 to 10 mM and a substantial fraction of methylmercury is reported to be bound to glutathione in brain (Thomas and Smith, 1979). For DMSA and cysteine, the addition of 2 mM sulfhydryl groups to the medium may have created a sulfhydryl gradient responsible for extracting methylmercury from the cells.

The role of added sulfhydryl groups in reducing cellular mercury and promoting reassembly of microtubules may be further complicated by the reentry into the cells of mercury bound to the sulfhydryl compound. Some evidence for this exists. In the neuroblastoma, glioma cells and fibroblasts pretreated with methylmercury, as described above, the addition of 100 mM DMSA for 2 hours reduced cell mercury levels and allowed reassembly of at least some microtubules (Table 6 and Sager and Syversen, 1984). However, if the incubations with DMSA were continued, cell-associated mercury increased to about half the initial concentration after methylmercury pretreatment. Surprisingly, although concentrations of mercury associated with the cells now were above the apparent threshold for disassembly, there was no evidence of further dissociation of microtubules (Fig. 8B,D). There are several possible explanations. One is that the mercury is reentered into the cells bound to DMSA and remain bound to it, unavailable to bind to microtubule proteins. Alternatively, microtubules formed during the earlier phase of recovery, such as at 2 hours when mercury concentrations are low, might be resistant to the subsequent disruptive action of methylmercury. Indeed the cellular network which reforms varied somewhat in appearance from that in normal cells. For example in fibroblasts treated with DMSA (Figs. 7B,C, and 8D) the microtubules appeared to be very long and coiled or curved within the cell, rather than projecting straight from the nucleating center to the periphery of the cell. The number of microtubules frequently appeared to be reduced in cells treated with methylmercury followed by DMSA. Furthermore, it is not known how microtubule reassembly or the addition of the sulfhydryl compounds affected other measures of cell function such as cell cyle or long-term culture growth.

DISRUPTION OF MITOSIS IN DEVELOPING BRAIN

Up to this point, the effect of methylmercury on microtubules had not been tested as a mechanism of injury in the developing nervous system. Since methylmercury had been shown to interact directly with microtubules in vitro, to disrupt cytoplasmic microtubule networks and to inhibit mitosis in cultured cells, we examined its effect on a microtubule-related function in vivo. Earlier human and animal data had suggested that cell proliferation was depressed after methylmercury exposure. Furthermore, during brain development, proliferation occurs in both spatially and temporally defined patterns, which are highly regulated. Thus, measures of mitotic activity in developing brain provide sensitive indicators of damage. While alterations in rates of proliferation may derive from a variety of mechanisms related to direct cell killing, DNA replication, or spindle function, our measures were chosen to distinguish between such mechanisms. Thus if methylmercury were to act by disrupting the microtubules of the mitotic spindle, presumably it would affect the progression of cells through mitosis (measured as percentage late mitotic figures) since without a proper spindle, the chromosomes cannot separate in anaphase. Classic microtubule poisons such as colchicine were first identified by their capacity to arrest mitosis at metaphase. The entrance of cells into mitosis (mitotic index) would be unaffected by interference with microtubules but would be decreased by failure of DNA replication.

The external granular layer of the cerebellum (Fig. 9) was chosen for analysis since it is an easily defined region of high mitotic activity in the neonatal mouse brain; it also produces neurons which migrate to specific locations within a few days (Fujita et al., 1966; Fujita, 1967). In the initial study, 2-day-old animals were administered methylmercury (8 mg Hg/kg per os) and sacrificed 24 hours later (Sager et al., 1982). Cell numbers and mitotic indices were determined for selected areas of the external granular layer (Fig. 9A). Total number of mitotic figures and the fraction of late mitotic figures (anaphase and telophase) were also counted. As shown in Table 7A, the measures of mitotic activity which were altered were the percentage late mitotic figures and the total number of mitotic figures. The decrease in cells completing mitosis is strongly suggestive of interference with mitotic spindle function, and hence microtubule integrity, since cells entered mitosis at a normal rate. Similar results were demonstrated in animals receiving 4 mg Hg/kg methylmercury (Table 7B and Sager et al., 1984). Surprisingly, the supression of mitotic activity was most pronounced in males. The molecular basis for this differential response is unknown and remains particularly intriguing since brain concentrations of methylmercury in the brains of males and females remained equal from postnatal day 3 through day 21, after treatment of day 2 (Sager et al., 1984).

These data strongly suggest that methylmercury treatment alters mitotic activity in proliferative areas of the brain by a mechanism related to arrest during mitosis. Indeed, in more recent studies by Rodier and coworkers, changes in proliferation have been demonstrated in several regions of developing brain when mice were exposed to methylmercury in utero in day 12 of gestation. In all cases, the changes in mitotic activity were consistent with the failure of cells to complete mitosis; either the total number of mitotic figures was decreased, fraction of late mitotic figures decreased, fraction of early mitotic figures increased or the ratio of early to late mitotic figures increased (Rodier et al., 1984). Again, in no case was there evidence of a reduction in the number of cells entering mitosis. Thus, it appears that the antimitotic action of methylmercury is a

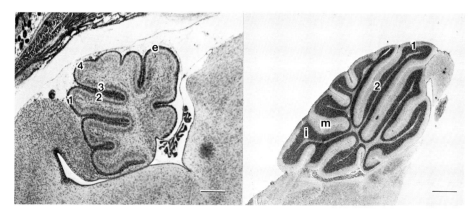

Fig. 9. Midsagittal sections from mouse cerebellum.
  A. 3-day-old mouse. The proliferative zone, the external granular layer (e) on the surface of the cerebellum, produces neurons which will migrate to other layers of the cerebellar cortex. Cell numbers and mitotic indices were determined in regions 1-4. Total mitotic figures were counted for the entire external granular layer in each matched section. Bar = 250 μm.
  B. 21-day-old mouse. The external granular layer has disappeared and the neurons it produced now reside in the molecular layer (m) and the internal granular layer (i) which are separated by a row of Purkinje neurons which form prenatally. Numbers of cells in the molecular layer and Purkinje cells were counted in regions 1 and 2 which correspond to the same areas as in Panel A. Bar = 500 μM. Measurements (for both A and B) were made in the defined layer over a fixed unit length.

Table 7. Mitotic Activity in External Granular Layer of Cerebellar Cortex*

A. 8 mg Hg/kg methylmercury

Measurement		Control (n=10)	Methylmercury treated animals (n=14)	Significance**
Mitotic Index	TOTAL	1.5(0.1)***	1.5(0.2)	N.S.
Total Mitotic Figures/Section	TOTAL	67(6)	56(4)	N.S.
	males	74(7)	52(6)	$p<0.05$
	females	59(9)	59(6)	N.S.
Late Mitotic Figures	TOTAL	4.6(1.1)	1.0(0.03)	$p<0.002$
	males	4.9(1.6)	1.0(0.05)	$p<0.05$
	females	4.3(1.0)	1.1(0.4)	$0.05<p<0.10$

B. 4 mg Hg/kg methylmercury

Measurement		Control (n=10)	Methylmercury treated animals (n=14)	Significance**
Mitotic Index	TOTAL	1.3(0.2)	1.0(0.1)	N.S.
Total Mitotic Figures/Section	TOTAL	70(3)	53(3)	$p<0.001$
	males	76(5)	53(4)	$p<0.005$
	females	65(3)	54(4)	$p<0.05$
Late Mitotic Figures	TOTAL	4.3(0.6)	2.0(0.5)	$p<0.025$
	males	4.7(0.9)	0.5(0.4)	$p<0.001$
	females	4.0(0.9)	3.3(0.7)	N.S.

*Two day old BALB/c mice received 4 or 8 mg Hg/kg methylmercury or 5 mM Na_2CO_3 (in a volume of 10 ml/Kg body weight) per os. Animals were sacrificed 24 hours later, and the heads fixed, embedded in paraffin and 6 μm sections prepared. One matched section from each animal was used for analysis (adapted from Sager et al., 1982, 1984).
**Groups were compared by analysis of variance.
***Mean (standard deviation).

generalized mechanism underlying damage in developing neural tissue; the evidence is also most consistent with disruption of the mitotic spindle, thus decreasing the number of cells which are able to complete mitosis.

Several consequences can be predicted from the effect of methylmercury on mitosis; the production of neurons would be expected to decrease, particularly since compensatory proliferation rarely, if ever, results in complete repair (Altman et al., 1969; Langman et al., 1972). In point of fact, cell numbers in selected areas of the external granular layer were reduced 24 hours after methylmercury treatment (Table 8). Furthermore, cell numbers in cerebellar cortical layers, such as the molecular layer, derived from the external granular layer remain depressed 19 days after treatment on day 2 (Table 8). This is in contrast to Purkinje cell numbers which remain unaltered by methylmercury treatment; these cells are produced prenatally and therefore were expected to be unaffected by a postnatal, antimitotic agent. Subtle reductions in cellularity, such as these in the molecular and internal granular layer, produced by other agents have been correlated with behavioral deficits (Langman et al., 1975; Rodier, 1977, 1980, 1983; Rodier et al., 1979).

Of course, the effect on mitosis cannot account for the lesions caused by methylmercury in adult brain since proliferation of neurons is thought to be restricted to the developmental period. It does not, however, require separate molecular targets for methylmercury in developing and mature nervous tissue. For example, disruption of microtubule structure in proliferative tissues might well be most noticeable as an effect on mitosis. On the other hand, in nondividing tissues, the manifestation of an interaction with tubulin would be quite different, affecting perhaps cell shape or intracellular movement.

Table 8. Cell Numbers in Areas of the Cerebellar Cortex After Methylmercury Treatment*

Measurement			Control	Methylmercury treated animals	Significance**
DAY 3					
External granular layer			(n=11)	(n=10)	
	Region 1	males	95(1)***	86(4)	0.05<p<0.10
		females	86(3)	78(4)	N.S.
	Region 2	males	95(2)	90(5)	N.S.
		females	97(4)	84(4)	p<0.05
DAY 21					
Molecular layer			(n=15)	(n=15)	
	Region 1	males	108(5)	91(5)	p<0.05
		females	98(3)	99(4)	N.S.
	Region 2	males	102(4)	88(3)	p<0.05
		females	99(5)	91(3)	N.S.
Purkinje cells					
	Region 1	males	17(1)	15(1)	N.S.
		females	17(1)	17(1)	N.S.
	Region 2	males	17(1)	16(1)	N.S.
		females	17(1)	17(1)	N.S.

*Two-day old mice received 4 mg Hg/kg methylmercury or carrier per os and were sacrificed 24 hr or 19 days later. Matched sections were prepared as for Table 7. External granular layer is the proliferation zone producing neurons which will migrate into the internal granular layer and molecular layer by day 21. Since Purkinje cells are produced prenatally and therefore their number should be unaffected by postnatal treatment with an antimitotic agent, they serve as a control for artifactual changes in tissue size (adapted from Sager et al., 1984).
**Groups compared by analysis of variance.
***Mean (standard deviation).

SUMMARY

The disruption of cytoplasmic and spindle microtubules by methylmercury has been fairly well established (Miura et al., 1978, 1984; Imura et al., 1980; Sager et al., 1983, Sager and Syversen, 1984). Interference with either of these cellular structures would be expected to have deleterious effects especially in developing tissues. Indeed, the antimitotic effect of methylmercury on the developing central nervous system (Sager et al., 1982; Rodier et al., 1984) appears to have persistent morphological consequences (Sager et al., 1984) which may correlate with functional deficits (Rodier, 1977, 1980, 1983).

The molecular mechanism underlying the dissociation of microtubules by methylmercury is most likely related to an interaction with tubulin sulfhydryl groups. A direct binding can be demonstrated in vitro (Abe et al., 1975; Vogel et al., 1982) where methylmercury acts similarly to other sulfhydryl-binding drugs. In cells, however, the action of methylmercury may be mediated through intracellular levels of reduced glutathione, which have been implicated in the regulation of microtubule polymerization (Nath and Rebhun, 1976; Oliver et al., 1976). While it is most likely that a methylmercury-sulfhydryl interaction is involved, it is by no means clear whether the disruption of microtubules in cells results from a direct action on tubulin or via some other mechanism regulating microtubule assembly.

ACKNOWLEDGMENTS

The authors are deeply grateful to T.W. Clarkson for his encouragement and support of our individual and collaborative research interests, J.B. Olmsted and P.M. Rodier for their invaluable expertise and advice, and R.A. Doherty for the use of lab facilities. We also thank D.G. Myles and W.F. Sunderman, Jr. for reviewing this manuscript, and J. Jannace for preparing it. The work reported here was completed, for the most part, while the authors were associated with the Environmental Health Sciences Center and the Division of Toxicology at the University of Rochester School of Medicine, and was supported in part by NIH Training Grant ES 07026, Center Grant ES 01247, Program Project Grant ES 01248.

REFERENCES

Abe, T., Haga, T. and Kurokawa, M., 1975, Blockage of axoplasmic transport and depolymerization of assembled microtubules by methylmercury, Brain. Res., 86:504-508.

Altman, J., Anderson, W.J. and Wright, K.A., 1969, Early effects of X-irradiation of the cerebellum of infant rats: decimation and reconstitution of the external granular layer, Exp. Neurol., 24:196-216.

Amin-Zaki, L., Elhassani, S., Majeed, M.A., Clarkson, T.W., Doherty, R.A. and Greenwood, M., 1974, Intrauterine methylmercury poisoning in Iraq, Pediatr., 54:587-595.

Amin-Zaki, L., Majeed, M.A., Elhassani, S.B., Clarkson, T.W., Greenwood, M.R. and Doherty, R.A., 1979, Prenatal methylmercury poisoning, Am. J. Dis. Child., 133:172-177.

Bamburg, J.R., Shooter, E.M., and Wilson, L., 1973, Developmental changes in microtubule protein of chick brain, Biochem., 12:1476-1482.

Chen, W.J., Body, R.L. and Mottet, N.K., 1979, Some effects of continuous low-dose congenital exposure to methylmercury on organ growth in the rat fetus, Teratol., 20:31-36.

Choi, B.H. and Lapham, L.W., 1981, Effects of meso-2,3-dimercaptosuccinic acid on methylmercury-injured human fetal astrocytes in vitro, Exp. and Mol. Pathol., 34:25-33.

Choi, B.H., Lapham, L.W., Amin-Zaki, L. and Saleem, T., 1978, Abnormal neuronal migration, deranged cerebral cortical organization, and diffuse white matter astrocytosis of human fetal brain: a major effect of methylmercury poisoning in utero, J. Neuropathol. Exp. Neurol., 37:719-733.

Clarkson, T.W., Cox, C., Marsh, D.O., Myers, G.J., Al-Tikriti, S.K., Amin-Zaki, L. and Dabbagh, A.R., 1981, Dose-response relationships for adult and prenatal exposures to methylmercury, in: "Measurements of Risks," G.G. Berg, and H.D. Maillie, eds., pp. 111-129, Plenum Press, New York.

Fellous, A., Francon, J., Virion, A. and Nunez, J., 1975, Microtubules and brain development, FEBS Lett., 37:5-8.

Fujita, S., 1967, Quantitative analysis of cell proliferation and differentiation in the cortex of the postnatal mouse cerebellum, J. Cell Biol., 32:277-287.

Fujita, S., Shimada, M. and Nakamura, T., 1966, H^3-thymidine autoradiographic studies of the cell proliferation and differentiation in the external and internal granular layers of the mouse cerebellum, J. Comp. Neurol., 128:191-208.

Gaskin, F., 1981, In vitro microtubule assembly regulation by divalent cations and nucleotides, Biochem., 20:1318-1322.

Gaskin, F. and Kress, Y., 1977, Zinc-ion induced assembly of tubulin, J. Biol. Chem., 252-6918-6924.

Harada, Y., 1977, Congenital Minamata disease, in: "Minamata Disease, Methylmercury Poisoning in Minamata and Niigatta, Japan," T. Tsubaki, and K. Irukayama, eds., pp. 209-239, Elsevier North Holland, Amsterdam.

Hughes, J.A. and Sparber, S.B., 1978, Reduction of methylmercury concentration in neonatal rat brains after administration of dimercaptosuccinic acid to dams while pregnant, Res. Comm. Chem. Path. Pharmacol., 22:357-363.

Hunter, D. and Russell, D.S., 1954, Focal cerebral and cerebellar atrophy in a human subject due to organic mercury compounds, Neurol. Neurosurg. Psychiat., 17:235-241.

Imura, N., Miura, K., Inokawa, M. and Nakada, S., 1980, Mechanism of methylmercury cytotoxicity: by biochemical and morphological experiments using cultured cells, Toxciol., 17:241-254.

Kuriyama, R., 1976, In vitro polymerization of flagellar and ciliary outer fiber tubulin into microtubules, J. Biochem., 80:153-165.

Kuriyama, R. and Sakai, H., 1974, Role of tubulin-SH groups in polymerization to microtubules: functional -SH groups in tubulin for polymerization, J. Biochem., 76:651-654.

Langman, J., Shimada, M. and Rodier, P.M., 1972, Floxuridine (5-FUdR) and its influence in postnatal cerebellar development, Pediatr. Res., 6:758-764.

Langman, J., Webster, W.S. and Rodier, P.M., 1975, Morphological and behavioral abnormalities caused by insults to the CNS in the perinatal period, in: "Teratology: Trends and Applications," C.L. Berry, and D.E. Posuille, eds., pp. 182-200, Springer-Verlag, New York.

Lennon, A.M., Francon, J., Fellous, A. and Nunez, J., 1980, Rat, mouse and guinea pig development and microtubule assembly, J. Neurochem., 35:304-813.

Lowry, O.H., Rosenbrough, N.J., Farr, A.L. and Randall, R.J., 1951, Protein measurement with the folin phenol reagent, J. Biol. Chem., 193:265-275.

Magos, L., 1976, The effects of dimercaptosuccinic acid on the excretion and distribution of mercury in rats and mice treated with mercuric chloride and methylmercury choloride, Br. J. Pharmacol., 56:479-484.

Magos, L., Peristianis, G.C. and Snowden, R.T., 1978, Postexposure preventive treatment of methylmercury intoxication in rats with dimercaptosuccinic acid, Toxicol. Appl. Pharmacol., 45:463-475.

Miura, K., Suzuki, K. and Imura, N., 1978, Effects of methylmercury on mitotic mouse glioma cells, Environ. Res., 17:453-471.

Miura, K., Inokawa, M. and Imura, N., 1984, Effects of methylmercury and some metal ions on microtubule networks in mouse glioma cells and in vitro tubulin polymerization, Toxicol. Appl. Pharmacol., 73:218-231.

Mottet, N.K., 1974, Effects of chronic low-dose exposure of rat fetuses to methylmercury hydroxide, Teratol., 10:173-190.

Nath, J. and Rebhun, L.I., 1976, Effects of caffeine and other methylxanthines on the development and metabolism of sea urchin eggs, involvement of $NADP^+$ and glutathione, J. Cell Biol., 68:440-450.

Nishida, E. and Kobayashi, T., 1977, Relationship between tubulin SH groups and bound guanine nucleotides, J. Biochem., 81:343-347.

Oliver, J.M., Albertini, D.F. and Berlin, R.D., 1976, Effects of glutathione-oxidizing agents on microtubule assembly and microtubule-dependent surface properties of human neutrophils, J. Cell Biol., 71:921-932.

Prasad, K.N. Nables, E. and Ramanujam, M., 1979, Differential sensitivities of glioma cells and neuroblastoma cells to methylmercury toxicity in cultures, Environ. Res., 19:189-201.

Ramel, C., 1967, Genetic effects of organic mercuric compounds, Hereditas 57:445-447.

Roderer, G. and Doenges, K.H, 1983, Influence of trimethyl lead and inorganic lead on the in vitro assembly of microtubules from mammalian brain, Neurotoxicol., 4:171-180.

Rodier, P.M., 1977, Correlations between prenatally-induced alterations in CNS cell populations and postnatal function, Teratol., 16:235-246.

Rodier, P.M., 1980, Chronology of neuron development: animal studies and their clinical implications, Develop. Med. Child Neurol., 22:525-545.

Rodier, P.M., 1983, Effects of metals: the developing nervous system, in: "Reproductive and Developmental Toxicity of Metals," T.W. Clarkson, G.F. Nordberg, and P.R. Sager, eds., pp. 127-150, Plenum Press, New York.

Rodier, P.M., Reynolds, S.S. and Roberts, W.N., 1979, Behavioral consequences of interference with CNS development in the early fetal period, Teratology, 19:327-336.

Rodier, P.M., Aschner, M. and Sager, P.R., 1984, Mitotic arrest in the developing CNS after prenatal exposure to methylmercury, Neurobehav. Toxical. Teratol., 6:379-385.

Sager, P.R. and Syversen, T.L.M., 1984, Differential responses to methylmercury exposure and recovery in neuroblastoma and glioma cells and fibroblasts, Exper. Neurol., 85:371-382.

Sager, P.R., Doherty, R.A. and Rodier, P.M., 1982, Effects of methylmercury on developing mouse cerebellar cortex, Exp. Neurol., 77:179-193.

Sager, P.R., Doherty, R.A. and Olmsted, J.B., 1983, Interaction of methylmercury with microtubules in cultured cells and in vitro, Exp. Cell Res., 146:127-137.

Sager, P.R., Aschner, M. and Rodier, P.M., 1984, Persistent, differential alterations in developing cerebellar cortex of male and female mice after methylmercury exposure, Devel. Brain Res., 12:1-11.

Spyker, J.M. and Smithberg, M., 1972, Effects of methylmercury on prenatal development in mice, Teratol., 5:181-190.

Su, M.M. and Okita, G.T., 1976, Embryocidal and teratogenic effects of methylmercury in mice, Toxicol. Appl. Pharmacol., 38:207-216.

Tan, L.P., Ng, M.L. and Das, V.G.K., 1978, The effect of trialkyltin compounds on tubulin polymerization, J. Neurochem., 31:1035-1041.

Thomas, D. and Smith, J.C., 1979, Partial characterization of a low-molecular weight methylmercury complex in rat cerebrum, Toxicol. Appl. Pharmacol., 47:547-556.

Vogel, D.G., Margolis, R.L. and Mottet, N.K., 1982, The effects of methylmercury binding to microtubules, The Toxicologist, 2:5 (abstract).

Van deWater, L., Guttman, S., Gorovsky, M.A. and Olmsted, J.B., 1982, Preparation of antisera and radioimmunoassay for tubulin, Methods Cell Biol., 24:79-96.

Webb, J.L., 1966, Mercurials, in: "Enzyme and Metabolic Inhibitors," Vol. II, p. 753, Academic Press, New York.

SESSION 2.   CYTOSKELETON OF THE NERVOUS SYSTEM

Chairperson: Tore L.M. Syversen
University of Trondheim, School of Medicine

Rapporteur: John B. Cavanagh
Institute of Neurology, Queen Square

PRIMARY AND SECONDARY CHANGES IN AXONAL TRANSPORT IN NEUROFIBRILLARY DISORDERS

Bruce G. Gold[a], John W. Griffin[b], Donald L. Price[c] and Paul N. Hoffman[d]

[a]Department of Environmental Health Sciences
[b]Departments of Neurology and Neuroscience
[c]Departments of Pathology, Neurology and Neuroscience,
[d]Department of Ophthalmology,
The Johns Hopkins University School of Medicine
Baltimore, Maryland, USA

INTRODUCTION

Neurofibrillary changes are pathological hallmarks of a variety of neurotoxic and degenerative disorders. Progress during the last five years in reconstructing the pathogenesis of neurofibrillary changes has resulted from correlations of ultrastructural findings with aberrations in the axonal transport of neurofilament proteins. The first neurotoxic agent studied in this fashion was β,β'-iminodipropionitrile (IDPN) (Griffin et al., 1978). Recently, comparable studies have become available on aluminum (Troncoso et al., 1983; Bizzi et al., 1984; Parhad et al., 1984), colchicine (Komiya and Kurokawa, 1980), 3,4-dimethyl-2,5-hexanedione (DMHD) (Griffin et al., 1984), acrylamide (Sidenius and Jakobsen, 1983; Gold et al., 1985), and 2,5-hexanedione (Gold, Griffin, Price, unpublished observation). From these studies, the following principles can be derived:

First, many types of neurofibrillary pathology result from maldistribution of neurofilaments within the neuron. To date, there is presumptive evidence for decreased turnover or breakdown in neurofilaments in only one model, leupeptin, an agent which inhibits proteolysis produces neurofilamentous axonal swellings when aplied to goldfish optic nerve terminals (Roots, 1983). No example of an increase in the synthesis of neurofilaments has been recognized.

Second, maldistribution of neurofilaments results in changes in caliber within affected neurons. Several lines of evidence suggest that one important role of neurofilaments in normal axons is the regulation of axonal caliber; the "volume" pathology, e.g., axonal swellings or shrinkage, seen in neurofibrillary disorders follows from this function of neurofilaments. Changes in caliber are easily quantitated within axons because of the parallel cylindrical geometry of nerve fibers and the simplicity of organization of axonal neurofilaments.

Third, the selectivity of transport abnormalities produced by different toxic agents varies widely. Some agents, including IDPN and DMHD, impair neurofilament transport in a relatively selective fashion. Other agents, such as acrylamide, produce alterations that affect all proteins equally within the slow components of axonal transport. In both of these situations, neurofilaments may accumulate in focal regions. The prominence of neurofibrillary pathology in many toxic disorders may reflect the variety of effects on axonal transport of neurofilaments and the limited capacity of the nonterminal axon to dispose of neurofilamentous masses once formed.

Finally, it is important to distinguish between the primary effect of toxins on neurofilament transport and secondary effects of reordering neuronal synthesis which might result from axonal disease. This distinction comprises the major theme of this chapter. Primary effects of toxins are most easily identified when single doses of an agent produce prompt abnormalities in transport. Secondary alterations in axonal transport can be suspected on the basis of similarities of changes in transport to those produced by axotomy or other types of direct axonal injury. To evaluate primary and secondary processes affecting axonal transport, we have found it useful to examine, in particular, the proximal regions of axons. Increases or decreases in neurofilament content in the proximal axon reflect a discrepancy between cell body synthesis and delivery of neurofilaments to the axon and the capacity of the axon to transport neurofilaments. Alterations in either the rate of neurofilament transport or the relative abundance of neurofilaments undergoing transport should be reflected in structural changes in the proximal axon. In addition, because slowly transported elements pass through the proximal axon at early times after synthesis, pulse-labeling studies of axonal transport can be performed with relatively short postlabeling intervals. Such short postlabeling intervals allow examination of the consequences of injection of single doses of toxic agents without confounding secondary responses.

This report first describes transport abnormalities produced by acrylamide and then compares these findings with those induced by axotomy and by other toxic agents, including 2,5-hexanedione. In the acrylamide model, a single high dose of toxin impairs slow axonal transport in a nonspecific fashion, resulting in neurofilament accumulations and in the development of axonal swelling in the proximal axon. However, after repeated doses, both transport and pathological changes resemble those produced by axotomy. At this stage, the proximal axon undergoes atrophy, which correlates with marked reductions in the abundance of neurofilaments undergoing axonal transport.

ACRYLAMIDE INTOXICATION

Primary Effect

To examine the primary effects of acrylamide on the axonal cytoskeleton, rats were given single doses of acrylamide (75 mg/kg, i.p.), and the slow component of axonal transport was labeled by intraspinal injections of [^{35}S] methionine. Sciatic nerves and spinal roots were removed six days later and divided into 3-mm segments. Each segment was run as a separate track on polyacrylamide slab gels, and the distribution of radioactivity in individual slow component proteins was assessed by gel fluorography. Quantification of the slow transport of the neurofilament proteins and tubulin was based on scintillation counting of these proteins in each gel track. Results demonstrated that acrylamide reduced the transport of both proteins by approximately 20 percent, compared to age-matched controls (Gold et al., 1985). Inspection of fluorograms confirmed that the slow transport defect retarded all slow component proteins equally. Seven days after acrylamide administration, the proximal L5 ventral root demonstrated a modest increase in caliber of large axons as well as an increase in neurofilament density. Together, increased size and increased neurofilament density resulted in a 70 percent increase in neurofilament numbers within the largest fibers. At this stage, some fibers showed typical giant axonal swellings (Fig. 1). This increase in neurofilament number was presumably the morphological correlate of the impairment in slow transport; it reflects a mismatch between delivery of neurofilaments to the axon from the nerve cell body and the ability of the axon to carry them centrifugally.

Secondary Effects

Animals receiving repeated smaller doses of acrylamide (30 mg/kg, i.p. for 24 days) showed a different alteration in slow transport and in morphology of proximal axons. These animals showed an increased prominence of the slow component b (SCb) proteins of Lasek and Hoffman (1976), including tubulin, and a marked reduction in the proportion of neurofilament proteins within slow component a (SCa). The morphological correlate of these changes was a striking decrease in axonal caliber in the proximal portion of fibers. These axons were smaller than in aged-matched controls, and lost circularity, resulting in a crenated appearance (Fig. 2).

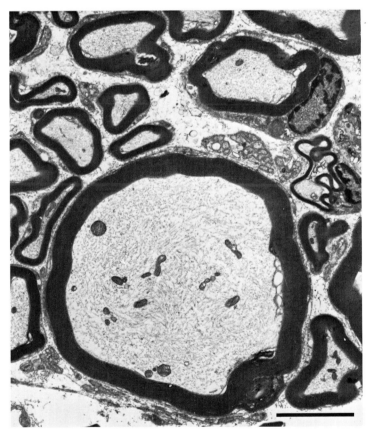

Fig. 1. Electron micrograph of a proximal giant axonal swelling from a dorsal root fiber within the dorsal root ganglia. A three-week-old rat was given a single injection of acrylamide (75 mg/kg, i.p.) followed by daily injections (30 mg/kg, i.p.) for a total of ten days. Note the large size of the axon, the relative thinness of the myelin sheath, and numerous neurofilaments within the axoplasm. Such giant axonal swellings were only occasionally observed; bar = 5 μm.

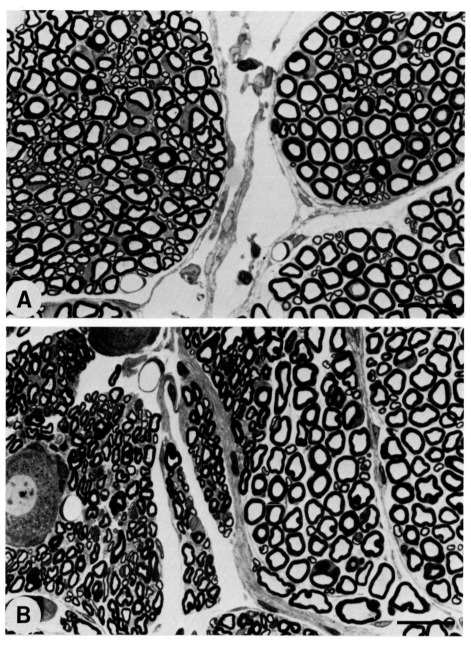

Fig. 2. Light micrographs of dorsal (left) and ventral (right) root fibers at the level of the L5 dorsal root ganglia of seven-week old rats. A, normal; B, chronic acrylamide intoxication. Note that dorsal root fibers from the acrylamide animal are markedly atrophic at this very proximal level of the nerve; dorsal root ganglia cell bodies are present among the fibers. Ventral root fibers appear normal. Epon sections (1 μm) stained with toluidine blue; bar = 25 μm.

Indirect evidence strongly suggests that changes in axonal caliber, related to slow transport, reflect a secondary response of the neuron to axonal disease. Changes produced by repeated acrylamide administration closely resemble alterations produced by axotomy. Since the late 19th century, it has been known that axons within the proximal stump of transected nerves undergo a reduction in caliber. Several lines of evidence indicate that this axonal atrophy is the result of a reduced amount of neurofilaments within the slow component, reflecting a reordering of delivery of slow component constituents to the axon.

Hoffman and Lasek (1980) showed that nerve section resulted in an increase in the relative abundance of tubulin and SCb components of the slow phase of transport and a striking reduction in the amount of label in neurofilament proteins, resulting in a decrease in neurofilament/tubulin ratio. This reordering occurred at the level of the perikaryon; slow component constituents already en route down the axon underwent comparatively little change. Further, Hoffman et al. (1984) examined the spatial and temporal evolution of axonal atrophy and correlated these findings with changes in slow transport. They found that axonal atrophy began within a few days after sciatic nerve transection and was localized initially in the most proximal region of lumbar root fibers. In fact, changes were apparent first in the intraparenchymal segment of motor fibers and appeared in the proximal ventral root 1-2 weeks after axotomy. A wave of axonal atrophy progressed down the root from proximal-to-distal regions at a rate of 1.7 mm/day, equivalent to the rate of slow transport. Atrophy in ventral roots persisted if the nerve was transected, and no successful reinnervation of muscle occurred. When axotomy was produced by crushing sciatic nerves (rather than transection), successful reinnervation occurred; in these nerves, the caliber of ventral root axons was eventually reestablished. Throughout this cycle of caliber change in motor fibers, neurofilament density remained constant at about $115/\mu m^2$, i.e., the normal linear correlation between the cross-sectional area of axons and neurofilament numbers was maintained. Thus, atrophy appears to have been a reflection of a selective decrease in delivery of neurofilaments to the axon.

In addition to profound effects on fast transport in the distal axon, acrylamide can retard slow axonal transport to a modest extent with high doses. This effect was sufficient to produce neurofilament accumulations in the proximal axon. However, in models of acrylamide neuropathy produced by repeated small doses, the direct effect of acrylamide on slow transport was obscured by prominent reductions in the perikaryal synthesis and/or delivery of neurofilament proteins to the axon. This reduction in the abundance of neurofilament proteins resulted in a decrease in axonal caliber which began proximally and moved distally with time. Since these alterations in slow transport and axonal caliber are very similar to those observed following axotomy, we suggest that they represent a secondary response and are not due to a direct effect of the toxin.

OTHER TOXIC AGENTS

Ongoing studies suggest that the same distinction between direct toxic effects on axonal transport and secondary changes applies to other disorders. Gold et al. (in preparation) have shown that very high doses of 2,5-hexanedione (1 gm/kg/day for three days) produced very proximal neurofilamentous axonal swellings; in sensory nerves, these early swellings were located within the dorsal root ganglion (Fig. 3). However, in animals given a lower dose (600 mg/kg/day for two days, with removal of nerves on day 7), axonal atrophy was already apparent within the dorsal root ganglion. In contrast, in the mid-portion of the dorsal root, neurofilamentous axonal swellings were present (Fig. 4). We interpret these findings to suggest that high doses of 2,5-hexanedione were capable of producing a selective impairment of neurofilament transport in the proximal

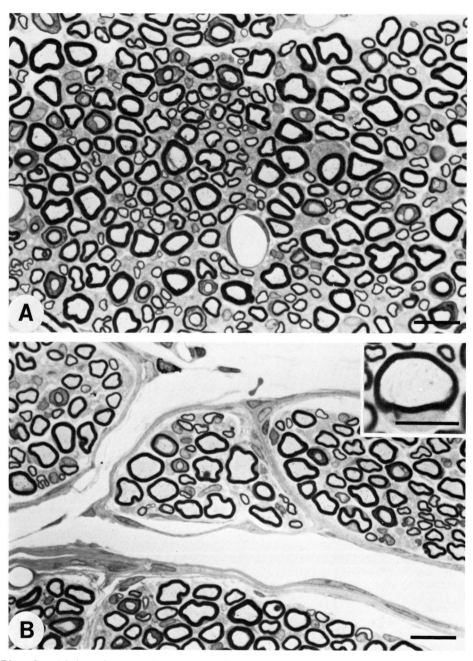

Fig. 3. Light micrographs of the L5 dorsal root just proximal to dorsal root ganglia from a normal rat (A) and a rat given high doses of 2,5-hexanedione (1 gm/kg/day) for three days (B). Note that the fibers from the 2,5-hexanedione-intoxicated animal are larger in caliber than those from the age-matched control. Insert: giant axonal swelling packed with neurofilaments from the proximal dorsal root of a similarly intoxicated rat. Epon sections (1 μm) stained with toluidine blue; bar = 15 μm.

portion of the axon in a fashion similar to changes produced by the potent analogue, DMHD (Griffin et al., 1984). However, at later times, even with

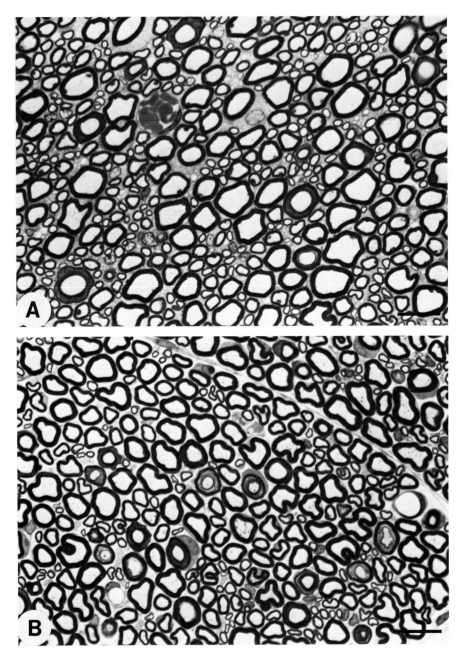

Fig. 4. Light micrographs of the L5 dorsal root from a rat given moderately high doses of 2,5-hexanedione (600 mg/kg/day) for two days; sections were taken at the mid-portion of the root (A) and immediately proximal to dorsal root ganglia (B). Note that fibers are slightly enlarged at mid-root (A). In the more proximal region (B), fibers are mildly atrophic; note the creanated appearance of many of these fibers. Epon sections (1 μm) stained with toluidine blue; bar = 15 μm.

more modest doses, the nonspecific secondary response of axonal atrophy dominated the pathological picture, and atrophy began in the intraganglionic portion of sensory axons.

It is noteworthy that the initiation of proximal axonal atrophy did not appear to require disconnection of the axon from its target; atrophy could be established by administration of botulinum toxin, an agent whose sole known action is to prevent release of miniature endplate potentials at the neuromuscular junction (Stanley et al., 1984). Thus, physiological abnormalities appear capable of inducing centrifugal axonal atrophy. It remains possible, based upon these observations, that secondary alterations in transport and caliber, observed following repeated exposure to a toxic agent (i.e., acrylamide, 2,5-hexanedione), may have arisen prior to axonal degeneration, perhaps as a consequence of loss of a trophic signal from the periphery due to an impairment in retrograde transport.

SUMMARY

These studies demonstrate that neurofibrillary changes in several models reflect an alteration in neurofilament proteins and that different alterations in axonal caliber and neurofilament content arise at different stages during the development of toxic neuropathies. The direct effect of these agents appears to be an impairment in the rate of slow transport. In contrast, under chronic conditions, axonal atrophy arises as a secondary consequence of nerve disease. At either stage, under appropriate conditions, alterations in the balance between delivery of neurofilaments from the cell body and transport along the axon will be reflected earliest in the most proximal regions of the axon.

ACKNOWLEDGMENTS

These studies were supported by grants from the U.S. Public Health Service (NIH NS 15080, ES 07094, NS 15721, and NS 07179). Drs. Hoffman and Griffin are the recipients of Research Career Development Awards (NS 00896 and NS 00450, respectively).

REFERENCES

Bizzi, A., Crane, R.C., Autilio-Gambetti, L. and Gambetti, P., 1984, Aluminum effect on slow axonal transport: a novel impairment of neurofilament transport, J. Neurosci., 4:722-731.

Gold, B.G., Griffin, J.W. and Price, D.L., 1985, Slow axoplasmic transport in acrylamide neuropathy: different abnormalities produced by single-dose and continuous administration, J. Neurosci., (in press).

Griffin, J.W., Anthony, D.C., Fahnestock, K.E., Hoffman, P.N. and Graham, D.G., 1984, 3,4-Dimethyl-2,5-hexanedione impairs the axonal transport of neurofilament proteins, J. Neurosci., 4:1516-1526.

Griffin, J.W., Hoffman, P.N., Clark, A.W., Carroll, P.T. and Price, D.L., 1978, Slow axonal transport of neurofilament proteins: impairment by β,β'-iminodipropionitrile administration, Science, 202:633-635.

Hoffman, P.N. and Lasek, R.J., 1980, Axonal transport of the cytoskeleton in regenerating motor neurons: constancy and change, Brain Res., 202:317-333.

Hoffman, P.N., Griffin, J.W. and Price, D.L., 1985, Control of axonal caliber by neurofilament transport, J. Cell Biol., 99:705-714.

Komiya, Y. and Kurokowa, M., 1980, Preferential blockade of the tubulin transport by colchicine, Brain Res. 190:505-516.

Lasek, R.J. and Hoffman, P.N., 1976, The neuronal cytoskeleton, axonal transport and axonal growth, Cold Spring Harbor Conferences on Cell Proliferation, 3:1021-1049.

Parhad, I.M., Griffin, J.W. and Koves, J.F., 1984, Aluminum intoxication in the visual system: Morphological and axonal transport studies, Neurology, 34(Suppl 1):197.

Roots, B.I., 1983, Neurofilament accumulation induced in synapses by leupeptin, Science, 221:971-972.

Sidenius, P. and Jakobsen, J., 1983, Anterograde axonal transport in rats during intoxication with acrylamide, J. Neurochem., 40:697-704.

Stanley, E.F., Hoffman, P.N., Griffin, J.W. and Price, D.L., 1984, Effect of axotomy on conduction in the proximal motor nerve, Neurology, 34(Suppl 1):182.

Troncoso, J.C., Griffin, J.W., Hoffman, P.N., Hess-Kozlow, K.M., Blum, J.R. and Price, D.L., 1983, Slow axonal transport of neurofilament proteins in aluminum intoxication, Soc. Neurosci. Abstr., 9:1190.

CHEMICAL NEUROTOXINS ACCELERATING AXONAL TRANSPORT OF NEUROFILAMENTS

Pierluigi Gambetti, Salvatore Monaco, Lucila Autilio-Gambetti, and Lawrence M. Sayre*

Division of Neuropathology, Institute of Pathology, and
*Department of Chemistry
Case Western Reserve University School of Medicine
Cleveland, Ohio, USA

INTRODUCTION

It has been known for years that chemical neurotoxins produce alterations of axonal morphology similar to those in some human diseases. Recently, it has been discovered that several of these agents have a specific effect on slow axonal transport and on the organization of the axonal cytoskeleton (Sayre et al., 1985). Therefore, they have become valuable probes for investigating functional and structural aspects of the normal axon as well as the pathogenesis of axonal lesions.

All neurotoxic agents known to affect specifically slow axonal transport produce focal axonal enlargements. The enlargements contain predominantly neurofilaments (NF) which are often centrally located whereas other organelles such as cisternae of smooth endoplasmic reticulum and mitochondria occupy the periphery of the axon (Spencer et al., 1977).

Neurofilament-containing axonal enlargements may form either at the distal, intermediate or proximal segments of the axon. For convenience, neurotoxic agents are classified according to the axonal location of the enlargements that they induce, although the mechanism of formation of the enlargements is probably the same regardless of their location (Fig. 1). Hence, in this chapter we examine the effects on slow axonal transport and on axonal cytoskeleton of: 2,5-hexanedione (2,5-HD) and carbon disulfide (CS_2), two agents producing distal enlargements; acrylamide which also causes distal axonal enlargements which are, however, morphologically different from those caused by 2,5-HD and CS_2; and 3-methyl-2,5-HD (3-M-2,5-HD), which produces enlargements located in the intermediate zones of the axon. The effects of these agents are compared with that of β,β'-iminodipropionitrile (IDPN) and 3,4-dimethyl-2,5-HD (3,4-DM-2,5-HD), two compounds that produce proximal enlargements. Before discussing the effects of the individual neurotoxic agents, general aspects of axonal transport and the method of slow axonal transport analysis used in this study are reviewed.

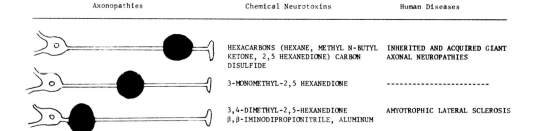

| Axonopathies | Chemical Neurotoxins | Human Diseases |

Fig. 1. Diagramatic representation of distal, intermediate and proximal giant axonopathies with the corresponding chemical neurotoxic agents and human diseases.

AXONAL TRANSPORT

One of the paradoxes in neurobiology is that the axon, which exceeds in volume the cell body by several hundred-fold (and occasionally several thousand-fold, is incapable of protein synthesis. Thus, all materials required for axonal growth and maintenance are actively transported from the perikaryon by axonal transport (Grafstein and Forman, 1980). Three principal components of axonal transport have been identified (Table 1): the fast component which moves at 250-400 mm/day; and the two subcomponents of slow transport, slow component b (SCb), which moves at 3-5 mm/day and slow component a (SCa), which moves at 0.2-1.5 mm/day (Lasek, 1980).

Each of these components has a distinct protein composition, and with few exceptions, proteins transported in one component are not transported in any other (Willard et al., 1974; Tytell et al., 1981). Membraneous organelles move with the fast transport. The microtubule (MT) proteins, MT-associated proteins (MAPs) and NF-proteins account for more than 80 percent of the proteins migrating with SCa. SCb consists of more than 100 different proteins including actin, clathrin, and fodrin (a myosin-like protein), as well as calmodulin and various metabolic enzymes (Black and Lasek, 1979; Brady and Lasek, 1981; Brady et al., 1981; Garner and

Table 1. Principal Components of the Anterograde Axonal Transport

Transport Group	Rate (mm/day)	Protein Composition	Type of Structure
Fast Comp. or I	250-400	glycoproteins, glycolipids, peptides, catecholamines, acetylcholinesterase	vesicles, granules smooth endoplasmic reticulum
Slow Comp.b or IV	2-5	actin, clathrin, enolase, calmodulin, actin-binding protein, fodrin	axoplasmic matrix
Slow Comp.a or V	0.2-2	tubulin, neurofilament proteins, tau proteins, fodrin	microtubule-neurofilament network

Lasek, 1981; Levine and Willard, 1981; Willard, 1977; Willard et al., 1979). It has been proposed that most proteins are transported down the axon as components of identifiable cytological structures (Lasek, 1982). Hence, SCa would comprise the MT-NF meshwork, a complex superstructure of NF and MT connected to each other by numerous side arms (Black and Lasek, 1981). The major portion of SCb would form a macromolecular structural complex comprising enzymes as well as cytoskeletal proteins such as actin and clathrin, organized in a morphologically ill-defined but stable structure referred to as the cytomatrix (Brady and Lasek, 1982). Slow axonal transport therefore is currently viewed as the movement of at least two separate structural complexes, or superstructures, quite different from each other in their protein composition.

Different methods can be applied to the study of axonal transport (Brady and Lasek, 1982). The most effective method to analyze the protein composition of the various components of the transport is polyacrylamide gel electrophoresis (PAGE), followed by fluorography, of consecutive axonal segments following radiolabeling of the neuronal cell body. In the present study, this method was applied to examine slow axonal transport in the primary visual system. Briefly, ^{35}S-methionine was injected into the vitreous to label the proteins of the retinal ganglion cells (Fig. 2). After a time interval sufficient to allow for separation of the labeled proteins migrating with SCa and SCb, the animals were sacrificed; the primary visual system was dissected out and cut into 3 mm consecutive segments which were individually processed for one- or two-dimensional PAGE (1-D- or 2-D-PAGE). Radioactive polypeptides present in each consecutive

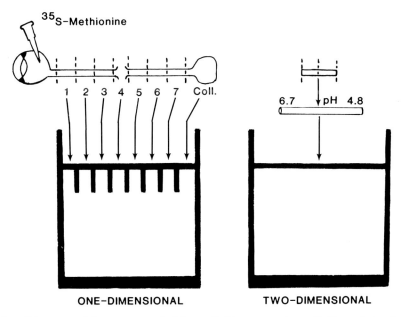

Fig. 2. Diagrammatic representation of the procedure followed to investigate slow axonal transport in rat primary visual system. Polypeptides from 3 mm segments of the system were separated according to molecular weight by one-dimensional gel electrophoresis. To characterize further some of the labeled polypeptides, selected segments were processed for two- dimensional gel electrophoresis to separate them also by their isoelectric point. An X-ray film was then applied on the gel slabs to visualize the radioactive polypeptides.

segment were detected by fluorography, thus allowing a determination of the distribution of each labeled polypeptide moving along the primary visual system with SCa and SCb.

2,5-HD

2,5-HD, a γ-diketone, is the common toxic metabolite of methyl-n-butylketone (MnBK) and n-hexane (nH). The latter two compounds, widely used as solvents, have caused outbreaks of polyneuropathies among industrial workers (Allen et al., 1975; Herskowitz et al., 1971) and individuals intentionally inhaling glue vapors (Korobin et al., 1975). Morphologically, this polyneuropathy is characterized by NF-containing focal enlargements in the preterminal region of the axon. In advanced stages, the enlargements extend proximally and the distal part of the involved axon often undergoes degeneration, both in central and peripheral systems. The axonal changes produced by 2,5-HD, MnBK and nH in experimental and clinical conditions are morphologically indistinguishable from those observed in the human inherited forms of giant axonal neuropathy (Durham et al., 1983).

Accumulation of NF in the distal region of the axon points to an abnormality in SCa transport. We selected the visual system to study slow axonal transport in 2,5-HD-treated and control rats because, despite the presence of numerous enlargements, this system has the advantage of not undergoing axonal degeneration (Cavanagh, 1981; Griffiths et al., 1981; Jones and Cavanagh, 1982; Monaco et al., 1985).

Fluorograms of 2,5-HD-treated rats 25 days after labeling showed that five polypeptides of m.w. 68,000, 145,000, 200,000, 64,000 and 62,000 were transported at a faster rate and were spread over virtually the entire length of the primary visual system, being identifiable up to the superior colliculus (Fig. 3A). In the controls, on the contrary, the same polypeptides migrated as components of the SCa and could be seen only up to 12 mm from the sclera (Fig. 3B). Fluorograms of 2-D-PAGE allowed the definitive identification of the 68,000, 145,000 and 200,000 m.w. polypeptides as NF subunits.

Because it can easily be identified in 1-D-PAGE, the 145,000 m.w. polypeptide was selected as representative of the five polypeptides transported at a higher rate. Quantitation of the radioactivity present in the 145,000 m.w. band as a function of distance along the primary visual system revealed an apparently bimodal distribution in the 2,5-HD-treated animals which was significantly different from the unimodal distribution found in the controls (Fig. 4). The bimodal distribution of the radioactivity suggests that a fraction of the transported NF moves coherently with SCa, as in controls, while another fraction moves at an increased rate approaching that of SCb. The distribution of a 30,000 m.w. polypeptide, a normal component of the SCb, was not significantly different from that of control animals. These results support the conclusion that the accelerated rate of transport involves preferentially the three NF subunits and the 64,000 and 62,000 m.w. polypeptides which are components of SCa, while components of SCb appear to be much less affected.

The nature of the 64,000 and 62,000 m.w. polypeptides is not known. Our findings (Monaco et al., 1985) that they a) do not copurify with tubulin following cycles of temperature-dependent assembly and disassembly, b) are resistant to Triton X-100 extraction, and c) react with a monoclonal antibody that recognizes all intermediate filaments (IF), suggest that the 64,000 and 62,000 m.w. polypeptides are not "tau factors" (Tytell et al., 1984) but are related to IF. Studies in control animals indicate that they are normally transported with the SCa in the visual system, probably as components of the MT-NF meshwork.

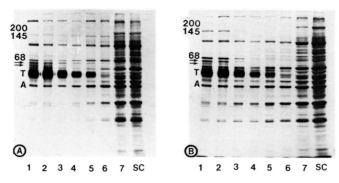

Fig. 3. Fluorograms of polyacrylamide gels following electrophoresis of 3 mm segments of primary visual system, 25 days after intraocular administration of ^{35}S-methionine. A:2,5-HD administered for 13 weeks; B:Control. 200:145; 68:NF subunits of corresponding molecular weight; T:tubulin; A:actin. In the 2,5-HD-treated rat the 68 Kd and 145 Kd subunits as well as two polypeptides of approximately 64 Kd and 62 Kd (arrows) can be detected only up to the superior colliculus (SC) whereas, in the control, they are visible up to approximately 12 mm from the sclera. The 200 Kd NF subunit is not sufficiently labeled to be clearly visible (from Monaco et al., 1985).

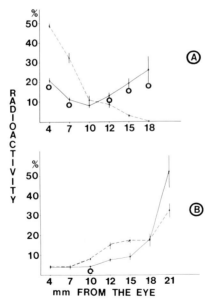

Fig. 4. Distribution percentage of the radioactivity migrating along the primary visual system with the 145 Kd NF subunit and with the 30 Kd polypeptide, a normal component of SCb; (-----) 2,5-HD treated rat; (———) control. While the distribution of the radioactivity migrating with the 145 Kd NF subunit in 2,5-HD-treated animal is apparently bimodal, that of the control is unimodal. No significant difference in distribution is seen for the 30 Kd polypeptide. (o): Significantly different from control ($p < 0.05-0.0001$) (from Monaco et al., 1985).

Acceleration of NF transport occurred both in advanced stages of intoxication (13th to 16th week of 2,5-HD administration), when the superior colliculus contained a very large number of axonal enlargements, as well as in early stages (5th week of intoxication), when enlargements were not yet present. Thus, acceleration of NF transport preceded, not followed, the formation of axonal enlargements. Moreover, NF transport was also accelerated when 2,5-HD was first administered ten days after intraocular injection of the precursor, at which time labeled NF had already entered the axon. This latter finding indicates that 2,5-HD affects NF which have already been assembled and are being transported along the axon.

Although the <u>distribution</u> of radiolabeled NF polypeptides differed between 2,5-HD-treated and control animals, the <u>total</u> NF-related radioactivity present in the primary visual system appeared to be similar. Therefore, we inferred that if the total number of NF in the system were the same in 2,5-HD and controls, a faster rate of transport should result in a decreased number of NF in regions proximal to the axonal enlargement. Ultrastructural examination indeed showed that the number of NF per optic axon in 2,5-HD-treated rats was reduced when compared to controls whereas the number of MT remained unchanged; the size of the 2,5-HD axons was also reduced (Fig. 5). These morphologic changes were confirmed by morphometric analysis (Monaco et al., 1985).

In conclusion, in the rat primary visual system 2,5-HD produced a selective acceleration of the transport rate of NF and two other polypeptides that normally migrate with the SCa, probably as constituents of the cytoskeleton. Acceleration of NF transport preceded the formation of axonal enlargements and resulted from an effect of 2,5-HD on the axon rather than on the cell body. Probably as a result of the accelerated transport, there was a longitudinal redistribution of NF whereby the accumulations in the distal enlargements were accompanied by a decreased density in more proximal axonal segments.

CS_2

As a solvent used in the manufacture of artificial fibers, CS_2 has been involved in several cases of occupational intoxication (Seppalainen and Haltia, 1980). Exposure to CS_2 leads to a chronic neurologic syndrome dominated by polyneuropathic symptoms. Available morphologic studies of axonal changes are limited to the sciatic system where numerous NF-containing axonal enlargements, identical in morphology and location to those produced by 2,5-HD (Jirmanova and Lukas, 1984), have been observed. Many of these axons undergo degeneration.

In a preliminary study we exposed rats to CS_2 by inhalation or by intraperitoneal (i.p.) injection (Pappolla et al., 1984). Light microscopic examination of the primary visual system revealed large numbers of axonal enlargements located in the layer of the optic fibers of the superior colliculus and in the distal segment of the optic tract. Except for a slightly more proximal distribution, the enlargements were similar to those following administration of 2,5-HD. In contrast to the sciatic axon, no evidence of axonal degeneration was observed in the primary visual system.

The aim of this study was to determine whether two compounds such as CS_2 and 2,5-HD which are chemically different, but produce similar morphologic changes in the axon, had a similar effect on slow axonal transport. This finding would suggest a direct correlation between certain morphologic changes, such as NF-containing distal axonal enlargements, and specific alterations of slow axonal transport, such as acceleration of NF transport rate.

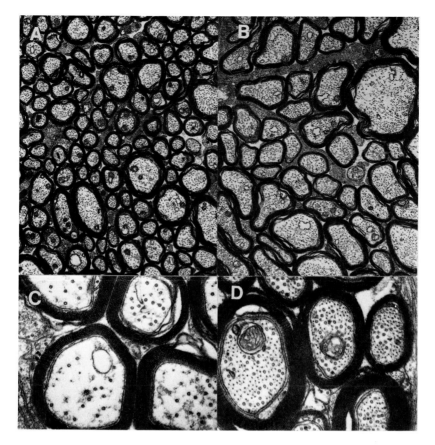

Fig. 5. Electron microphotographs of optic axons in cross-section, 5 mm from the sclera. A, C:Rat treated with 2,5-HD for 11 weeks; B, D:Controls. Axons of 2,5-HD-treated rats are reduced in size and also show a noticeable reduction in number of NF but not of microtubules (from Monaco et al., 1985).

Fluorograms of samples from rats exposed to CS_2 showed a marked increase in the rate of NF polypeptides and of the 64,000 and 62,000 m.w. polypeptides (Pappolla et al., 1984). This finding was confirmed by the analysis of the distribution percentage of the radioactivity moving with various bands which showed that the NF subunits and the 64,000 m.w. polypeptides (the radioactivity of the 62,000 m.w. was too little to be measured) were selectively transported at an increased rate, whereas the rate of transport of other polypeptides, including polypeptides in the 55,000 m.w. band comprising mostly tubulin, were much less affected (Pappolla et al., 1984).

Preliminary morphometric studies of the segments of the optic axons proximal to the NF-containing enlargements indicate decreases in the number of NF and in axonal size, as in the case of 2,5-HD.

In conclusion, following CS_2 exposure, the transport of NF and two other polypeptides of 64,000 and 62,000 m.w. was preferentially accelerated. In addition, there was a longitudinal rearrangement of the axonal cytoskeleton which was characterized by a decrease of NF in the proximal part of the axons and distal focal accumulation. Therefore, the changes in the slow axonal transport and in the organization of the axonal cytoskeleton produced by CS_2 were qualitatively identical to those occurring following administration of 2,5-HD.

3-M-2,5-HD and 3,4-DM-2,5-HD

Dimethylation of 2,5-HD, giving 3,4-DM-2,5-HD, increases neurotoxicity by twenty- to thirty-fold (Anthony et al., 1983) and, in contrast to 2,5-HD, the NF-containing enlargements form in the proximal regions of the axon (Anthony et al., 1983). In the 3,4-DM-2,5-HD model, axonal transport of NF has been reported to be preferentially impaired at the level of the axonal enlargements (Griffin et al., 1984). In these studies, however, the transport rate in the segments of the axon proximal to the enlargements could not be determined because of the proximity of the enlargements to the cell body. We postulated that monomethylation of 2,5-HD, giving 3-M-2,5-HD, would not result in as marked a proximal shift of the NF-containing enlargements as observed for dimethylation. Thus, we synthesized this compound with the aim of investigating whether methylation alters the effect of 2,5-HD on the rate of slow axonal transport.

Following chronic administration of 3-M-2,5-HD, rats developed NF-containing axonal enlargements ultrastructurally indistinguishable from those produced by the 2,5-HD but located more proximally in both the sciatic and primary visual systems (Monaco et al., 1984). In the sciatic system, most of the enlargements were located approximately 10 cm from the spinal cord; in the primary visual system, they first appeared in the medial regions of the optic tract and later extended proximally and distally to involve the entire optic tract and the superior colliculus (Monaco et al., 1984). As with 2,5-HD and CS_2, optic axons showed no evidence of degeneration (Monaco et al., 1984).

Analysis of the slow axonal transport showed again that the transport rate of both NF and the 64,000 and 62,000 m.w. polypeptides was increased (Monaco et al., 1984). A study of the distribution of radioactivity migrating with the 145,000 m.w. NF subunit and with various other bands confirmed that the accelerated transport involved preferentially NF and the 64,000 and 62,000 m.w. polypeptides. Comparison of the distribution of radioactivity related to the 145,000 m.w. NF subunit in 3-M-2,5-HD intoxication to that in 2,5-HD intoxication, indicates that a greater fraction of radioactivity migrates to the distal part of the optic tract following 3-M-2,5-HD. However, the data did not suggest that NF had reached the distal regions of the optic pathway in a shorter time. Although a more detailed study is needed, these findings suggest that 3-M-2,5-HD differs from 2,5-HD in that it affects the transport of a greater number of NF, rather than producing a greater acceleration of the same number of NF.

In conclusion, the monomethyl derivative of 2,5-HD, 3-M-2,5-HD produced NF containing enlargements located more proximally along the central and peripheral axons than those produced by 2,5-HD. It also produced acceleration of the transport rate of NF and of the 64,000 and 62,000 m.w. polypeptides. The effect on NF transport, therefore, was qualitatively similar to that produced by 2,5-HD, but 3-M-2,5-HD appeared to affect a greater number of NF.

Acrylamide

Acrylamide administration is known to produce distal axonal enlargements in central and peripheral axons (Tilson, 1981). Morphologically, however, these enlargements differ from those produced by γ-diketones because they contain a large number of dense bodies, membraneous bodies and cisternae along with NF (Prineas, 1969; Suzuki and Pfaff, 1973). We produced a significant number of such enlargements in the optic axons at the level of the superior colliculus by administering the highest acrylamide doses compatible with survival (50 mg/kg/day, up to a total of 1,200 mg/kg) (Liepold et al., 1983). These enlargements, however, were not nearly as numerous as those present following γ-diketone administration. In addition, some of these axons were degenerating (Liepold et al., 1983). A study of slow axonal transport showed a possible increase in the rate of transport of NF polypeptides that, however, did not reach statistical significance (Liepold et al., 1983).

In conclusion, we could not reproduce with acrylamide the selective acceleration of NF transport that occurs following 2,5-HD, 3-M-2,5-HD and CS_2. Acrylamide may impair fast transport leaving the slow transport unaffected, as it has been suggested in previous studies (Souyri et al., 1981); in that case it would have a mechanism of action completely different from that of 2,5-HD, CS_2 and 3-M-2,5-HD. However, in view of a) the excessive number of NF in the distal axonal enlargements, and b) the consistent, though slight acceleration of NF transport, we favor the possibility that acrylamide has an accelerating effect on NF transport similar to that of the above neurotoxic agents, but that this effect is weaker and perhaps concealed by other toxic effects such as that on fast transport.

MECHANISMS OF ACTION OF CHEMICAL NEUROTOXINS AND PATHOGENESIS OF NF-CONTAINING AXONAL ENLARGEMENTS

Three mechanisms of action have been proposed for the production of NF-containing axonal enlargements by chemical neurotoxins. According to Sabri et al. (1979), neurotoxic agents impair one or more pathways of the intermediary metabolism supplying the energy needed for NF transport (Spencer et al., 1979). As a result, NF are transported at a lower rate, eventually stop, and accumulate focally within the axon. This mechanism is unlikely for at least two reasons. First, if the activity of a metabolic enzyme were impaired, one would expect an entire component of slow axonal transport to be affected, not only NF and two additional polypeptides. Second, much higher and longer term dosages of these agents are needed to produce neurotoxic effects than would be expected if they acted as enzyme inhibitors.

Two other proposed mechanisms are based on the finding that neurotoxic γ-diketone compounds react with primary amines to form pyrroles (Anthony et al., 1983; DeCaprio et al., 1983; Graham et al., 1982). Evidence that such pyrrole formation is a crucial pathogenic step is that only those diketones which can form pyrroles (i.e., those with a γ-spacing) are neurotoxic. In addition, we observed that the 3,3-dimethyl-2,5-HD, a γ-diketone which cannot form pyrroles, lacks neurotoxicity even when administered in high doses (Sayre et al., unpublished data). Moreover, the degree of toxicity of the neurotoxic γ-diketones is directly related to the rate of pyrrole formation. Graham et al. (1982) proposed that pyrrole oxidation results in intra- and intermolecular cross-link formation. Cross-link formation would cause polymerization of NF and would result in a progressive slow-down and eventual accumulation of NF (Graham et al., 1982). These workers recently reported that a high concentration of 3,4-DM-2,5-HD produces covalent cross-linking of NF <u>in vitro</u> (Graham et al., 1984). However, Decaprio et al. (1983) found no evidence for cross-linking of NF following 2,5-HD treatment and proposed that pyrrole formation alters the physicochemical properties of proteins by converting a positively charged $-NH_3^+$ group to a neutral pyrrole group. According to these workers, this charge neutralization would lower the NF "solubility" and would eventually induce aggregation and precipitation of NF bundles.

The above hypotheses, which postulate polymerization and/or loss of solubility as a necessary pathogenetic step, imply that formation of the NF-containing axonal enlargements results from a slow-down of NF transport. However, this is difficult to reconcile with our findings: a) of an increased rate of transport of NF, regardless of the distal or intermediate location of the NF-containing axonal enlargments; and b) of the lack of insoluble protein aggregates in the axon. We propose that chemical neurotoxins which produce intermediate or distal NF-containing axonal enlargements alter the physicochemical properties of NF and/or related cytoskeletal proteins as originally suggested by Decaprio et al. (1983), but that this alteration results in the acceleration rather than the slowing down of NF transport.

A distinctive feature of NF proteins are unique sequences at the carboxyl-terminus which contain a high percentage of charged amino acids, and which may serve as electrostatic scaffolds for interaction with other NF or possibly other cytoskeletal components such as MT (Geisler et al., 1983). Neutralization of these charges would disrupt the NF-MT meshwork that constitutes the SCa and would free individual NF from the connection with other NF and/or MT. Under these conditions, NF may be transported at a higher rate. A comparable situation exists during development, when few inter-NF connections are seen in the visual system (Hirokawa et al., 1984) and when the rate of transport of NF polypeptides is severalfold that of control animals (Willard and Simon, 1983). Support for an effect of these neurotoxic agents on the NF-MT connections is provided by the finding that separation of NF from MT follows local or systemic administration of 2,5-HD and IDPN (Papasozomenos et al., 1981; Griffin et al., 1982).

In a recent study, we showed that all the neurotoxic agents examined in this review can theoretically neutralize the charges at the carboxyl-terminal region of NF proteins under physiologic conditions (Sayre et al., 1985). If correct, this charge-neutralization hypothesis would provide a unified mechanism of action by which all these different neurotoxic agents produce similar effects on the morphology of the axon and on axonal transport.

The aforementioned observation that the dimethyl derivative of 2,5-HD, the 3,4-DM-2,5-HD is 25 times more potent than the 2,5-HD, and produces proximal enlargements similar to those produced by IDPN indicates that a simple structural change within a single chemical class of compounds can produce a switch in the location of the axonal enlargements (Anthony et al., 1983). The location of the axonal enlargements produced by a given agent can also be altered by changing doses or mode of administration (Griffin et al., 1982). These findings, together with our observation that the monomethyl derivative 3-M-2,5-HD, which is intermediate in neurotoxicity between 3,4-DM-2,5-HD and 2,5-HD, produces axonal enlargements in the intermediate regions of the axon, strongly indicate that the basic mechanism of formation of NF-containing axonal enlargements is the same regardless of their location, proximal or distal.

As shown in the present study, the main compounds forming intermediate or distal NF-containing axonal enlargements cause an acceleration in the rate of NF transport from the cell body to the site of the axonal enlargement. However, since the rate of NF transport between cell body and axonal enlargements cannot be measured when the enlargements are proximal, as in the case of 3,4-DM-2,5-HD and IDPN, we do not know the real effect on transport rate of these two compounds. It is possible that they also produce acceleration of NF; in this case, the proximal location of the enlargements may result from the effect on a larger number of NF, causing accumulation to occur at nodes of Ranvier in the proximal axonal segments. An alternative is that the various neurotoxic agents produce different degrees of alteration of NF (or of other cytoskeletal proteins), thus resulting in a threshold effect on the rate of NF transport: the agents of lower toxicity, such as 2,5-HD, produce an NF alteration that results in acceleration of NF transport; those of higher toxicity, such as 3,4-DM-2,5-HD, alter NF further to the point that they become untransportable. Further study is needed to resolve these two possibilities.

How acceleration of NF transport results in the accumulation of NF at discrete axonal sites can only be conjectured. In view of our finding that acceleration appears to precede rather than follow the formation of the distal enlargements, it is likely that acceleration of the transport is the cause, and not the result, of these enlargements. One may postulate that the faster rate of transport is incompatible with the normal passage of NF through nodes of Ranvier, where the axon is narrower and the cytoskeleton probably undergoes reorganization (Berthold, 1978; Ishikawa and Tsukita, 1982; Raine, 1982). Disruption of transport could result in accumulation of NF at these sites.

In conclusion, chemical neurotoxins producing distal and intermediate NF-containing axonal enlargements cause acceleration in the rate of transport of NF and two other cytoskeletal polypeptides. The mechanism responsible for these changes remains to be determined. We propose that these agents act directly by neutralizing charges in specific domains of NF and/or related cytoskeletal proteins. This, in turn, would free NF from the NF-MT meshwork and, as a result, NF would be transported at an increased rate. Accelerated transport may impair the coursing of NF through nodes of Ranvier, where axons are thinner and the cytoskeleton probably undergoes reorganization, resulting in formation of NF-containing enlargements. Further study is needed to determine whether those neurotoxins which produce proximal enlargements also accelerate NF transport between neuronal cell body and enlargements. Should this be the case, one may postulate that the proximal or distal location of the enlargements is related to the number of NF per axon transported at a higher rate: a larger number resulting in a more proximal location.

Neurotoxic agents producing NF-containing axonal enlargements have provided valuable information on slow axonal transport and on organization of the normal axonal cytoskeleton as well as on the pathogenetic mechanisms of diseases such as inherited giant axonal neuropathies. If these neurotoxic agents do indeed modify a specific domain of NF and/or related cytoskeletal proteins, further study may provide relevant information concerning molecular interactions responsible for the coherent axonal transport of the various axonal components and the molecular events involved in the initial pathogenetic steps of the human diseases they mimic.

ACKNOWLEDGMENT

This work was supported by Grants NS 14509 and NS 18714 from the National Institutes of Health.

REFERENCES

Allen, N., Mendell, J.R., Billmaier, D.J., Fontaine, R.E. and O'Neill, J., 1975, Toxic polyneuropathy due to methyl-n-butyl ketone. An industrial outbreak, Arch. Neurol., 32:209-218.
Anthony, D.C., Boekelheide, K., Anderson, C.W. and Graham, D.G., 1983, The effect of 3,4-dimethyl substitution on the neurotoxicity of 2,5-hexanedione. II. Dimethyl substitution accelerates pyrrole formation and protein crosslinking, Toxicol. Appl. Pharmacol., 71:372-382.

Anthony, D.C., Boekelheide, K. and Graham, D.G., 1983, The effect of 3,4-dimethyl substitution on the neurotoxicity of 2,5-hexanedione. I. Accelerated clinical neuropathy is accompanied by more proximal axonal swellings, Toxicol. Appl. Pharmacol., 71:362-371.

Anthony, D.C., Giangaspero, F. and Graham, D.G., 1983, The spatio-temporal pattern of the axonopathy associated with the neurotoxicity of 3,4-dimethyl-2,5-hexanedione in the rat, J. Neuropathol. Exp. Neurol., 42:548-560.

Berthold, C.-H., 1978, Morphology of normal peripheral axons, in: "Physiology and Pathobiology of Axons," S.G. Waxman, ed., pp. 3-63, Raven Press, New York.

Black, M.M. and Lasek, R.J., 1979, Axonal transport of actin: slow component b is the principal source of actin for the axon, Brain Res., 171:410-413.

Black, M.M. and Lasek, R.J., 1980, Slow components of axonal transport: Two cytoskeletal networks, J. Cell Biol., 86:616-623.

Brady, S.T. and Lasek, R.J., 1981, Nerve specific enolase and creatin phosphokinase in axonal transport: soluble proteins and the axoplasmic matrix, Cell, 23:523-531.

Brady, S.T. and Lasek, R.J., 1982, The slow components of axonal transport: Movements, compositions and organization, in: "Axoplasmic Transport," D.G. White, ed., pp. 206-217, Springer-Verlag, Berlin-Heidelberg-New York.

Brady, S.T. and Lasek, R.J., 1982, Axonal transport: A cell-biological method for studying proteins that associate with the cytoskeleton, in: "Methods in Cell Biology," Vol. 25, D.M. Prescott, ed., pp. 365-395, Academic Press, New York.

Brady, S.T., Tytell, M., Heriot, K. and Lasek, R.J., 1981, Axonal transport of calmodulin: A physiologic approach to identification of long term associations between proteins, J. Cell Biol., 89:607-614.

Cavanagh, J.B., 1981, The pattern of recovery of axons in the nervous system of rats following 2,5-hexanediol intoxication: a question of rheology, Neuropathol. Appl. Neurobiol., 8:19-34.

Decaprio, A.P., Strominger, N.L. and Weber, P., 1983, Neurotoxicity and protein binding of 2,5-hexanedione in the hen, Toxicol. Appl. Pharmacol., 68:297-307.

Durham, H.D., Pena, S.D.J. and Carpenter, S., 1983, The neurotoxins 2,5-hexanedione and acrylamide promote aggregation of intermediate filaments in cultured fibroblasts, Muscle Nerve, 6:631-637.

Garner, J.A. and Lasek, R.J., 1981, Clathrin is axonally transported as part of slow component b: The axoplasmic matrix, J. Cell Biol., 88:172-178.

Geisler, N., Kaufmann, E., Fischer, S., Pressmann, U. and Weber, K., 1983, Neurofilament architecture combines structural principals of intermediate filaments with carboxy-terminals extensions increasing in size between triplet proteins, EMBO J., 2:1295-1302.

Grafstein, B. and Forman, D.S., 1980, Intracellular transport in neurons, Physiol. Rev., 60:1167-1283.

Graham, D.G., Anthony, D.C., Boekelheide, K., Maschmann, N.A., Richards, R.G., Wolfram, J.W. and Shaw, B.R., 1982, Studies of the molecular pathogenesis of hexane neuropathy. II. Evidence that pyrrole derivatization of lysyl residues leads to protein crosslinking, Toxicol. Appl. Pharmacol., 64:415-422.

Graham, D.G., Szakal-Quin, G., Priest, J.W. and Anthony, D.C., 1984, In vitro evidence that covalent crosslinking of neurofilament occurs in γ-diketone neuropathy, Proc. Natl. Acad. Sci., 81:4979-4982.

Griffin, J.W., Anthony, D.C., Fahnestock, K.E., Hoffman, P.N. and Graham, D.G., 1984, 3,4-dimethyl-2-hexanedione impairs the axonal transport of neurofilament proteins, J. Neurosci., 4:1516-1526.

Griffin, J.W., Fahnestock, K.E., Price, D.L. and Cork, L.C., 1983, Cytoskeletal disorganization induced by local application of IDPN and 2,5-hexanedione, Ann. Neurol., 14:55-61.

Griffin, J.W., Gold, B.G., Cork, L.C., Price, D.L. and Lowndes, H.E., 1982, IDPN neuropathy in the cat, coexistence of proximal or distal axonal swellings, Neuropathol. Appl. Neurobiol., 8:351-364.

Griffiths, I.R., Kelly, P.A.T., Carmichael, S., McCulloch, M. and Waterson, M., 1981, The relationship of glucose utilization and morphological change in the visual system in hexacarbon neuropathy, Brain Res., 22:447-451.

Herskowitz, A., Ishii, N. and Schaumberg, H.H., 1971, n-Hexane neuropathy. A syndrome occurring as a result of industrial exposure, New Engl. J. Med., 285:82-85.

Hirokawa, N., Glicksman, M.A. and Willard, M.B., 1984, Organization of mammalian neurofilament polypeptides within the neuronal cytoskeleton, J. Cell Biol., 98:1523-1536.

Ishikawa, H. and Tsukita, S., 1982, Morphological and functional correlates of axoplasmic transport, in: "Axoplasmic Transport," D.G. Weiss, ed., pp. 251-259, Springer-Verlag, Berlin-Heidelberg.

Jirmanova, I. and Lukas, E., 1984, Ultrastructure of carbon disulfide neuropathy, Acta Neuropathol. (Berl.) 63:255-263.

Jones, H.B. and Cavanagh, J.B., 1982, The early evolution of neurofilamentous accumulations due to 2,5-hexanediol in the optic pathways of the rat, Neuropathol. Appl. Neurobiol., 8:289-301.

Korobkin, R., Asbury, A.K., Summer, A.J. and Nielsen, S.L., 1975, Glue in sniffing neuropathy, Arch. Neurol. 32:158-162.

Lasek, R.J., 1980, A dynamic view of neuronal structure. Trends Neurosci., 3:87-91.

Lasek, J.J., 1982, Translocation of neuronal cytoskeleton and axonal locomotion, Philos. Trans. R. Soc. London (Biol.), 299:313-327.

Levine, J. and Willard, M., 1981, Axonally transported polypeptides associated with internal periphery of many cells, J. Cell Biol., 90:643-652.

Liepold, T.A., Monaco, S. and Gambetti, P., 1983, Optic axons in acrylamide intoxication, J. Neuropathol. Exp. Neurol., 42:331 (Abstract).

Monaco, S., Wongmongkolrit, T., Sayre, L., Autilio-Gambett, L. and Gambetti, P., 1984, Giant axonopathy in 3-methyl-2,5-hexanedione (3-M-2,5-HD), J. Neuropathol. Exp. Neurol. 43:304 (Abstract).

Monaco, S., Autilio-Gambetti, L., Zabel, D. and Gambetti, P., 1985, Giant axonal neuropathy. Acceleration of neurofilament transport in optic axons, Proc. Natl. Acad. Sci., 82:920-924.

Papsozomenos, S.Ch., Autilio-Gambetti, L. and Gambetti, P., 1981, Reorganization of axoplasmic organelles following β,β'-iminodipropionitrile administration, J. Cell Biol., 91:866-871.

Pappolla, M., Monaco, S., Weiss, H., Miller, C., Sahenk, Z., Autilio-Gambetti, L. and Gambetti, P., 1984, Slow axonal transport in carbon disulfide (CS_2) giant axonopathy, J. Neuropathol. Exp. Neurol., 43:305 (Abstract).

Prineas, J., 1969, The paghogenesis of dying-back polyneuropathies. Part II. An ultrastructural study of experimental intoxication in the cat, J. Neuropathol. Exp. Ther., 28:598-621.

Raine, C.S., 1982, Differences between the nodes of Ranvier of large and small diameter fibers in the PNS, J. Neurocytol., 11:935-947.

Sabri, M.I., Ederle, K., Holdsworth, C.E. and Spencer, C.S., 1979, Studies on the biochemical basis of distal axonopathies. II. Specific inhibition of fructose-6-phosphate kinase by 2,5-hexanedione and methyl-n-butyl ketone, Neurotoxicol., 1:285-298.

Sabri, M.I., Moore, C.L. and Spencer, C.S., 1979, Studies on the biochemical basis of distal axonopathies. I. Inhibition of glyceraldehyde-3-phosphate dehydrogenase by neurotoxic hexacarbon components, J. Neurochem., 32:683-689.

Sayre, L.M., Autilio-Gambetti, L. and Gambetti, P., 1985, Experimental giant axonopathies: A unified hypothesis of neurofilament modification of neurotoxic chemicals, Brain Res. Rev., in press.

Seppalainen, A.M. and Haltia, M., 1980, Carbon disulfide, in: "Experimental and Clinical Neurotoxicology," P.S. Spencer, H.H. Schaumberg, eds., pp. 356-373, Williams and Wilkins Co., Baltimore, MD.

Souyri, F., Chretien, M. and Droz, B., 1981, Acrylamide-induced neuropathy and impairment of axonal transport of proteins. I. Multifocal retention of fast transported proteins at the periphery of axons as revealed by light microscope radioautography, Brain Res., 205:1-13.

Spencer, P.S., Sabri, M.I., Schaumberg, H.H. and Moore, C.L., 1977, Does a defect of energy metabolism in the nerve fiber underlie axonal degeneration in polyneuropathies, Ann. Neurol., 5:501-507.

Spencer, P.S. and Schaumburg, H.H., 1977, Ultrastructural studies of the dying back process. III. The evolution of experimental peripheral giant axonal degeneration, J. Neuropathol. Exp. Neurol., 36:276-299.

Suzuki, K. and Pfaff, L., 1973, Acrylamide neuropathy in rats: An electron microscopy study of degeneration and regeneration, Acta Neuropathol. (Berl.), 24:197-213.

Tilson, H.A., 1981, The neurotoxicity of acrylamide: An overview, Neurobehav. Toxicol. Teratol., 3:445-461.

Tytell, M., Black, M., Garner, J. and Lasek, R.J., 1981, Axonal transport: each oif the major component consists of distinct macromolecular complex, Science, 214:179-181.

Tytell, M., Brady, S.T. and Lasek, R.J., 1984, Axonal transport of a subclass of T proteins: Evidence for the regional differentiation of microtubules in neurons, Proc. Natl. Acad. Sci., 81:1570-1574.

Willard, M., 1977, The identification of two intra-axonally transported polypeptides resembling myosin in some respect in the rabbit visual system, J. Cell Biol., 75:1-11.

Willard, M., Cowan, W.M. and Vegelos, P.R., 1974, The polypeptides composition of intraaxonally transported proteins: evidence for four transport velocities, Proc. Natl. Acad. Sci., 71:2183-2187.

Willard, M. and Simon, M., 1983, Modulations of neurofilament axonal transport during the development of rabbit retinal ganglion cells, Cell, 35:551-559.

Willard, M., Wiseman, M., Levine, J. and Skene, P., 1979, Axonal transport of actin in rabbit retinal ganglion cells, J. Cell Biol., 81:581-591.

THE EFFECT OF SOME DITHIOCARBAMATES, DISULFIRAM AND 2,5-HEXANEDIONE ON THE CYTOSKELETON OF NEURONAL CELLS IN VIVO AND IN VITRO

Veli-Pekka Lehto, Ismo Virtanen, and Kai Savolainen[*]

Department of Pathology
University of Helsinki
Helsinki, Finland

[*]Department of Environmental Hygiene and Toxicology
National Public Health Institute
Kuopio, Finland

INTRODUCTION

Dithiocarbamates

Dithiocarbamates are widely used fungicides because of their low toxicity to mammalian species and because of their rather rapid biodegradation in the soil and by plants. Their fungicidal action is based on interference with the citric acid cycle of several fungi (Schlagbauer and Schlagbauer, 1972). Several dithiocarbamates show neurotoxic effects in central and peripheral nervous system (Seppalainen and Linnoila, 1976; Freudenthal et al., 1977; Chernoff et al., 1979), but the mechanisms of the neurotoxicity of dithiocarbamates are largely unexplained.

Disulfiram

Disulfiram (tetraethylthiuram disulfide) is metabolized in mammals largely to diethyldithiocarbamate, a close structural analogue of the dithiocarbamates. DSF is used in the therapy of alcoholism by virtue of its aldehyde dehydrogenase antagonizing capacity. It is also used in large quantities as an accelerator in the rubber industry. Observations in humans and rodents have shown that long-term treatment with DSF can lead to peripheral neuropathy (Anzil, 1980). The mechanism of this neurotoxic effect is also not known.

2,5-Hexanedione

2,5-hexanedione is a metabolite of commonly used solvents n-hexane and methyl-n-butyl ketone (Di Vincenzo et al., 1976) which are known to induce distal axonopathies in the peripheral and central nervous system (Herskowitz et al., 1971; Allen et al., 1975). Administration of 2,5-hexanedione has been used widely as a experimental model system to study the molecular neurotoxic mechanism of this type of axonopathy (Graham, 1980; Spencer et al., 1980).

In the present studies we have investigated: 1) the effect of some dithiocarbamates, thiurams (e.g. thiram), and their metabolic products, e.g., ethylenethiourea (ETU), on the cytoskeleton of neuronal cells in vivo and in vitro; 2) the effect of disulfiram (belongs also to thiurams) on the cytoskeleton of neuronal cells in vivo and in vitro, and 3) the effect of 2,5-hexanedione on cultured mouse neuroblastoma cells (C-1300) in vitro.

MATERIALS AND METHODS

Animals and Treatments

The in vivo experiments were divided into two parts. In the first experiment two groups of male Wistar rats weighing about 140 g (WI rats) and 200 g (WII rats), and one group of Sprague-Dawley rats weighing 140 g (SD rats) (Tuohilampi Farm and Animal Centre, Finland) were used. Disulfiram, suspended in 1 percent gum acacia in water, was given daily by intragastric intubation to the rats for the duration of the experiment, namely 1 or 3 weeks. The WI and SD rats received 2.5 ml of the suspension twice daily for the first week, and those exposed thereafter 3.5 ml once daily. The daily doses were approximately 470 mg/kg in the first week, 245 mg/kg in the second week, and 220 mg/kg in the third week.

The WII rats were treated similarly except that they were given 5 ml of the suspension twice daily for the first week and then once daily. The daily doses of the WII rats were 580 mg/kg for the first week, 280 mg/kg for the second week, and 250 mg/kg for the third week. Controls received corresponding volumes of 1 percent gum acacia in water. Standard pellet diet (Hankkija, Finland) and tap water were offered ad libitum, and the animals were kept on artificial light/dark rhythm of 12 h. At the end of exposure, the animals were sacrificed and the left sciatic nerve removed as described later.

In the second series of in vivo experiments the chemicals were administered to male Wistar (Kuo/Mol) strain rats, weighing 160-220 g, by intraperitoneal injection either in physiological saline (ETU, nabam), or in physiological saline containing 1 percent gum acacia (other agents). The dosing schedule is given in Table 1. Control rats were given the respective vehicle. The animals were kept in plastic animal cages where no pesticides were used. They received standard pellet diet (Hankkija, Finland) and tap water ad libitum; and were kept in an artificial light/dark rhythm of 14/10 h. The agents were injected as 4 daily doses on succeeding days, and rats were decapitated in the morning of the 5th day. The left sciatic nerve was immediately removed and divided into different samples for further analyses.

Chemicals

Analytical grade disulfiram (tetraethylthiuram disulfide, DSF) and N,N-ethylenethiourea (ETU) were from Fluka AG (Switzerland), and maneb (Ma), nabam (Na), zineb (Zin), ferbam (Fer), thiram (Thi), and ziram (Zir) were obtained from Ehrenstorfer (FRG).

Cells and Treatments

Murine C1300 neuroblastoma cells (clone $NB_{42}B$), (Augusti-Tocco and Sato, 1969) were kindly provided by Professor N. Ringertz (Stockholm, Sweden). The cells were cultured in Eagle's minimal essential medium (MEM) supplemented with 10 percent fetal calf serum (Flow Laboratories, Irvine, Scotland) and antibiotics. In order to induce morphologic differentiation, the neuroblastoma cells were cultured for up to 7 days in the presence of 1 mM N_5, O^8-dibutyryl adenosine 3'5'-cyclic monophosphate (db-cAMP, Sigma

Table 1. Cumulative Doses of Chemicals Given Intraperitoneally, Once Daily, for 4 Succeeding Days.

Chemical	Cumulative dose (mg/kg)	Dosage regimen
Control	-	(4 x saline i.p.)
Disulfiram	800	(200+200+200+200)
Ethylenethiourea	800	(200+200+200+200)
Ferbam[+]	240	(100+ 80+ 30+ 30)
Thiram[+]	80	(30+ 30+ 10+ 10)
Ziram[+]	20	(10+ 10+ 0+ 0)
Maneb[+]	600	(200+200+100+100)
Nabam	60	(15+ 15+ 15+ 15)
Zineb	800	(200+200+200+200)

Each dose was given in the morning, and the rats were decapitated in the morning of the 5th day. All the intraperitoneal doses were calculated as 4 percent of the peroral LD50-values (mg/kg) given in the literature. [+]The dose was decreased or the number of doses was decreased (ziram) because of overt toxicity of the chemical or because of the different p.o./i.p. ratios of different agents.

Chemical Co., St. Louis, USA), (Prasad and Hsie, 1971). Some cultures were exposed to 10 µg/ml of vinblastine sulphate (Eli Lilly Co., Indianapolis, USA) for 2 hours. For cold treatment, the cells were kept in $0°C$ medium for 60 min, washed in cold phosphate buffered saline (PBS, pH 7.2) and then fixed in methanol, cooled to $-20°C$. In order to expose the cells to the chemicals, the cells were cultured in the presence of 1 µg/ml or 10 µg/ml of ETU, DSF and the various dithiocarbamates (see above) dissolved in dimethyl sulfoxide (Merck AG, Darmstadt, FRG) or in the presence of 100 µg/mg of 2,5-hexanedione (Merck).

Electron Microscopy

Animals were sacrificed by decapitation after one or three weeks of the treatment. The popliteal segment of the left sciatic nerve was carefully removed and immersed in fresh 2.5 percent glutaraldehyde (TAAB, Reading, England) in 0.1 M sodium cacodylate buffer, pH 7.2, dehydrated in ethanol, immersed in propylene oxide and embedded in Epon 812. Sections were cut and stained with toluidine blue and examined by light microscopy. On the basis of this preliminary screening, representative fields were chosen, of which ulrathin sections were made, stained with uranyl acetate and lead citrate, and examined in a Jeol 100CX electron microscope at the acclerating voltage of 60 kV.

Antibodies

Production and specificity of polyclonal antibodies to vimentin (pc anti-vim) and monoclonal antibodies to vimentin (mc anti-vim) have been described previously (Virtanen et al., 1981; Lehtonen et al., 1983). Monoclonal antibodies against nonerythroid spectrin were raised against bovine lens spectrin (Lehto et al., 1983). Monoclonal antibodies to neurofilaments (mc anti-nf) were produced by immunizing mice with neurofilament polypeptides isolated from bovine spinal cord (Virtanen et al., 1984). Hybridomas were screened using a solid phase immunoassay with purified neurofilament polypeptides, immunofluorescence microscopy and western blotting technique. The monoclonal antibodies selected for this study recognized the 200,000 kilodalton subunit protein of the neurofilament triplet (Fig. 1). Monoclonal antibodies to α-tubulin and β-tubulin (anti-tub) were purchased from Amersham Int. (Amersham, England) and were used in immunofluorescence microscopy in a 1:1 mixture.

Immunofluorescence Microscopy (IFL)

For indirect IFL small pieces of the left sciatic nerve of the exposed animals were fixed in ethanol, cooled to -20°C, dehydrated, and then embedded in paraffin in a routine manner. This procedure gives good preservation of the cytoskeletal antigens (Altmannsberger et al., 1981). Five micron thick sections were allowed to attach firmly to the glass slides by incubating them at 37°C over night. After deparaffinization, the sections were exposed to anti-tub or mc-anti-nf antibodies followed by fluorescein isothiocyanate (FITC)-conjugated goat anti-mouse IgG (Cappel Labs, Cochraneville, PA).

For indirect IFL of the cultured cells, the cells were fixed in methanol, at -20°C, for 15 min. Thereafter, the cells were exposed either to anti-vim, anti-spectrin or anti-tub antibodies as described above. The tissue sections and the cells were examined in a Zeiss Universal microscope equipped with an epi-illuminator IIIRS and appropriate filters for FITC-fluorescence.

RESULTS

The Effect of Dithiocarbamates on Neuronal Cells In Vivo

The effects of ETU and the dithiocarbamates were studied using light microscopy of the toluidine blue-stained Epon sections of the sciatic nerve

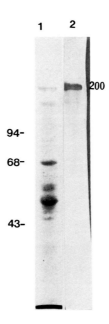

Fig. 1. Western blot of the polypeptides of isolated bovine spinal cord reacted with monoclonal anti-neurofilament antibodies. Note the distinct reaction with a polypeptide of 200,000 daltons.
Lane 1: protein staining. Lane 2: Immunoblot.

specimens. On the basis of the light microscopic evaluation, representative fields of the nerve specimens were chosen for electron microscopic investigation. The results of the light and electron microscopic observations are compiled in Table 2. Only minor alterations were observed in experimental animals when compared with the controls. The alterations were infrequent and are suggestive of nonspecific degenerative processes (Fig. 2).

No major morphological alterations were seen by immunofluorescence microscopy, with anti-nf and anti-tub antibodies, of specimens taken from the experimental animals (Fig. 3).

The Effect of DSF on Neuronal Cells In Vivo

Occasional morphological alterations suggestive of early degenerative changes could be seen by light microscopy of the Epon sections of the sciatic nerve specimens; these included slight swelling of the axons, some organelle loss and slight degeneration, probably artefactual, of the myelin sheaths (Fig. 4). By electron microscopy, a slight decrease in the filament density, and more consistently, a clear decrease in the number of microtubules could be seen (Fig. 5). In order to obtain an estimate of the extent of the axonal changes, a quantitative analysis of the cross sections of the axons was performed. Three randomly selected fields of four sections of a sciatic nerve were photographed and the number of microtubules, neurofilaments and all filament cross sections was counted in four 0.8 cm^2 randomly selected areas of three to four axons per photograph. The areas were selected from both the peripheral and central portions of the axons. Only axons with similar diameter were included in the analysis (cf. Pannese et al., 1984). Neurofilaments and microtubules could be easily distinguished by their characteristic morphology and diameter.

There was a significant decrease in the microtubule/neurofilament ratio in the treated rat as compared to the control rat (Table 3). The reason for this was apparently a decrease in the number of microtubules, since the density of neurofilaments seemed to be about the same in the treated and the control rat.

The Effect of Dithiocarbamates on Cultured C1300 Cells

Mouse C1300 neuroblastoma cells were allowed to grow to subconfluence. Thereafter db-cAMP was added to some culture dishes and the cells were incubated for an additional 48 h, by which time a distinct morphologic alteration to neural-like cells appeared: the cells extended long bipolar or multipolar neurites while the uninduced cells remained roundish (Fig. 6) (See also Fig. 8 A-F).

Table 2. Light and Electron Microscopic Observations on the Sciatic Nerves of Rats Exposed to ETU and Various Dithiocarbamates.

Morphologic alteration	ETU	Fer	Thi	Chemical Zir	Ma	Na	Zin
Axonal swelling	--	slight	slight	--	--	--	--
Myelin figures	some	some	--	some	--	--	--
Atrophy	--	--	--	--	--	--	--
Number of neurofilaments	normal	normal	normal	normal	normal	normal	normal
Number of microtubules	normal	normal	normal	normal	normal	normal	normal

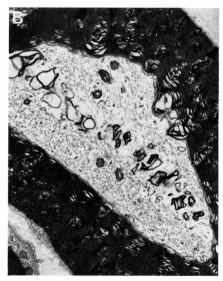

Fig. 2. Electron micrograph of sciatic nerves of rats exposed to ziram (A) and ferbam (B). Note the myelin figures in the axons (A, x 18,000; B, x 20,000; original magnifications).

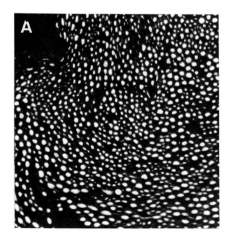
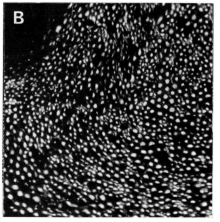

Fig. 3. IFL micrographs of sciatic nerves of rats exposed to ziram and stained for neurofilaments (A) and tubulin (B) (A,B x 120; original magnification).

At confluence, ETU and the various dithiocarbamates were added to the culture dishes of uninduced and induced cells at concentrations 1 µg/ml and 10 µg/ml in DMSO. As a control, DMSO alone was added to some cultures. The growth of the cells was monitored at 6 h intervals. The effect of the chemicals on the cytoskeletal organization was evaluated after 12 h by fixing the cells and staining them with anti-tub, anti-vim and anti-spectrin antibodies for IFL-microscopy.

In uninduced cultures, the chemicals that cause most prominent morphological alterations (rounding, cell detachment) were ferbam, thiram, zineb and ziram. With these agents, the effects were seen already at the

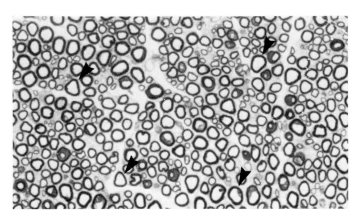

Fig. 4. Light micrograph (Epon-section stained with toluidine blue) of sciatic nerve from a rat exposed to DSF. Note the slight swelling of some axonal profiles (x 400; original magnification).

lower concentration. On the other hand, maneb and nabam seemed to have no or only a slight effect; ETU seemed to have an intermediate effect. When the control cells (DMSO-exposed) were compared with the ferbam-, zineb- and ziram-exposed cells by IFL microscopy after anti-tub staining, a reorganization of the microtubule network could be seen in the latter cells; microtubules showed a tendency to form dense, perinuclear accumulations (not shown).

A similar kind of perinuclear accumulation of intermediate filaments was seen in uninduced cells treated with ferbam, thiram, zineb and ziram. Also in ETU-treated cells some perinuclear accumulation of intermediate filaments could be seen (not shown). No change was seen in the organization of the structures reactive with anti-spectrin antibodies (not shown).

The effect of various dithiocarbamates on the db-cAMP-induced cells was to a large extent similar to that on uninduced cells. Also in the case of induced cells, ferbam, thiram, zineb and ziram seemed to have the most pronounced effect. All these chemicals brought about the retraction of the axon-like extensions and accumulation of both microtubules and intermediate filaments either to perinuclear masses or coils (Fig. 6 C-F). In addition, ziram seemed to detach most of the cells even at the lower concentration. Maneb and nabam had no pronounced effects while ETU had an intermediate effect.

The Effect of DSF on C1300 Cells

In db-cAMP cells, DSF caused rounding up of the cells and led to a distinct accumulation of tubulin in discrete clumps, which were often seen as cellular protrusions, and to a perinuclear coiling of the intermediate filaments (Fig. 7).

The Effect of 2,5-Hexanedione on Cultured C1300 Cells

Exposure of the db-cAMP-induced C1300 cells to 2,5-hexanedione led to a rapid reorganization of intermediate filaments. Already after 2 h of exposure the vimentin-specific fluorescence could been seen to be retracted from the long neurite-like extensions of control cells to distinct juxtanuclear aggregates (Fig. 8 G,H). The neurite-like extension were preserved, however, and the staining of the db-cAMP-induced cells with anti-tub antibodies demonstrated that the microtubule system of these extensions remained intact (Fig. 8 I,J).

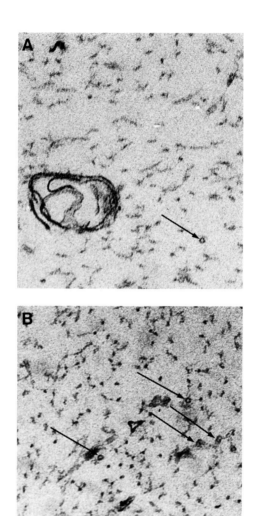

Fig. 5. Electron micrographs of sciatic nerves from a rat exposed to DSF (A) and a control rat (B). Note the decreased number of microtubules (arrows) in (A) (A,B x 80,000; original magnification).

Table 3. Microtubule: Neurofilament Ratio (100x mt/nf) in the Sciatic Nerve of DSF-Treated and Control WI-Rats.

DSF-treated	Exposed
2.4 ± 0.9% (n = 6)	6.5 ± 2.6% (n = 4)

n = number of sections

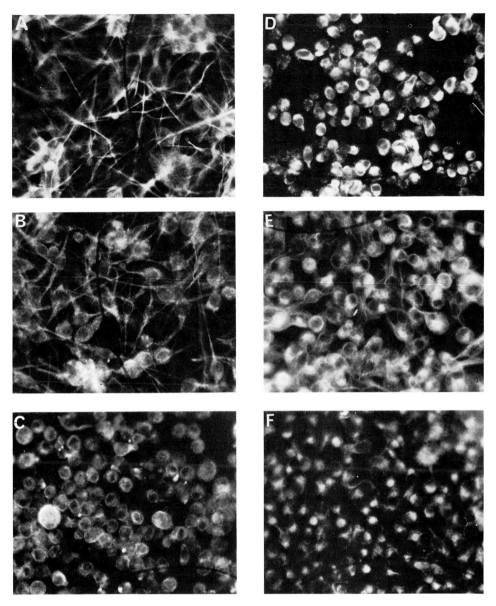

Fig. 6. IFL-micrographs of db-cAMP-induced C1300 neuroblastoma cells cultured in the presence of DMSO alone (A,B), ferbam (C,D), and thiram (E,F) and stained for tubulin (A,C,E) and vimentin (B,D,F) (A-F, x 500; original magnification).

DISCUSSION

Dithiocarbamate

The toxicity profile of different types of dithiocarbamates closely resemble each other. The acute toxicity of these chemicals appears to be rather low except for nabam, ziram and thiram (Autio, 1982). The targets of the long-term effect seems to be the thyroid gland, neural tissue and testes (Hodge et al., 1952; Graham and Hansen, 1972; Fishbein, 1976; Lee et al.,

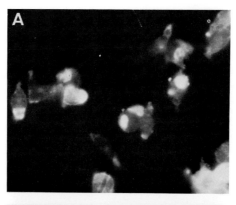
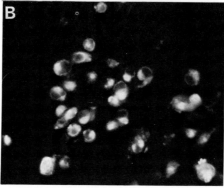

Fig. 7. IFL-micrographs of db-cAMP-induced C1300 neuroblastoma cells cultured in the presence of DSF and stained for tubulin (A) and vimentin (B) (A, x 600; B, x 400; original magnifications).

1978). We have not, however, any coherent view of the mechanisms of these toxic effects largely due to our lack of knowledge of the kinetics and various metabolites of these chemicals.

The results of the present study show that intraperitoneal administration of various dithiocarbamates and ETU had only a slight effect on peripheral nerves as judged by electron microscopy and IFL microscopy. With IFL microscopy no segregation of microtubule structures to the center of the axonal profiles was seen (cf. Griffin et al., 1983). With electron microscopy of ferbam- and thiram-exposed animals a slight swelling of the axons and some degenerative alterations was seen. Preliminary quantitative evaluation did not reveal any pronounced alteration in the number of neurofilaments and microtubules per cross section. In this regard, the results are different from those obtained with disulfiram, a structural analogue of dithiocarbanates, which seems to bring about a decrease in the number of microtubules in peripheral nerves of the exposed animals (Savolainen et al., 1983).

With IFL microscopy of the cells exposed to dithiocarbamates and ETU, nonspecific cytopathic changes were seen. In addition to these, especially ferbam and thiram seemed to have a pronounced effect on the morphology and structure of both microtubules and intermediate filaments which may be of pathogenetic significance. Further studies are under way to elucidate the effect of these agents on microtubules and intermediate filaments also at a biochemical level. It also will be of interest to determine whether the metals carried as complexes with some dithiocarbamates contribute the adverse effects on the cytoskeleton.

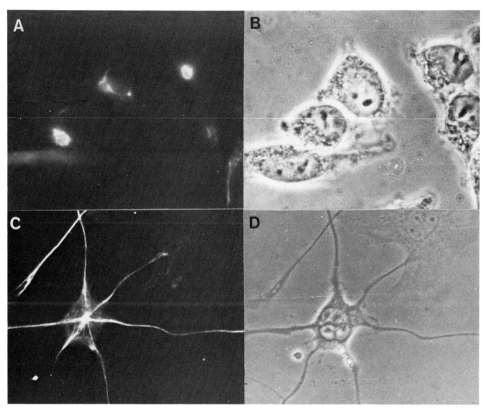

Fig. 8. Undifferentiated mouse C1300 neuroblastoma cells were stained for vimentin (A) and tubulin (E); differentiated cells were stained for vimentin (C) and tubulin (F). (B) and (D) are phase contrast micrographs of the same fields as (A) and (C), respectively. Differentiated cells cultured in the presence of 2,5-hexanedione were stained for vimentin (G) and tubulin (I). (H) and (J) are phase contrast micrographs of the same fields as (G) and (I), respectively (A-J, x 600; original magnifications).

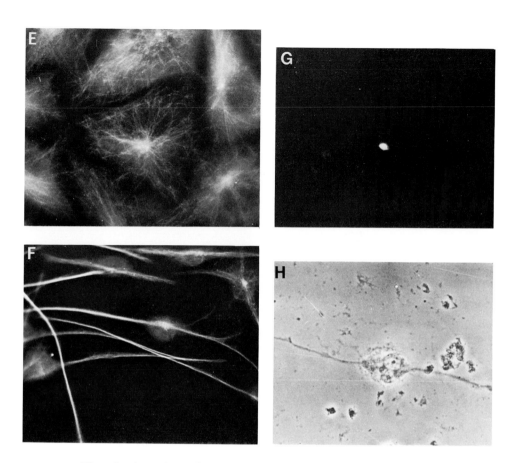

Fig. 8. (continued)

Disulfiram

Disulfiram intoxication induces peripheral neuropathy (Sterman and Schaumburg, 1980; Moddel et al., 1978; Mokri et al., 1981). Disulfiram, and obviously thiram, ferbam and ziram are metabolized to diethyldithiocarbamate, further to CS_2, and then to carbonyl sulfide (COS) (Faiman et al., 1980; Sauter and von Wartburg, 1977; Hodgson et al., 1975). CS_2 and COS are very reactive and bind readily to cellular macromolecules possibily leading to lipid peroxidation and membrane damage (Heikkila et al., 1976; Kappus and Sies, 1981; Chengelis and Neal, 1980; Savolainen et al., 1980). Hydrogen sulfide (H_2S) has been claimed to be responsibile for the high toxicity of COS (Dalvi et al., 1975).

Rainey (1977) has emphasized the similarities between DSF and CS_2 neuropathy, and they have been reported to induce a similar neuropathy in experimental animals and man (Seppalainen and Haltia, 1980). The mechanism of the CS_2 neurotoxicity has been suggested to be mediated via its interference with neurofilaments (Savolainen et al., 1977; Seppalainen and Haltia, 1980; Sterman and Schaumburg, 1980).

The present results show, for the most part, DSF-induced axonal changes in the peripheral nerves of the exposed animals. A notable feature of these morphological alterations was a decrease in the number of microtubules. No major change in neurofilament number or morphology was seen. This finding is at variance with the suggestion that CS_2 was the causative agent in DSF

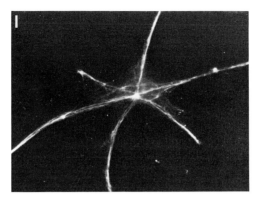

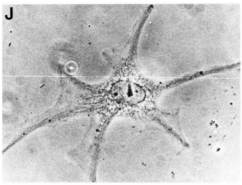

Fig. 8. (continued)

intoxication (Ausbacher et al., 1982). Our findings are in agreement with the findings of Anzil (1980) and Bouldin et al., (1980) on impoverishment of microtubules in the axons of myelinated peripheral nerves of DSF-exposed rats.

The molecular mechanisms leading to a decrease in microtubules in DSF exposed animals is not known. To some extent this phenomenon is reminiscent of the effect of β,β'-iminodipropionitrile (IDPN) toxicity, which is characterized by segregation of microtubules from neurofilaments (cf. Griffin et al., 1983). In DSF neurotoxicity there are, however, no signs of microtubule segregation. It may be hypothesized that disulfiram or its metabolites may have a direct effect, either depolymerizing or inhibitory, that could interfere with the microtubule component of the axonal flow and, in this way, affect the normal physiology of both peripheral and autonomic nerves (Savolainen et al., 1983). This view is supported by the IFL-data from cultured cells exposed to DSF that showed microtubule disorganization that goes beyond the alterations that can be normally seen upon cell rounding and detachment.

2,5-Hexanedione

Neuropathies caused by n-hexane, methyl-n-butyl ketone or by their metabolite 2,5-hexanedione, are characterized by axonal swellings with large neurofilament aggregates (Spencer et al., 1980; Griffin et al., 1983). Usually the swellings can be seen in the distal axons and 2,5-hexanedione has been considered as a model system for studying the mechanisms of hexacarbon neuropathy and other distal axonopathies (cf. Griffin et al., 1983). Recently, a common pathogenetic pathway involving interference with the interaction between microtubules and neurofilaments have been suggested for neuropathies caused by 2,5-hexanedione and β,β'-iminopropionitrile (Griffin et al., 1983).

In the present study we examined the effect of 2,5-hexanedione on cultured mouse neuroblastoma cells which were induced to differentiate by using db-cAMP. These cells express, in culture conditions, only vimentin-type of intermediate filaments but not neurofilaments (Virtanen et al., 1984). They express, however, neuronal properties and have been widely used as a model system to study the growth and other phenotypic properties of neuronal cells (Prasad, 1975).

The results of the present study show that 2,5-hexanedione brought about a perinuclear redistribution of the vimentin filaments without affecting the integrity of the microtubule framework or without causing the retraction of the axon-like extensions. The results therefore suggest that the reorganization of the intermediate filaments can take place without morphological alterations in microtubules. Thus, the primary target of the drug may be intermediate filament themselves or some cytoskeletal components that interact with both intermediate filaments and microtubules. In the light of these results, it is of interest that normal human fibroblasts seem to respond in a similar fashion to 2,5-hexanedione exposure (Durham et al., 1983).

REFERENCES

Allen, N., Mendell, J.R., Billmaier, D.J., Fontaine, R.E., and O'Neill, J., 1975, Toxic polyneuropathy produced by industrial solvent methyl-n-butyl ketone, Arch. Neurol., 32:209-218.
Altmannsberger, M., Osborn, M., Schauer, A., and Weber, K., 1981, Antibodies to different intermediate filament proteins. Cell type-specific markers on paraffin-embedded human tissues, Lab. Invest., 45:427-434.
Ausbacher, L.E., Bosch, E.P., and Cancilla, P.A., 1982, Disulfiram neuropathy: a neurofilamentous distal axonopathy, Neurology, 32:424-428.
Anzil, A.P., 1980, Selected aspects of experimental disulfiram neuromyopathy, in: "Advances in Neurotoxicology," L. Manzo, I. Laxasse and L. Roche, eds., pp. 359-366, Pergamon Press, New York.
Augusti-Tocco, G., and Sato, G., 1969, Establishment of functional clonal lines of neurons from mouse neuroblastoma, Proc. Natl. Acad. Sci., 64:311-315.
Autio, K., 1982, Ethylenethiourea: metabolism, analysis and aspects of toxicity, Dissertation, Technical Research Centre of Finland, Research Report 91, Espoo.
Bouldin, T.W., Hall, C.D., and Krigman, M.R., 1980, Pathology of disulfiram neuropathy, Neuropath. Appl. Neurobiol., 6:155-160.
Chengelis, C.P., and Neal, R.A., 1980, Studies of carbonyl sulfide toxicity: metabolism by carbonic anhydrase, Toxicol. Appl. Pharmacol., 55:198-204.
Chernoff, N., Karlock, R.I., Rogers, E.H., Carver, B.D., and Murray, S., 1979, Perinatal toxicity of maneb, ethylene thiourea and ethylenebisisothiocyanate sulfide in rodents, J. Toxicol. Environ. Health, 5:821-834.
Dalvi, R.R., Hunter, A., and Neal, R.A., 1975, Toxicological implications of the mixed function oxidase catalyzed metabolism of carbon disulfide, Chem.-Biol. Interact., 10:347-361.
Di Vincenzo, G., Kaplan, C.J., and Dedinas, J., 1976, Characterization of the metabolites of methyl-n-butyl ketone, metyl isobutyl ketone and metyl ethyl ketone in guinea-pig serum and their clearance, Toxicol. Appl. Pharmacol., 36:511-522.
Durham, H.D., Pena, S.D.J., and Carpenter, S., 1983, The neurotoxins 2,5-hexanedione and acrylamide promote aggregation of intermediate filaments in cultured fibroblasts, Muscle and Nerve, 6:631-637.

Faiman, M.D., Artman, L., and Haya, K., 1980, Disulfiram distribution and elimination in the rat after oral and intraperitoneal administration, Alcoholism: Clin. Exp. Res., 4:412-419.
Fishbein, L., 1976, Environmental health aspects of fungicides, I. Dithiocarbamates, J. Toxicol. Environm. Health., 1:713-735.
Freudenthal, R.I., Kerchner, G.A., Persing, R.L., Baumell, I., and Baron, R.L., 1977, Subacute toxicity of ethylenbisisothiocyanate sulfide in the laboratory rat, J. Toxicol. Environ. Health, 2:1067-1078.
Graham, D.G., 1980, Hexane neuropathy: a proposal for pathogenesis of a hazard of occupational exposure and inhalant abuse, Chem.-Biol. Interactions, 32:339-345.
Graham, S.L., and Hansen, W.H., 1972, Effects of short-term administration of ethylenethiourea upon thyroid function of the rat, Bull. Environ. Contam. Toxicol. (U.S.), 7:19-25.
Griffin, J.W., Price, D.L., and Hoffman, P.N., 1983, Neurotoxic probes of the axonal cytoskeleton, Trends in Neurol. Sci., 6:490-495.
Heikkila, R.E., Cabbat, F.S., and Cohen, G., 1976, In vivo inhibition of superoxide dismutase in mice by diethyldithiocarbamate, J. Biol. Chem., 251:2182-2185.
Herskowitz, A., Ishii, N., and Schaumburg, H.H., 1971, n-Hexane neuropathy: A syndrome occurring as a result of industrial exposure, N. Engl. J. Med., 285:82-85.
Hodge, H.C., Maynard, E.A., Downs, W., Blanchet, H.J., and Jones, C.K., 1952, Acute and short-term oral toxicity tests of ferbam and ziram, J. Am. Pharm. Assoc., 41:662-666.
Hodgson, J.R., Hoch, J.C., Castles, T.R., Helton, D.O., and Lee, C.-C., 1975, Metabolism and disposition of febram in the rat, Toxicol. Appl. Pharmacol., 33:505-513.
Kappus, H., and Sies, H., 1981, Toxic drug effects associated with oxygen metabolism: Redox cycling and lipid peroxidation, Experientia, 37:1233-1241.
Lee, C.-C., Russell, J.Q., and Minor, J.L., 1978, Oral toxicity of ferric dimethyldithiocarbamate (ferbam) and tetramethylthiourea disulfide (thiram) in rodents, J. Toxicol. Environm. Health, 4:93-106.
Lehto, V.-P., and Virtanen, I., 1983, Immunolocalization of a novel, cytoskeleton-associated polypeptide of Mr 230,000 daltons (p230), J. Cell Biol., 96:703-716.
Lehtonen, E., Lehto, V.-P., and Virtanen, I., 1983, Parietal and visceral endoderm differ in their expression of intermediate filaments, EMBO J., 2:1023-1028.
Moddel, G., Bilbao, J.M., Payne, D., and Ashby, P., 1978, Disulfiram neuropathy, Arch. Neurol., 35:658-660.
Mokri, B., Ohnishi, A., and Dyck, P.J., 1981, Disulfiram neuropathy, Neurology, 31:730-735.
Pannese, E., Procacci, P., Ledda, M., Arcidiacono, G., and Rigamonti, L., 1984, A quantitative study of microtubules in motor and sensory axons, Acta. Anat., 118:193-200.
Prasad, K.N., 1975, Differentiation of neuroblastoma cells in culture, Biol. Rev., 50:129-165.
Prasad, K.N., and Hsie, A.W., 1971, Morphologic differentiation of mouse neuroblastoma cells induced in vitro by dibutyryl adenosine 3'5'-cyclomonophosphate, Nature New Biol., 233:141-142.
Rainey, J.M., 1977, Disulfiram toxicity and carbon disulfide poisoning, Am. J. Psychiatry, 134:371-378.
Sauter, M., and von Wartburg, J.P., 1977, Quantitative analysis of disulfiram and its metabolites in human blood by gas-liquid chromatography, J. Chromatogr., 133:167-172.
Savolainen, K., Hervonen, H., Lehto, V.-P., and Mattila, M.J., 1983, Neurotoxic effects of disulfiram on autonomic and peripheral nervous system in rat, Toxicol. Lett. Suppl., 1:34.

Savolainen, H., Lehtonen, E. and Vainio, H., 1977, CS_2 binding to rat spinal neurofilaments, Acta Neuropathol., 37:219-224.

Savolainen, H., Tenhunen, R., Elovaara, E., and Tossavainen, A., 1980, Cumulative biochemical effects of repeated subclinical hydrogen sulfide intoxication in mouse brain, Int. Arch. Environ. Health., 46:87-92.

Schlagbauer, B.G.L., and Schlagbauer, A.W.J., 1972, The metabolism of carbamate pesticides, Residue Rev., 42:1-84.

Seppalainen, A.M., and Linnoila, I., 1976, Electrophysiological findings in rats with experimental carbon disulfide neuropathy, Neuropath. Appl. Neurobiol., 2:209-215.

Seppalainen, A.M., and Haltia, M., 1980, Carbon disulfide, in: "Experimental and Clinical Neurotoxicology," P.S. Spencer and H.H. Schaumburg, eds., pp. 356-373, Williams and Wilkins, Baltimore.

Spencer, P.S., Schaumburg, H.H., Sabri, M.I., and Veronesi, B., 1980, The enlarging view of hexacarbon neurotoxicity, CRC Crit. Rev. Toxicol., 7:279-350.

Sterman, A.B., and Schaumburg, H.H., 1980, Neurotoxicity of selected drugs, in: "Experimental and Clinical Neurotoxicity," P.S. Spencer and H.H. Schaumburg, eds., pp. 593-612, Williams and Wilkins, Baltimore.

Virtanen, I., Lehto, V.-P., Lehtonen, E., Vartio, T., Stenman, S., Kurki, P., Wager, O., Small, J.V., Dahl, D., and Badley, R.A., 1981, Expression of intermediate filaments in cultured cells, J. Cell Sci., 50:45-63.

Virtanen, I., Miettinen, M., Lehto, V.-P., Kariniemi, A.-L., and Paasivuo, R., 1984, Diagnostic application of monoclonal antibodies to intermediate filaments, Ann. N.Y. Acad. Sci., in press.

NEUROFIBRILLARY DEGENERATION: THE ROLE OF ALUMINUM

Michael L. Shelanski

Department of Pharmacology
New York University School of Medicine
New York, New York, USA

INTRODUCTION

Studies on the effects of drugs and toxic agents on the cytoskeleton have their origin in experimental neuropathology. Pathologists of the nervous system have long recognized the effects of a variety of agents on the argentophilic neurofibrils and classified them under the rubric of "neurofibrillary proliferation."

The role of cytoskeletal rearrangements in nonneural cells was not appreciated until the electron microscopic studies of Robbins and Gonatas (1964) showed bundling of intermediate filaments as well as loss of microtubules in Hela cells arrested in metaphase by treatment with anti-mitotic drugs. At about the same time, it became clear that the neurofibrils seen with the light microscope were the counterparts of neurofilaments seen with the electron microscope.

Aluminum has not been commonly considered to be a neurotoxin. Even now, it is debatable whether dietary aluminum, regardless of source, is involved in either the etiology or pathogenesis of neurological disease. However, aluminum applied directly to the surface of the brain in the form of alumina paste causes the formation of a focus of epileptic activity. This "aluminum model" is widely employed in experimental studies of the physiology and pharmacology of seizure disorders.

When brains which have been exposed to topical or intrathecal aluminum are examined histologically, the neurons show displacement of the Nissl substance and intensely argentophilic fibrillary bundles in their cytoplasm and initial axonal segment (Klatzo et al., 1965). Viewed with the light microscope, these fibrillar changes are quite similar to the neurofibrillary tangles seen in Alzheimer's disease. However, upon electron microscopic observation, it is clear that the aluminum-induced filaments are of the unpaired 10 nm diameter type, apparently identical to normal neurofilaments in appearance, while the filaments in Alzheimer's disease are paired structures with a "twist" or node visible every 80 nm (Kidd, 1963; Terry, 1963). In spite of the difference in structure at the electron microscopic level, the similarity in appearance at the light microscopic level has led to numerous investigations of aluminum in the brains of Alzheimer patients and brought aluminum under suspicion of being an etiological agent in this disease.

Ultrastructrual studies in the late 1960s and early 1970s revealed bundling and apparent proliferation of 10 nm filaments in neuronal cells in response to any of a large number of chemical agents including colchicine, vinca alkaloids, acrylamide, lathyrogens and others. Similar morphologic changes were seen in Wallerian degeneration, amyotrophic lateral sclerosis, sporadic motor neuron disease, and in the neuropathy that occurred as a side effect of vincristine or vinblastine therapy of leukemias (Wisniewski et al., 1971).

These changes did not always affect the cell body and often showed some specificity as to cell type. Certain agents, such as the antimitotics, cause cytoskeletal rearrangements in most cell types, while others, typified by aluminum, cause changes only in neurons.

In order to understand the diverse pathways by which these agents cause cytoskeletal rearrangements it is necessary to understand the structure of the cytoskeleton, the interactions between its components, and the mechanisms by which these interactions are mediated. While our knowledge of these factors is far from complete, enough data is available to enable us to begin such an analysis.

FILAMENTS, TUBULES AND CROSSBRIDGES

Though there was an awareness of a fibrillar matrix in the neuron in the 19th century, it was not until the introduction of electron microscopy that the richness and complexity of cytoplasmic structure became apparent. Initial studies concentrated on the morphologically well defined and abundant neurofilaments and microtubules, but even these images showed a less constant network of fibers of varying size which appeared to connect the linear elements of the cytoplasm with one another.

As confidence in electron microscopic images increased, links were described between adjacent filaments (Hirokawa, 1982), between adjacent tubules (Kohno, 1964; Paley et al., 1968), and between filaments and tubules (Hinkley, 1973; Tsukita and Ishikawa, 1980). Microtubules were also seen to be linked to mitochondria by distinct bridges (Raine et al., 1971; Smith et al., 1977) and to synaptic vesicles (Smith, 1971; Gray, 1975). More recently links between the cytoskeleton and the membrane have been observed (Westrum and Gray, 1977; Fifkova and Delay, 1982). These observations led to the proposal of functional analogies with both muscle and the microtubule-dynein system in cilia. In the latter, the radial arms projecting from the axonemal microtubules are composed of an ATPase, dynein, which generates the force for ciliary bending by binding to the adjacent pair of microtubules and undergoing a conformational change (Summers and Gibbons, 1973).

While the totality of these crosslinks has been referred to as the microtrabeculae or "microtrabecular lattice" (Wolosewick and Porter, 1979), it has become clear that at least a large fraction of the links are composed of proteins which are familiar to us from biochemical studies. These include the microtubule accessory proteins (Kim et al., 1979; Sloboda and Rosenbaum, 1979) and the higher molecular weight components of the neurofilament triplet as well as others.

Biochemical studies have reflected the diversity of morphological structures and have begun to reveal their composition. Microtubules purified by cycles of assembly and disassembly (Weisenberg, 1972; Shelanski et al., 1973) co-purify with high molecular weight (Murphy et al., 1977) proteins which lack apparent ATPase activity (Gaskin et al., 1974; Burns and Pollard, 1974), but do form lateral projections or arms on the microbubule (Kim et al., 1979; Sloboda and Rosenbaum, 1979). The 200,000 dalton protein of the neurofilament triplet is laterally located on the filament (Willard and Simon, 1981; Sharp et al., 1982) and appears to bridge one filament to another. The high molecular weight microtubule associated proteins (MAPs) are capable of binding to neurofilaments as well as microtubules (Leterrier et al., 1982) and appear to mediate filament-tubule interactions in vitro (Runge et al., 1981). Similar MAP-mediated interactions between actin microtubules have also been reported (Griffith and Pollard, 1978; Satillaro et al., 1981). Immunological cross-reactivity has also been reported between MAPs and the α subunit of spectrin suggesting common features in these cytoplasmic "linkers" (Davis and Bennett, 1982).

The diversity of links could function in the formation of stable, metastable, and labile interconnections between cytoskeletal elements. These could be regulated by phosphorylation (Leterrier et al., 1981) or some of the links could possess a force-generating ATPase.

POTENTIAL SITES FOR TOXIC ACTION ON THE NEURONAL CYTOSKELETON

The multiplicity of structures and proteins in the neuronal cytoskeleton suggest numerous sites at which toxic substances may act. To date, most toxic effects on the cytoskeleton have caused morphological changes in the cytoskeleton, but it is clear that "freezing" the cytoskeleton in a morphologically perfect state would also result in the disruption of the dynamic processes subserved by it.

Since early observations showed the formation of bundles of filaments in the cytoplasm in response to agents which induced microtubule depolymerization such as colchicine or vinblastine (Robbins and Gonatas, 1964; Wisniewski et al., 1968), it was clear that removal of microtubules caused a rearrangement of filaments either directly by releasing an elastic meshwork from its constraints or indirectly as the entire cytoskeleton reorganized. Since the loss of microtubules is rapid while the bundling of filaments or capping of preexisting filament bundles is not complete for one or more days (Blose et al., 1976), it is likely that the rearrangements are more complex than a "snapping" together of rubber band-like filaments. Nonetheless, the loss of the linkage site normally found on the intact microtubule may play an important initiating role.

However attractive the colchicine model is, it is not generalizable to all toxic neurofibrillary changes since in many of these, including experimental aluminum encephalopathy, microtubules remain intact in the presence of the toxic agent. Furthermore, the characteristic filamentous accumulations are limited to neurons, in many such cases, while the colchicine effect is found in a vast variety of cell types. Therefore, we must look for effects directly on the linker molecules or their binding sites rather than on the microtubule itself. For example, the recent work by Dr. Doyle Graham (Graham et al., this volume) indicated that covalent crosslinking of cytoskeletal proteins might play a role in the n-hexane neuropathies.

Because the neuron can synthesize proteins only in the cell body and in the dendritic cytoplasm, elaborate transport mechanisms exist to convey these products down the axon and to the nerve ending. Cytoskeletal elements are both transported and involved in the transport machinery. Thus, a physical blockade of axonal transport could, if filament synthesis persisted, lead to an accumulation of filaments in the cell body and the axon proximal to the blockage.

A toxic agent could also lead to a selective overproduction of the filament proteins; when this production exceeded the ability of the transport system to remove it from the cell soma, an accumulation would occur.

Thus, neurofilamentous bundling could be due to: a disruption of the normal relations between filaments and other organelles; to a blockade of transport; to overproduction of filaments; or combinations of these factors. Failure to activate neurofilament degrading enzymes or inhibition of such enzymes could also play important roles.

BIOCHEMISTRY OF ALUMINUM-INDUCED NEUROFIBRILLARY PROLIFERATION

The large neurons of animals exposed to intrathecal aluminum salts (Klatzo et al., 1965) develop very prominent bundles or tangles of closely packed intermediate filaments. The neurons are distended and the animals show a profound paresis. Microchemical determination of the protein content of affected neurons of aluminum-treated animals reflect the swollen appearance of the cells with an increased protein content in the treated cells (Embree and Hess, 1970).

To ascertain if the filaments of the treated neurons were of normal composition, we separated the neurons from spinal cords of both treated and untreated rabbits, and then compared their protein compositions and content. The treated neurons showed a constant 20 percent increase in total protein as compared to controls.

On SDS-polyacrylamide gel electrophoresis, the aluminum treated cells showed significant differences from control cells only in proteins of molecular weight corresponding to the components of the neurofilament triplet (Selkoe et al., 1978). Immunocytochemical studies also showed specific staining of the tangles with antineurofilament antibodies (Selkoe et al., 1979; Dahl et al., 1980, 1982).

These studies do not answer the question of whether the increase in neurofilaments in the cell body is due to increased synthesis, decreased degradation or decreased export from the soma. Studies on axonal transport of neurofilaments in aluminum treated animals (Bizzi et al., 1984) find a selective inhibition of neurofilament transport suggesting a major role for the last of these possibilities.

Unlike the mitotic spindle poisons, aluminum seems to require the presence of neurofilaments for the development of characteristic morphologic changes. This would also explain the preponderance of changes in filament-rich larger neurons. This specificity focuses attention on components of the neurofilament lacking in other intermediate filaments, specifically the high molecular weight proteins of the triplet, as attractive possibilities for the site of aluminum action. However, unique regions of the P70 "core" protein such as those involved in MAP binding (Heimann et al., 1983) must also be considered. It is also possible that the primary effect of aluminum is not on the cytoskeleton (Fornell et al., 1982; McLachlan et al., 1983) and that the neurofibrillary changes are secondary events.

ALUMINUM AND HUMAN NEUROLOGICAL DISEASE

In considering the possible toxic role of aluminum in human neuropathology, it is useful to separate the disorders characterized by neurofibrillary changes into one group and the "dialysis dementias" into another group.

"Dialysis dementia" is an unwanted medical complication of chronic hemodialysis. This progressive encephalopathy is characterized by disturbed speech patterns, facial grimacing, and dementia, either singly or together. Aluminum in the dialysis buffer has been implicated as the agent responsible, at least in part, for this condition (Arieff et al., 1979; Berlyne, 1980; Alfrey et al., 1980). Neurofibrillary changes have not been found to be characteristic of this disease.

The human neurofibrillary diseases include numerous idiopathic, toxic, and iatragenic disorders characterized by the presence of bundles of 10 nm neurofilaments of normal appearance (Wisniewski et al., 1971). Biochemical studies on these disorders in man have been limited and little is known about the association of metallic elements with the filaments. However, the situation is much more complicated in the case of Alzheimer's disease which is a major cause of human senile dementia. Though the paired helical filaments, characteristic of this disease, differ markedly from normal neurofilaments when viewed with the electron microscope, the similarity of the light microscopic appearance to the experimental lesions induced by aluminum led to the analysis of post-mortem specimens of Alzheimer and control brains for aluminum. Initial studies showed significantly higher levels of aluminum in the brains of patients dying of Alzheimer's disease than in controls (Crapper et al., 1976). These findings agreed well with earlier work that showed aluminum localization in the neurons of aluminum treated animals (Terry and Pena, 1965) and heightened concerns about a possible etiologic role for aluminum in Alzheimer's disease and senile dementia. A subsequent study (Markesbery et al., 1981) failed to find differences between Alzheimer's and control specimens. The earlier study was done in Toronto, Canada, while the latter was in Lexington, Kentucky. Since the analyses appear to be valid in both studies, it is likely that aluminum is not a necessary factor in these diseases. The Markesbery report points out that alumina is used in the clarification of drinking water in Toronto, while it is not in Lexington, and that the brains of patients with Alzheimer's might preferentially accumulate aluminum if it is available.

These studies have led to more refined approaches using x-ray dispersive analysis under the electron microscope. This technique allows quantitative elemental analysis of small areas of cells which have been identified precisely by electron microscopy. Analyses of cases of Alzheimer's disease (Perl and Brody, 1980) and Guam-Parkinsonism dementia (Perl et al., 1982; Garruto et al., 1984) show markedly elevated levels of aluminum in the area of the neurofibrillary tangles which in both cases are composed of PHF. The lesions also showed elevation of calcium concentrations as compared to controls. Since Guam Parkinsonism-dementia affects only Chamorro people, initial emphasis was on genetic factors. However, decline in the incidence of the disease in the past 30 years has shifted emphasis to the role of environmental factors, especially the fact that the soil of Guam is deficient in both calcium and magnesium and that this might be the cause of the secondary hyperparathyroidism seen in this population. This, in turn, could lead to abnormal deposition of calcium and aluminum in the central nervous system (Garruto et al., 1984).

As in the Alzheimer cases, these data do not differentiate between a predilection for tangle-bearing neurons to accumulate aluminum and calcium and an etiologic role for one or both of these ions in these diseases. The aformentioned occurrance of Alzheimer's disease without aluminum accumulation would favor the former conclusion while the decrease in the incidence of the Guamanian P-D as more calcium rich diets were available would favor the latter.

SUMMARY

Aluminum is an abundant and useful metal which can cause neurological changes in several experimental conditions. Whether it is also a significant environmental neurotoxin is not clear though available evidence suggests that further investigation of this possibility is urgently required.

REFERENCES

Alfrey, A.C., Hess, A. and Craswell, P., 1980, Metabolism and toxicity of aluminum in renal failure, Am. J. Clin. Nutr., 33:1509-1516.

Arieff, A.I., Cooper, J.C., Armstrong, D. and Lazarowitz, V.C., 1979, Dementia, renal failure and brain aluminum, Ann. Intern. Med., 90:741-747.

Berlyne, G.M., 1980, Aluminum toxicity in renal failure, Int. J. Artif. Organs., 3:60-61.

Bizzi, A., Crane, R.C., Autilio-Gambetti, L. and Gambetti, P., 1984, Aluminum effect on slow axonal transport: a novel impairment of neurofilament transport, J. Neurosci., 4:722-731.

Blose, S.H., Shelanski, M.L. and Chacko, S., 1977, Localization to intermediate filaments in guinea pig vascular endothelial cells and chick cardiac muscle cells of antibody prepared against bovine brain filaments, Proc. Natl. Acad. Sci., 74:662-665.

Crapper, D.R., Krishan, S.S. and Quittkat, S., 1976, Aluminum, neurofibrillary degeneration and Alzheimer's disease, Brain, 99:67-80.

Dahl, D., Bignami, A., Bich, N.T. and Chi, N.H., 1980, Immunohistochemical characterization of neurofibrillary tangles induced by mitotic spindle inhibitors, Acta, Neuropathol. (Berl)., 51:165-168.

Dahl, D., Nsuyen, B.T. and Bignami, A., 1982, Ultrastructural localization of neurofilament proteins in aluminum-induced neurofibrillary tangles and rat cerebellum by immunoperoxidase labeling, Dev. Neurosci., 5:54-63.

Davis, J. and Bennett, V., 1982, Microtubule-associated protein 2, a microtubule-associated protein from brain, is immunologically related to the α-subunit of crythrocyte spectrin, J. Biol. Chem., 257:5816-5820.

Embre, L.J. and Hess, H.H., 1970, Microchemistry of ATPase in normal and Alzheimer's disease cortex, J. Neuropath. Exp. Neurol., 29:136-137.

Farnell, B.J., De-Boni, U. and Crapper-McLachlan, D.R., 1982, Aluminum neurotoxicity in the absence of neurofibrillary degeneration in CA1 hippocampal pyramidal neurons in vitro, Exp. Neurol., 78:241-258.

Fifkova, E. and Delay, R.J., 1982, Cytoplasmic actin in neuronal processes as a possible mediator of synaptic plasticity, J. Cell Biol., 95:345-350.

Garruto, R.M., Fukatsu, R., Yanasinara, R., Gajdusek, D.C., Hook, G. and Fiore, C.E., 1984, Imaging of calcium and aluminum in neurofibrillary tangle-bearing neurons in parkinsonism-dementia of Guam, Proc. Natl. Acad. Sci., 81:1875-1879.

Gaskin, F., Kramer, S.B., Cantor, C.R., Adelstein, R. and Shelanski, M.L., 1974, A dynein-like protein associated with neurotubules, FEBS Lett., 49:281.

Gray, E.G., 1975, Presynaptic microtubules and their association with synaptic vesicles, Proc. Royal Soc. London, Ser. B., 190:360-372.

Griffith, L.M. and Pollard, T.D., 1978, Evidence for actin filament-microtubule interaction mediated by microtubule-associated proteins, J. Cell. Biol., 78:958-965.

Heimann, R., Shelanski, M.L. and Liem, R.K.H., 1983, Specific binding of MAPs to the 68,000 dalton neurofilament protein, J. Cell Biol., 97:1079.

Hinkley, R.E., Jr., 1973, Axonal microtubules and associated filaments stained by alcian blue, J. Cell Sci., 13:753-761.

Hirokawa, N., 1982, Crosslinker system between neurofilaments, microtubules and membranous organelles in frog axons by the quick-freeze, deep-etching method, J. Cell Biol., 94:129-142.

Kidd, M., 1963, Paired helical filaments in electromicroscopy of Alzheimer's disease, Nature, 197:192-193.

Kim, H., Binder, L.I. and Rosenbaum, J.L., 1979, The periodic association of MAP-2 with brain microtubules "in vitro", J. Cell Biol., 80:266-276.

Klatzo, I., Wisniewski, H. and Streicher, E., 1965, Experimental production of neurofibrillary degeneration, I. Light microscopic observations, J. Neuropath. Exp. Neurol., 24:187.

Kohno, K., 1964, Neurotubules contained within the dendrite and axon of Purkinje cell of frog, Bull. Tokyo Med. Dent. Univ., 11:411-442.

Leterrier, J.-F., Liem, R.K.H. and Shelanski, M.L., 1981, Preferential phosphorylation of the 150,000 molecular weight component of neurofilament by a cyclic AMP dependent microtubule associated protein kinase, J. Cell Biol., 90:755-760.

Leterrier, J.-F., Liem, R.K.H. and Shelanski, M.L., 1982, Interactions between neurofilaments and microtubule associated proteins. A possible mechanism for intraorganellar bridging, J. Cell Biol., 95:982-986.

Markesbery, W.R., Ehmann, W.D., Hossain, T.I., Aluddin, M. and Goodin, D.T., 1981, Instrumental neutron activation analysis of brain aluminum in Alzheimer's disease and aging, Ann. Neurol., 10:511-516.

McLechlan, D.R., Dam, T.V., Farnell, B.J. and Lewis, P.N., 1983, Aluminum inhibition of ADP-ribosylation in vivo and in vitro, Neurobehav. Toxicol. Teratol., 5:645-647.

Murphy, D.B., Vallee, R.B. and Borisy, G.G., 1977, Identity and polymerization stimulatory activity of nontubulin proteins associated with microtubules, Biochem., 16:2598-2605.

Pelay, S.L., Sotelo, C., Peters, A. and Orkand, P.M., 1968, The axon hillock and the initial segment, J. Cell Biol., 38:193-201.

Perl, D.P. and Brady, A.R., 1980, Alzheimer's disease: X-Ray spectrometric evidence of aluminum accumulation in neurofibrillary tangle-bearing neurons, Science, 208:297-299.

Perl, D.P., Gajdusek, D.C., Garruto, R.M., Yanigahara, R.T. and Gibbs, C.J., Jr., 1982, Intranuronal aluminum accumulation in amyotrophic lateral sclerosis and parkinsonism-dementia of Guam, Science, 217:1053-1055.

Raine, C.S., Ghetti, G. and Shelanski, M.L., 1971, On the association between tubules and mitochondria in axons, Brain Res., 34:389-393.

Robbins, E. and Gonatas, N.K., 1964, Histochemical and ultrastructural studies on HeLa cell cultures exposed to spindle inhibitors and special reference to the interphase cell, J. Histochem. Cytochem., 12:704.

Runge, M.S., Lane, T.M., Yphantis, D.A., Lifsics, M.R., Saito, A., Altin, M., Reinke, K. and Williams, R.C., Jr., 1981, ATP-induced formation of associated complex between microtubules and neurofilaments, Proc. Natl. Acad. Sci., 78:1431-1435.

Samson, F.E., 1976, Pharmacology of drugs that affect intracellular movement, Ann. Rev. Pharmacol. Toxicol., 16:143-159.

Satillaro, R.F., Dentler, W.L. and LeCluyse, E.L., 1981, Microtubule associated proteins (MAPs) and the organization of actin filaments in vitro, J. Cell Biol., 90:467-473.

Selkoe, D.J., Leim, R.K.H., Yen, S.-H. and Shelanski, M.L., 1979, Biochemical and immunological characterization of neurofilaments in experimental neurofibrillary degeneration induced by aluminum, Brain. Res., 163:235-252.

Sharp, G.A., Shaw, G. and Weber, K., 1982, Immunoelectronmicroscopical localization of the three neurofilament triplet proteins along neurofilaments of cultured dorsal root ganglion neurons, Exp. Cell Res., 137:403-413.

Shelanski, M.L., Gaskin, F. and Cantor, C.R., 1973, Microtubule assembly in the absence of added nucleotides, Proc. Natl. Acad. Sci., 70:765-768.

Sloboda, R.D. and Rosenbaum, J.L., 1979, Decoration and stabilization of intact, smooth-walled microtubules with microtubule associated proteins, Biochem., 18:48-55.

Smith, D.S., 1971, On the significance of cross-bridges between microtubules and synaptic vesicles, Philos. Trans. R. Soc. London. Ser. B., 261:395-405.

Smith, D.S., Jarlfors, U. and Cayer, M.L., 1977, Structural cross-bridges between microtubules and mitochondria in central axons of an insect (Periplaneta americana), J. Cell Sci., 27:235-272.

Summers, K.E. and Gibbons, I.R., 1971, ATP-induced sliding of tubules in trypsin-treated flagella of sea urchin sperm, Proc. Natl. Acad. Sci., 68:3092-3096.

Terry, R.D., 1963, The fine structure of neurofibrillary tangles in Alzheimer's disease, J. Neuropathol. Exp. Neurol., 22:629-642.

Terry, R.D. and Pena, C., 1965, Experimental production of neurofibrillary degeneration 2: Electron microscopy, phosphatase histochemistry, and electron probe analysis, J. Neuropath. Exp. Neurol., 24:200-210.

Troncoso, J.C., Price, D.L., Griffin, J.W. and Parhad, I.M., 1982, Neurofibrillary axonal pathology in aluminum intoxication, Ann. Neurol., 12:278-283.

Tsukita, S. and Ishikawa, H., 1980a, The movement of membranous organelles in axons. Electron microscopic identification of anterogradely and retrogradely transported organelles, J. Cell Biol., 84:513-530.

Weisenberg, R.C., 1972, Microtubule formation in vitro in solutions containing low calcium concentrations, Science, 177:1104-1105.

Westrum, L.E. and Gray, E.G., 1977, Microtubules associated with postsynaptic "thickenings", J. Neurocytol., 6:505-518.

Willard, M. and Simon, C., 1981, Antibody decoration of neurofilaments, J. Cell Biol., 89:198-205.

Wisniewski, H., Shelanski, M.L. and Terry, R.D., 1968, Effects of mitotic spindle inhibitor on neurotubules and neurofilaments in anterior horn cells, J. Cell Biol., 38:224-229.

Wisniewski, H., Terry, R.D. and Hirano, A., 1971, Neurofibrillary pathology, J. Neuropath. Exp. Neurol., 29:173-181.

Wolosewick, J.J. and Porter, K.R., 1979, Microtrabecular lattice of the cytoplasmic ground substance: Artifact or reality, J. Cell Biol., 82:114-139.

PATHOGENETIC STUDIES OF THE NEUROFILAMENTOUS NEUROPATHIES

Doyle G. Graham, Gyöngyi Szakàl-Quin, Leslie Milam*,
Marcia R. Gottfried and D. Carter Anthony

Department of Pathology
Duke University Medical Center
Durham, North Carolina, USA

*Department of Anatomy
University of North Carolina
Chapel Hill, North Carolina, USA

INTRODUCTION: γ-DIKETONE NEUROPATHY

Exposure to both n-hexane and methyl n-butylketone (2-hexanone) have resulted in distal sensorimotor neuropathies in industrial settings (Yamamura, 1969; Allen et al., 1975), and n-hexane induced neuropathies have been seen in glue sniffers (Korokobin et al., 1975). The resulting neuropathy was characterized by paranodal axonal swellings filled with neurofilaments and was readily reproduced in experimental animals (Schaumburg and Spencer, 1976; Krasavage et al., 1980). Degeneration of the axon distal to the swellings has been observed in large myelinated axons in the peripheral nervous system (PNS), but less often in small myelinated or unmyelinated PNS axons (Spencer and Schaumburg, 1977) or in central nervous system (CNS) axons, in which the masses of neurofilaments travel to the end of the axon where they undergo eventual digestion by a calcium-activated protease (Cavanagh and Bennetts, 1981; Cavanagh, 1982).

That the ultimate toxic metabolite of n-hexane and methyl n-butylketone is the γ-diketone metabolite, 2,5-hexanedione (HD) is well established (Krasavage et al., 1980). What has been controversial has been the pathogenesis of the neurofilament-filled axonal swellings, and the relationship between these swellings and distal axonal degeneration. Hypotheses to account for neurofilament accumulations have included an inhibition of energy metabolism in the axon (Spencer et al., 1979), impaired neurofilament catabolism (Cavanagh and Bennetts, 1981), and increased hydrophobic interactions between neurofilaments (DeCaprio et al., 1982; 1983).

Studies in our laboratory have shown that HD reacts with lysyl residues of proteins to form an imine, which then cyclizes to a pyrrole. Autooxidation of the pyrrole results in orange chromophore development and protein crosslinking. Because the neurofilament appears to be the most stable and most slowly moving component of axoplasm, we proposed that neurofilaments were vulnerable to progressive covalent crosslinking during anterograde flow of axoplasm, resulting in an ever-growing aggregate (Graham et al., 1982b). The reduction in axonal diameter at nodes of Ranvier could be envisioned to present obstructions to the distal progression of these masses (Graham et al., 1982a,b; Cavanagh, 1982; Jones and Cavanagh, 1983). Morphological evidence for the stability of the aggregates of neurofilaments and for the impairment to their movement at nodes of Ranvier has been provided by Jones and Cavanagh (1983).

DIMETHYL SUBSTITUTION OF 2,5-HEXANEDIONE ENHANCES TOXICITY

The neurotoxicity of 3,4-dimethyl-2,5-hexanedione (DMHD) was investigated in order to test whether pyrrole formation and auto-oxidation were necessary steps in the pathogenesis of γ-diketone neuropathy. It was reasoned that dimethyl substitution would enhance the neurotoxicity of HD because alkyl substitution accelerates both the rate of ring formation (Eliel, 1962) and the subsequent auto-oxidation of pyrroles (Jones and Bean, 1977). The cyclization of DMHD with primary amines was found to be eight times faster than the reaction between HD and amines. Further, the tetramethylpyrrole formed from DMHD oxidized more readily than the dimethylpyrrole product of HD. The combined effect of these two factors led to an acceleration of the rate of protein crosslinking that was 40 times greater for DMHD than for HD (Anthony et al., 1983a).

As would be predicted from these findings, DMHD was found to be 20-30 times more potent than HD as measured by paralysis in treated rats. Further, while in HD intoxication, hindlimb weakness is far greater than that of forelimbs, DMHD affected hind and forelimbs equally. The resulting neurofilament-filled axonal swellings were in the proximal axon in the DMHD-treated rats, as opposed to the distal location of the axonal swellings after HD intoxication; severe degeneration was seen in the distal axon in the DMHD-treated rats (Anthony et al., 1983b). Axonal swellings were also observed proximally in cranial nerves (Anthony et al., 1983c). Another interesting observation was that the position of the neurofilamentous swellings varied with the rate of intoxication; more proximal swellings occurred near the neuronal perikaryon with rapid intoxication by DMHD (0.25 mmoles/kg every 8 hours), and beyond the anterior roots to the proximal sciatic nerve with less rapid intoxication (0.125 mmoles/kg/day) (Anthony et al., 1983b; Graham et al., 1982a).

The location of the neurofilamentous axonal swellings in the proximal axon in the DMHD model paralleled the findings seen after intoxication with β,β'-iminodipropionitrile (IDPN) (Clark et al., 1980); both compounds resulted in a marked retardation of neurofilament transport (Griffin et al., 1984). This led us to propose that DMHD represented a heretofore unavailable "missing link" between IDPN and those toxicants which result in neurofilament-filled swellings of the distal axon - HD, carbon disulfide, and acrylamide. We suggested that the neurofilamentous axonopathies represented a continuum in which the underlying principle was covalent crosslinking of neurofilaments and the anterograde location of the axonal swellings determined by the rate at which the neurofilaments aggregated to form masses too large to pass through nodes of Ranvier (Anthony et al., 1982, 1983b; Graham et al., 1982a).

EVIDENCE THAT NEUROFILAMENTS CAN BE CROSSLINKED BY γ-DIKETONES

When rat neurofilament preparations were reacted with DMHD there was a progressive loss of neurofilament bands on Coomassie Blue stained polyacrylamide gels, with the production of large molecular weight species that failed to enter the stacking gel. This observation was similar to those of Selkoe et al. (1982a,b) in their analysis of neurofibrillary masses in Alzheimer's disease or when neurofilaments were crosslinked with transglutaminase. Thus, we turned to nitrocellulose filters with different pore sizes to estimate the range of molecular sizes resulting from polymerization (Graham et al., 1984).

As shown in Fig. 1A, incubation of 20 mM [1,2-^{14}C]-2,5- hexanedione (synthesized in our laboratory) with desheathed rat peripheral nerve resulted in progressive alkylation of nerve proteins at 37°C and pH 7.4. When the labeled nerve protein was freed of lipids by acetone/ether extraction, denatured and reduced by boiling in 1 percent sodium dodecyl sulfate (NaDodSO$_4$) 5 percent 2-mercaptoethanol, and 8 M urea, there was retention of radiolabel by nitrocellulose filters. The percentage retained by 0.025 mm nitrocellular filters rose from 28.5 percent to 41.4 percent between 2 and 11 days of incubation. The corresponding percentages of retention for 0.2 μm and 12 μm filters were 19.5 to 32.4 percent and 2.6 to 8.7 percent, respectively. Thus, over time, there was progressive formation of polymers too large to pass through nitrocellulose filters with pore sizes as large as 12 μm, indicating that large molecular masses had been formed.

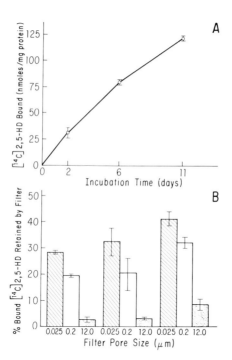

Fig. 1. Covalent crosslinking of nerve proteins by HD. Desheathed peripheral nerve was incubated in phosphate buffered saline (PBS), pH 7.4 at 37°C, with 20 mM 1,2-[^{14}C]-2,5-hexanedione. After 2, 6 or 11 days the nerves were washed with PBS, sonicated in 1 percent NaDodSO$_4$, and boiled for 10 min. After cooling, proteins were precipitated with 9 volumes of acetone, collected by centrifugation, delipidated with two extractions with acetone/ether (1:1, v:v), extracted with ether, dried, and then solubilized by boiling in 8 M urea, 5 percent 2-mercaptoethanol and 1 percent NaDodSO$_4$. Fig. 1A illustrates the quantity of [^{14}C] HD bound to protein after this treatment. In Fig. 1B the proportion of labeled protein retained by nitrocellulose filters with given pore sizes is shown for the three time points (mean ± SE, n=3 for each time point).

These data parallel those from experiments in which nerve was reacted with [2,5-^{14}C]-DMHD (Graham et al., 1984). In these studies, we also showed that DMHD resulted in massive polymerization of SCa (Hoffman and Lasek, 1975), the neurofilament-rich, slowest component of axonal transport, which had been pulse-labeled with [^{35}S] methionine.

EXPERIMENTS TO TEST WHETHER NEUROFILAMENT CROSSLINKING OCCURS IN VIVO

John Cavanagh (personal communication, 1983) once noted that the erythrocyte is a good model for the axon in the sense that, in both, proteins are separated from the nucleus--in the erythrocytes by time, and in the axon, by distance. This is an elegant expression of the rationale we employed when we demonstrated that the lysyl residues of globin are progressively derivatized to form pyrroles during chronic intoxication with HD or DMHD (Anthony et al., 1983a). In these experiments we also obtained the first evidence that protein crosslinking occurs in vivo with HD and DMHD; polyacrylamide gel electrophoresis of spectrin, eluted from washed erythrocyte membranes in 0.1 mM EDTA, revealed a 400,000 dalton protein corresponding in molecular weight to that of a spectrin dimer (Anthony et al., 1983a).

Recently we have turned to the marine worm Myxicola infundibulum (Fig. 2), as a species in which demonstration of crosslinking of neurofilaments may be possible in vivo. Myxicola survive in aerated seawater for over 4 days at 4°C in the presence of 0.25 mM DMHD. As shown in Fig. 3 neurofilaments can be readily isolated on glycerol gradients.

Neurofilament preparations from Myxicola exposed to 0.25 mM [2,5-^{14}C] DMHD (synthesized in our laboratory) demonstrate progressive alkylation (Fig. 4). Experiments to determine whether covalent crosslinking occurs in this setting are in progress.

INTERSPECIES DIFFERENCES IN VULNERABILITY TO γ-DIKETONE NEUROPATHY REFLECT THE INFLUENCE OF AXONAL LENGTH AND DIAMETER

Our hypothesis predicts that increasing axonal length and diameter should increase the chance for a given axon to develop prenodal neurofilament-filled axonal swellings and subsequent degeneration of the distal axon during chronic HD intoxication. The longer the axonal length,

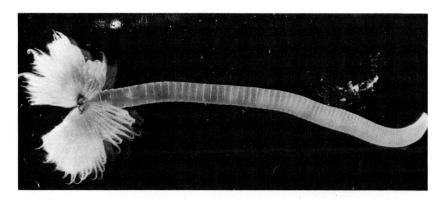

Fig. 2. Myxicola infundibulum. A relaxed worm, shown here with two-fold magnification (obtained from Ocean Products, P.O. Box 263, Eastport, Maine).

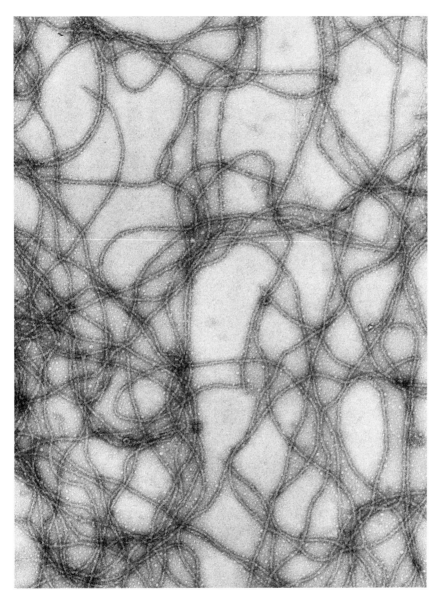

Fig. 3. Neurofilaments from <u>Myxicola</u> infundibulum. <u>Myxicola</u> axoplasm was removed by micropipette from control worms, homogenized in 0.45 ml buffer (100 mM NaF, 10 mM Tris, 5 mM EGTA, 0.8 mM 2-mercaptoethanol, pH 7.43) then layered over a discontinuous glycerol gradient (0.1 ml each of 7.7 percent, 25.8 percent, and 64.5 percent glycerol in the same buffer) and centrifuged at 40,000 rpm for 2 hrs in a SW 50.1 rotor. The 25.8 - 64.5 percent interface and 64.5 percent layer were mixed, diluted 10-fold in 64.5 percent glycerol, and negatively stained with 1 percent uranyl acetate on carbon coated grids. 80,000x.

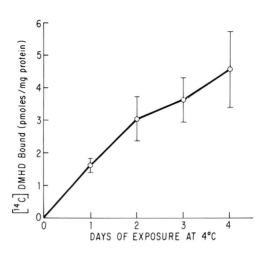

Fig. 4. In vivo intoxication of Myxicola with [^{14}C] DMHD. Myxicola were kept in aerated seawater at 4°C. After 1 to 4 days of exposure to 0.25 mM [2,5-^{14}C]-DMHD, neurofilaments were isolated as described in Fig. 3; binding was quantified by scintillation counting in Aquasol II (New England Nuclear). Protein concentrations were determined as described by Hartree (1972).

the greater the time a given length of neurofilament would have to become crosslinked. As pointed out by Waxman et al. (1976), the probability of axonal degeneration increases markedly with axonal length in multiple-hit models of axonal injury. In addition, axons with larger diameters are associated with greater proportional constrictions of axonal diameter at nodes of Ranvier (Hess and Young, 1952; Cavanagh and Bennetts, 1981; Cavanagh, 1982; Anthony et al., 1983c; Jones and Cavanagh, 1983), which we have proposed as necessary for the development of axonal swellings proximal, and of degeneration distal, to nodes of Ranvier. Our recent studies addressed these issues by comparing the vulnerability of rats and mice to γ-diketone neuropathy (Gottfried et al., in preparation). Mice, which have shorter axons with smaller diameters (Friede et al., 1984), did not reach the end-point of hind limb paralysis when treated with either HD or DMHD in drinking water, despite prolonged periods of intoxication or near-lethal doses. HD-treated mice exhibited some weakness, while those given DMHD remained normal. While neurofilaments became disordered and segregated from the other axoplasmic components in mice as well as rats during chronic HD intoxication, mice developed neurofilament-filled axonal swellings much less often than rats, and distal axonal degeneration was rare. On the other hand, mice treated with DMHD developed massive neurofilamentous swellings of the proximal axons similar to those seen in the rat. However, the DMHD-treated mice did not develop degeneration of the distal axon. Rather, they resembled IDPN-treated rats, which had proximal swellings without distal degeneration (Clark et al., 1980).

Thus, these experiments suggest that the relative insensitivity of mice to γ-diketone neuropathy is a product of having axons too short to develop large aggregates of neurofilaments and too small in diameter to have significant constrictions at nodes of Ranvier. As one can also conclude from the IDPN model in rats (Clark et al., 1980), axonal swellings in themselves do not result in weakness; rather, significant degeneration of distal axons appears to be necessary for this clinical end point to be reached.

REFERENCES

Allen, N., Mendell, J.R., Billmaier, D.J., Fontaine, R.E. and O'Neill, J., 1975, Toxic polyneuropathy produced by the industrial solvent, methyl n-butyl ketone, Arch. Neurol., 32:209-218.

Anthony, D.C., Boekelheide, K., Giangaspero, F., Allen, J.C., Jr., Parks, H., Priest, J.W., Webster, D. and Graham, D.G., 1982, The neurofilament neuropathies: A unifying hypothesis, J. Neuropathol. Exp. Neurol., 41:371.

Anthony, D.C., Boekelheide, K., Anderson, C.W. and Graham, D.G., 1983a, The effect of 3,4-dimethyl substitution on the neurotoxicity of 2,5-hexanedione. II. Dimethyl substitution accelerated pyrrole formation and protein crosslinking, Toxicol. Appl. Pharmacol., 71:372-382.

Anthony, D.C., Boekelheide, K. and Graham, D.G., 1983b, The effect of 3,4-dimethyl substitution on the neurotoxicity of 2,5-hexanedione. I. Accelerated clinical neuropathy is accompanied by more proximal axonal swellings, Toxicol. Appl. Pharmacol., 71:362-371.

Anthony, D.C., Giangaspero, F. and Graham, D.G., 1983c, The spatio-temporal pattern of the axonopathy associated with the neurotoxicity of 3,4-dimethyl-2,5-hexanedione in the rat, J. Neuropathol. Exp. Neurol., 42:548-560.

Cavanagh, J.B., 1982, The pattern of recovery of axons in the nervous system of rats following 2,5-hexanediol intoxication: A question of rheology, Neuropathol. Appl. Neurobiol., 8:19-34.

Cavanagh, J.B. and Bennetts, R.J., 1981, On the pattern of changes in the rat nervous system produced by 2,5-hexanediol. A topographical study by light microscopy, Brain, 104:297-318.

Clark, A.W., Griffin, J.W. and Price, D.L., 1980, The axonal pathology in chronic IDPN intoxication, J. Neuropathol. Exp. Neurol., 39:42-55.

DeCaprio, A.P., Olajos, E.S. and Weber, P., 1982, Covalent binding of a neurotoxic n-hexane metabolite: Conversion of primary amines to substituted pyrrole adducts by 2,5-hexanedione, Toxicol. Appl. Pharmacol., 65:440-450.

DeCaprio, A.P., Strominger, N.L. and Weber, P., 1983, Neurotoxicity and protein binding of 2,5-hexanedione in the hen, Toxicol. Appl. Pharmacol., 68:297-307.

Eliel, E.L., 1962, Stereochemistry of ring systems, in: "Stereochemistry of Carbon Compounds," E.L. Eliel (ed.), pp. 196-203, McGraw-Hill, New York.

Friede, R.L., Benda, M., Dewitz, A. and Stoll, P., 1984, Relations between axon length and axon caliber. Is maximum conduction velocity the factor controlling the evolution of nerve structure, J. Neurol. Sci., 63:369-380.

Graham, D.G., Anthony, D.C. and Boekelheide, K., 1982a, In vitro and in vivo studies of the molecular pathogenesis in n-hexane neuropathy, Neurobehav. Toxicol. Teratol., 4:629-634.

Graham, D.G., Anthony, D.C., Boekelheide, K., Maschmann, N.A. Richards, R.G., Wolfram, J.W. and Shaw, B.R., 1982b, Studies of the molecular pathogenesis of hexane neuropathy. II. Evidence that pyrrole derivatization of lysyl residues leads to protein crosslinking, Toxicol. Appl. Pharmacol., 64:415-422.

Graham, D.G., Szakal-Quin, Gy, Priest, J.W. and Anthony, D.C., 1984, In vitro evidence that covalent crosslinking of neurofilaments occurs in γ-diketone neuropathy, Proc. Natl. Acad. Sci., 81:4979-4982.

Griffin, J.W., Anthony, D.G., Fahnestock, K.E., Hoffman, P.N., and Graham, D.C., 1984, 3,4-dimethyl-2,5-hexanedione impairs the axonal transport of neurofilament proteins, J. Neurosci., 4:1516-1526.

Hartree, E.F., 1972, Determination of protein: A modification of the Lowry method that gives a linear photometric response, Anal. Biochem., 48:422-427.

Hess, A. and Young, J.Z., 1952, The nodes of Ranvier, Proc. R. Soc. Lond., 140:301-320.

Hoffman, P.H. and Lasek, R.J., 1975, The slow component of axonal transport, J. Cell. Biol., 66:351-366.
Jones, H.B. and Cavanagh, J.B., 1983, Distortions of the nodes of Ranvier from axonal distention by filamentous masses in hexacarbon intoxication, J. Neurocytol., 12:439-458.
Jones, R.A. and Bean, G.P., 1977, The chemistry of pyrroles, in: "Organic Chemistry," A.T. Blomquist and H.H. Wasserman (eds.), pp. 209-247, Academic Press, New York.
Korokobin, R., Asbury, A.K., Summer, A.J. and Nielsen, S.L., 1975, Glue-sniffing neuropathy, Arch. Neurol., 32:158-162.
Krasavage, W.J., O'Donoghue, J.L., DiVencenzo, G.D. and Terhaar, C.J., 1980, The relative neurotoxicity of MnBK, n-hexane, and their metabolites, Toxicol. Appl. Pharmacol., 52:433-441.
Schaumburg, H.H. and Spencer, P.S., 1976, Degeneration in the central and peripheral nervous systems produced by pure n-hexane: An experimental study, Brain, 99:183-192.
Spencer, P.S. and Schaumburg, H.H., 1977, Ultrastructural studies of the dying-back process. IV. Differential vulnerability of PNS and CNS fibers in experimental central-peripheral distal axonopathies, J. Neuropathol., 36:300-320.
Spencer, P.S., Sabri, M.I. and Moore, C.L., 1979, Does a defect in energy metabolism in the nerve fiber underlie axonal degeneration in polyneuropathies, Ann. Neurol., 5:501-507.
Waxman, S.G., Brill, M.H., Geschwind, N., Sabin, T.D. and Lettvin, J.Y., 1976, Probability of conduction deficit as related to fiber length in random-distribution models of peripheral neuropathies, J. Neurol. Sci., 29:39-53.
Yamamura, Y., 1969, n-Hexane polyneuropathy, Folia Psychiat. Neurol. Jpn., 23:45-57.

SESSION 3.   CYTOSKELETON AND MEMBRANE RELATED EVENTS

    Chairpersons: Paul L. La Celle
                             University of Rochester School of Medicine

                             Angelo C. Notides
                             University of Rochester School of Medicine

    Rapporteurs: Marshall A. Lichtman
                             University of Rochester School of Medicine

                             Arnjolt Elgsaeter
                             University of Trondheim School of Medicine

EFFECTS OF CALCIUM ON STRUCTURE AND FUNCTION OF THE HUMAN RED BLOOD CELL MEMBRANE

Hermann Passow, Melanie Shields, Paul La Celle*,
Ryszard Grygorczyk, Wolfgang Schwarz and Reiner Peters

Max-Planck-Institut fur Biophysik
Frankfurt/Main
Federal Republic of Germany

*Department of Radiation Biology and Biophysics
University of Rochester School of Medicine
Rochester, New York, USA

INTRODUCTION

The intracellular activity of ionized Ca^{++} in the red cell is below 0.4 µmoles/l (Schatzmann, 1973; Simons, 1982; Lew et al., 1982b) and hence much lower than calculated from intracellular Ca^{++} content divided by red cell volume (Lichtman and Weed, 1973). This indicates that much of the Ca^{++} is bound to the cell membrane (Lichtman and Weed, 1973; La Celle et al., 1973; Porzig and Stoffel, 1978) and intracellular constituents. The latter include the phosphoric acid esters and hemoglobin (Ferreira and Lew, 1977), all of which act as Ca^{++} buffers. The low intracellular Ca^{++} activity is maintained although the membrane is leaky for Ca^{++} (Ferreira and Lew, 1977) and the concentration of free Ca^{++} in blood plasma (about 1200 µmoles/l) exceeds that in cytosol by about four orders of magnitude. The enormous gradient is balanced by a powerful Ca^{++} pump (for review see Schatzmann, 1983). At 37°C, the maximal rate of pumping is about 10 mmoles/l cells/h. The half saturation concentration ($K_{1/2}$) of the pump, and hence one of the essential factors that determines the steady state Ca^{++} concentration, depends on the conditions inside the cell: the concentration of the energy supplying substrate ATP, the concentration of the activating calmodulin, and the concentration of Mg^{++} (calmodulin activates maximally when 1 Mg^{++} and 3 Ca^{++} ions are complexed). Under physiological conditions, $K_{1/2}$ seems to be about 0.3 µmole/l (Downes and Michell, 1981).

In red cells that undergo their physiological aging process in the circulation, the intracellular Ca^{++} level increased. In 80 day old cells, a value of 10 µmoles/l was found, of which an unknown fraction was bound to the Ca^{++} buffers. The increase occurred in spite of a slight reduction of Ca^{++} leakage and seemed to be related to a decrease of Ca^{++} buffer capacity and Ca^{++} pump activity. Both changes were due, at least in part, to a reduction of intracellular ATP (La Celle et al., 1973). The Ca^{++} content and permeability of human red cells are known to be altered also under certain pathological conditions, e.g., in sickle cell anemia (Eaton et al., 1973; Palek, 1977; Lew et al., 1980), congenital hemolytic anemias

(Wiley and Gill, 1976) and hereditary spherocytosis (Feig and Bassilan, 1974). After experimentally induced substrate depletion, the intracellular Ca^{++} concentration in cells suspended in their autologous plasma may increase up to 0.2-0.6 mmoles/l (total Ca^{++}) (Lichtman and Weed, 1973).

CALCIUM EFFECTS ON THE RED CELL MEMBRANE

The enormous transport activity of the Ca^{++} pump indicates that the maintenance of a low electrochemical Ca^{++} activity in the cytosol is a requirement for the red cell to remain in its physiological state. In fact, natural, pathological or experimental augmentation of the intracellular Ca^{++} activity above the steady state level leads to dramatic alterations of the cells, most of which are related to disturbances of membrane structure and function.

These alterations, summarized in Table 1, are listed in order of increasing intracellular Ca^{++} concentration, covering a range from about 0.1 to 1000 μmoles/l, i.e., roughly from the situation existing under physiological conditions up to complete equilibration of the cytosol with the free extracellular Ca^{++}. The concentrations indicated should only be viewed as approximate. They do not always refer to precise determinations

Table 1. Effects of internal Ca^{++} on the red blood cell membrane. The Ca^{++} concentrations indicated provide only a rough guideline since the effects may be modulated by factors which could not be considered for the purpose of this schematic representation.

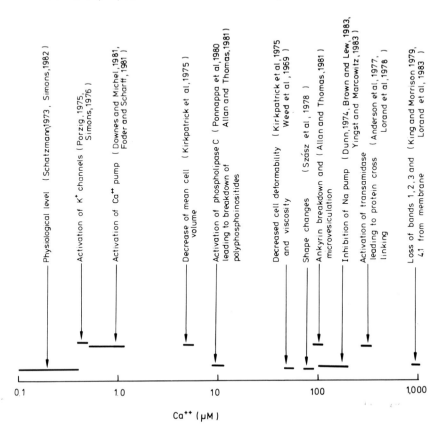

of free Ca^{++} since their toxic actions may be influenced by inhibiting or potentiating factors, such as the concentration of Mg^{++}, specific phosphoric acid esters, and calmodulin. Moreover, subtleties of the experimental conditions that are not yet properly understood may further add to the uncertainties.

Nevertheless, with these limitations in mind, the figure shows that a slight increase in free Ca^{++} leads to an activation of the Ca^{++} pump and the opening of K^+ selective channels. This latter event gives rise to the so-called Gardos effect (for review see Schwarz and Passow, 1983). The net K^+ efflux is accompanied by the efflux of an equivalent number of Cl^- ions and the cells shrink. In the cell membranes, the breakdown of polyphosphoinositides is stimulated, which should reduce the activity of the Ca^{++} pump. Phosphatidic acid accumulates, affecting the domain structure of the membrane lipids and changing their susceptibility to phospholipase activity.

At even higher Ca^{++} concentrations - in the range of 50 μmoles/l - the deformability of the red cells is decreased and as a consequence, blood viscosity increases. Shape changes ensue which further affect the rheological properties of the blood. When a Ca^{++} concentration of 100 μmoles/l is exceeded, the cytoskeleton shows signs of disintegration. Ankyrin breaks down and microvesiculation becomes apparent. At Ca^{++} concentrations above 300 μmoles/l, a transamidase becomes activated which leads to protein crosslinking. Overlapping with this effect, a proteolytic destruction of band 3 protein and of glycophorin takes place which, at 1000 μmoles/l includes degradation of the two spectrin promoters and band 4.1. This brief description of some of many effects of Ca^{++} illustrates clearly why the red cell needs to control the intracellular Ca^{++} level precisely.

EXPERIMENTAL: Ca^{++} MODIFICATION OF K^+ EFFLUX AND MEMBRANE VISCOELASTICITY

The experimental work summarized below focuses on the effects produced by Ca^{++} at the low concentration end on the scale presented in Table 1. It summarizes the results of parallel studies on the stimulation of K^+ efflux and the mechanical properties of the red cell membrane.

Ca^{++}-Stimulated K^+ Efflux in Untreated Red Cell Ghosts

The calcium-stimulated K^+ efflux is the consequence of the opening of K^+-selective aqueous channels in the red cell membrane (Hamill, 1981; Grygorczyk and Schwarz, 1983; Grygorczyk et al., 1984). This is demonstrated in Figs. 1 and 2, showing that the probabilities of the open state of the channels, as measured at a range of Ca^{++} concentrations by the patch clamp technique (Fig. 1), coincide quite well with measurements of the rate constant of K^+ efflux, determined from cellular K^+ content (Fig. 2). A comparison of the single channel conductance, with the conductance of the whole cell calculated from the isotopically measured K^+ movements, enables one to estimate the number of channels per cell that were activated by Ca^{++}. At the concentrations of Ca^{++} used to elicit the response, two populations of cells were observed (described by Hoffman et al., 1980; Yingst and Hoffman, 1984) with different rates of K^+ movements. One of the populations (75 percent of the cells in the suspension) showed 1-5 Ca^{++} activated channels; the other (i.e., the remaining 25 percent of the cells in the suspension) contained 11-55 channels per cell (Grygorczyk et al., 1984).

These numbers differ from those reported by Lew and associates (1982a). These authors had prepared inside-out vesicles and found that only a fraction of them responded to Ca^{++} with an increase of K^+ permeability.

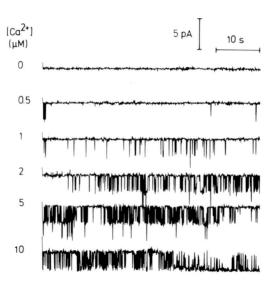

Fig. 1. Single-channel currents of Ca^{++}-activated K^+ channels, measured in an inside-out cell-free membrane patch, as a function of Ca^{++} concentration. After formation of a gigaseal, the cell was ruptured by briefly exposing it to the flow of distilled water. The solution in the pipette contained 140 mM KCl and no Ca^{++}. The bath solution contained 150 mM KCl and the Ca^{++} concentrations indicated in the figure. Single channel currents were recorded at a holding potential of -100 mV and a sampling rate of 20 Hz (Grygorczyk and Schwarz, 1983). Reproduced from Grygorczyk et al. (1984) with copyright permission of the Biophysical Society.

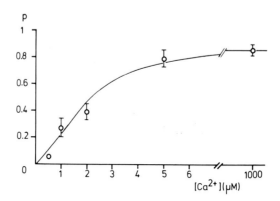

Fig. 2. Comparison of Ca^{++}-activated K^+ permeabilities derived from patch clamp measurements of single channel currents in individual cells with K^+ fluxes measured in cell suspensions. Symbols represent the probability of the open state ($\underline{p}$ ± SEM) of unitary currents determined from cell-free membrane patches during channel activity as in Fig. 1; the membrane potential was -100 mV. The solid line represents the rate constants of K^+ loss from red cell calculated from data of Heinz and Passow (1980; Fig 9a), and scaled to match the $\underline{p}$ values at saturating Ca^{2+} concentration. Reproduced from Grygorczyk et al. (1984) with copyright permission of the Biophysical Society.

On the assumption that the number of channels per cell was smaller than the number of vesicles derived from each cell, they inferred that there were about 100 - 200 channels per cell. This number is significantly higher than the estimate from our laboratory. The estimate of Lew et al. (1982a) depends on the assumption that the vesiculation procedure did not affect the transport system. While we determined the number of channels that were actually activated in the intact cell, the vesicles could contain channels that become activated by the vesiculation procedure. This raised the question of whether, in the intact red cell membrane, some of the channels could exist in a "dormant" form. Such channels could become activated by Ca^{++} after disruption of the cytoskeleton that could conceivably occur during the formation of the inside-out vesicles.

This question prompted us (1) to look for parallel effects of Ca^{++} on the cytoskeleton and the activation of K^+ selective channels and (2) to investigate whether tryptic digestion of the cytoskeleton would affect the susceptibility of the K^+ selective channels to activation by Ca^{++}.

Ca^{++}-Stimulated K^+ Efflux After Tryptic Degradation of the Cytoskeleton

Effects of internal Ca^{++} and trypsin on membrane viscosity and elasticity. As a measure of an effect of Ca^{++} on the cytoskeleton, we determined both the elasticity and the viscosity of the membrane. Figure 3 shows that in fact both quantities increased in ghosts that contained Ca^{++} at concentrations that activated the K^+ selective channels. However, this does not prove a causal relationship. This was seen when the cytoskeleton is subjected to controlled degradation by incorporation of trypsin into ghosts. Under these conditions, in the absence of Ca^{++}, both elasticity and viscosity remained essentially unaltered or, if anything, increased slightly (Table 2).

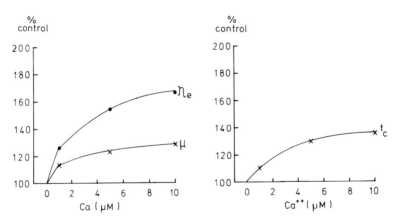

Fig. 3. Effect of Ca^{++} on elasticity (μ) and viscosity (η_e) of human red blood cells. Red cells were washed three times in medium containing 100 mM KCL, 46 mM NaCl, 10 mM Na citrate, 10 mM Tris, pH 7.6 with $CaCl_2$ added to give the final concentrations of free Ca^{++} indicated. Cells were suspended at 25 percent haematocrit in these media, A23187 was then added to a final concentration of 1 μM, except to the cells suspended in the absence of Ca^{++}. The cells were incubated for 10 min at 37°C. μ was calculated from measurements of the deformation of the red cells in a parallel plate flow chamber as described by Hochmuth et al. (1973). η_e was determined by measuring after elastic deformation of the cell, the rate of return of the cell to its undeformed shape and multiplying the relaxation time (t_c) by μ (see Hochmuth, 1982) (from Shields et al., 1985).

Table 2. Effect of Internal Trypsin on Elasticity and Viscosity of the Ghost Membrane.

Trypsin (ng ml^{-1})	(n)	μ (dynes cm^{-1}) x 10^{-3}	t_c (sec)	ηe (dynes·s cm^{-2})
0	9	2.12 ± 0.06	0.085 ± 0.002	0.18
5	5	2.50 ± 0.13	0.100 ± 0.003	0.25
10	4	2.52 ± 0.12	0.102 ± 0.007	0.26
20	1	2.40	0.110	0.26

Trypsin was added to an aliquot of 50 percent cell suspension in isotonic NaCl solution so that at haemolysis (at a ratio of cell:medium of 1:40) the final trypsin concentration was as indicated. The haemolysis medium was 2.5 mM KCl, 4 mM MgSO$_4$, 1.1 mM acetic acid, pH 4.5 at 0°C. After 10 min at 0°C, isotonicity was restored to final concentrations of 120 mM KCl, 20 mM Tris, 10 mM citrate, pH 7.6. Ghosts then resealed during a 1 hr incubation at 37°C. Measurements of μ and t_c were made at room temperature as indicated in the legend to Fig. 2 (from Shields et al., 1985).

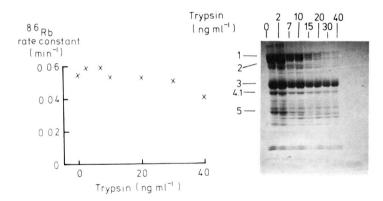

Fig. 4. left side. Rate constants for the Ca^{++} stimulated flux through the K$^+$ selective channel as a function of internal trypsin concentration. Measurements made by means of ^{86}Rb at equal K$^+$ concentrations on both surfaces of the red cell membrane (equilibrium exchange). Red cell ghosts were prepared as described in the legend to Table 2 except tracer ^{86}Rb was added to the hemolysis medium and incubated in 120 mM KCl, 20 mM Tris, 10 mM citrate, pH 7.6 with 0.04 mM CaCl$_2$ (free Ca^{++}, 1 µM). Rate constants were calculated from the time course of ^{86}Rb release.

Fig. 4. right side. SDS polyacrylamide gel electropherograms of human red cell ghosts that contained trypsin at the concentrations indicated. The ghosts were incubated for 1 hr at 37°C. Experimental conditions as in Fig. 4, left side. In gels showing greater resolution it could be seen that band 4.1, but not 4.2, disappeared at trypsin concentrations above 10 ng/ml. Thus, the decreased density of the band designated 4.1 in the figure reflects an essentially complete disappearance of band 4.1 and survival of 4.2 (from Shields et al., 1985).

Effect of internal trypsin on the cytoskeleton as revealed by SDS polyacrylamide gel electrophoresis. In spite of the absence of major changes of the viscoelastic properties of the ghost membrane, trypsination was correlated with a considerable degradation of the cytoskeleton. Over the concentration range used in the study of the mechanical properties of the membrane, trypsin produced the following effects (see Fig. 4, right side): the α and β chains of spectrin are largely digested although even at the highest concentrations some residual spectrin could still be detected. Ankyrin (band 2.1) was, perhaps, somewhat more resistant to proteolysis than spectrin, but the results were not quite clear since the rather faint band 2.1 became obscured by the appearance of a degradation product of one or both of the spectrin chains. At 20 ng/ml trypsin, however, degradation became clearly visible. Both the band 4.1 protein and actin (band 5) disappeared. This was seen only when the trypsin-containing ghosts, which had originally been resealed, were rendered permeable to peptides by washing in saponin-containing medium as in the ghosts used for the gel electropherogram shown in Fig. 4, right side. If the washes were performed in isotonic salt solutions without added saponin, the ghosts remained impermeable to peptides and the bands 4.1 and 5 were still demonstrable on the gel at their original locations (not shown). Thus, the proteins that form the cytoskeleton as well as the binding proteins that attach the mesh-work to the membrane were modified.

Effect of internal trypsin on the cytoskeleton as revealed by measuring the lateral diffusion of the band 3 protein. Trypsination in situ at the inner surface of the red cell membrane, lead to the proteolytic cleavage of the band 3 protein (MW 96 kd). The hydrophilic 42 kd domain becomes detached from the membrane while the hydrophobic 55 kd domain (which gives rise to a diffuse band that is not easily recognizable on the gels) remains attached to the bilayer (Steck, 1974). However, at the low trypsin concentrations

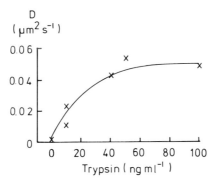

Fig. 5. Lateral diffusion of band 3 protein as measured by fluorescence microphotolysis in red cell ghosts after exposure to varying concentrations of internal trypsin. Cells were suspended at 10 percent haematocrit in a medium containing 30 mM surcose, 20 mM HEPES, 130 mM KCl, pH 8.5. Band 3 protein was labeled with either fluorescein isothiocyanate (FITC) or dichlorotriazinyl fluorescein (DCTAF). For FITC labeling, cells were incubated with 3.8 mg FITC/g cells for 24 hrs at 0°C. For DCTAF labeling, incubation was with 15 mg DCTAF/g cells for 30 min at 0°C. Both FITC- and DCTAF-labeled cells were washed twice in medium containing 0.5 percent bovine serum albumin (BSA), and then washed three times in medium without BSA. Ghosts were prepared as in the legend to Table 1. Lateral diffusion was measured at room temperature and the diffusion coefficient, D, was calculated as described by Peters (1983). The mean fluorescence recovery was 55 ± 3 percent (n = 5) (from Shields et al., 1985).

used in our experiments, most of band 3 survived and can still be demonstrated to migrate at 96 kd on SDS polyacrylamide gel electropherograms (Lepke and Passow, 1976).

The significance of the degradation of the cytoskeleton by trypsin was determined by measuring the rate of lateral diffusion of the band 3 protein. In the intact red cell membrane, the cytoskeletal mesh-work underneath the bilayer prevents the bulky protein from diffusing. The treatment with intracellular trypsin increased the lateral mobility. At 40 ng/ml the effect reached a plateau (Fig. 5). Nevertheless, some 40-50 percent of the band 3 molecules remained virtually immobile (as inferred from fluorescence recovery after photobleaching). This indicated that in spite of the considerable degradation of many cytoskeletal proteins, some structures survived that restricted the mobility of a sizeable fraction of the band 3 molecules. We believe that this is responsible for the survival of the elastic properties after treatment with internal trypsin.

The SDS polyacrylamide gel electropherograms in Fig. 4 show that the Commassie's blue stain in the band 3 region decreased with increasing trypsin concentration. This indicates a proteolytic removal of the cytoplasmic 42 kd domain, which left the poorly stained transmembrane fragment of 55 kd in the gel. It would seem possible that in the intact red cell membrane, some of the band 3 protein is attached to the cytoskeleton and, for this reason, remains inaccessible to digestion by low concentrations of intracellular trypsin. The remaining band 3 molecules could be mobile, although restricted in their diffusion to a narrow range by the meshes of the cytoskeleton. These molecules could possibly be the ones that are split by trypsin and whose lateral mobility increases as a consequence thereof, or as a consequence of the proteolytic opening of the cytoskeletal mesh-work. Regardless of whether this or some other interpretation will turn out to be correct, it is clear that these findings show the survival of structures that are essentially responsible for the mechanical properties of the red cell membrane.

<u>Absence of correlation between effects of trypsin on the cytoskeleton and Ca^{++}-activated K^+ channels.</u> At least up to a trypsin concentration of 40 ng/ml, the susceptibility of the K^+ selective channels to activation by Ca^{++} remained virtually unaltered (Fig. 4, left side). Thus, work with ghosts that underwent extensive degradation of the cytoskeleton lends no support for the existance of "dormant" K^+ selective channels whose activities are controlled by the cytoskeleton. However, the possibility cannot be excluded that the surviving structures suffice to inhibit activation of such "dormant" channels.

CONCLUSIONS

Our results extend the review of the actions of Ca^{++} on the red cell membrane summarized in Table 1 by demonstrating that micromolar levels of Ca^{++} caused changes in the mechanical properties of the cell. The underlying effect of Ca^{++} on the cytoskeleton is probably not responsible for the stimulation of K^+ efflux seen at the same micromolar levels of Ca^{++}, since major alterations of the cytoskeleton caused by trypsin had no effect on the stimulation of K^+ efflux by Ca^{++}. These major alterations did not affect the mechanical properties of the cell, which suggests that a small fraction of the cytoskeletal protein suffices for the maintenance of essentially normal mechanical properties of the red cell membrane.

ACKNOWLEDGEMENTS

P.L. La Celle was the recipient of a Humboldt Foundation Award. We thank Drs. V. Rudloff and D. Schubert for their comments on the manuscript.

REFERENCES

Allan, D. and Thomas, P., 1981, The effects of Ca^{2+} and Sr^{2+} on Ca^{2+}-sensitive biochemical changes in human erythrocytes and their membranes, Biochem. J., 198:441-445.

Anderson, D.R., Davis, J.L. and Carraway, K.L., 1977, Calcium-promoted changes of the human erythrocyte membrane, J. Biol. Chem., 252:6617-6623.

Brown, A.M. and Lew, V.L., 1983, The effect of intracellular calcium on the sodium pump of human red cells, J. Physiol., 343:455-493.

Downes, P. and Michell, R.H., 1981, Human erythrocyte membranes exhibit a cooperative calmodulin-dependent Ca^{2+}-ATPase of high calcium sensitivity, Nature, 290:270-271.

Dunn, M.J., 1974, Red blood cell calcium and magnesium: effects upon sodium and potassium transport and cellular morphology, Biochem. Biophys. Acta, 352:97-116.

Eaton, J.W., Skelton, T.D., Swofford, H.S., Kolpin, C.E. and Jacobs, H.S., 1973, Elevated erythrocyte calcium in sickle cell disease, Nature, 246:105-106.

Feig, S.A. and Bassilan, S., 1974, Abnormal RBC Ca^{2+} metabolism in hereditary spherocytosis (H.S.), Blood, 44:937.

Ferreira, H.G. and Lew, V.L., 1977, Passive Ca transport and cytoplasmic buffering in intact red cells, in: "Membrane Transport in Red Cells," J.C. Ellory and V.L. Lew, eds., pp. 53-91, Academic Press, London.

Foder, B. and Scharff, O., 1981, Decrease of apparent calmodulin affinity of erythrocyte (Ca^{2+} + Mg^{2+})-ATPase at low Ca^{2+} concentrations, Biochem. Biophys. Acta, 649:367-376.

Grygorczyk, R. and Schwarz, W., 1983, Properties of the Ca^{2+}-activated K^+ conductance of human red cells as revealed by the patch clamp technique, Cell Calcium, 4:499-510.

Grygorczyk, R., Schwarz, W. and Passow, H., 1984, Ca^{2+}-activated K^+ channels in human red cells, Biophys. J., 45:693-698.

Hamill, O.P., 1981, Potassium channel currents in human red blood cells, J. Physiol., 319:97P-98P.

Heinz, A. and Passow, H., 1980, Role of external potassium in the calcium-induced potassium efflux from human red blood cell ghosts, J. Memb. Biol., 57:119-131.

Hochmuth, R.M., 1982, Solid and liquid behaviour of red cell membrane, Ann. Rev. Biophys. Bioeng., 11:43-55.

Hochmuth, R.M., Mohandas, N., Blackshear, Jr., P.L., 1973, Measurement of the elastic modulus for red cell membrane using a fluid mechanical technique, Biophys. J., 13:747-762.

Hoffman, J.F., Yingst, D.R., Goldinger, R.M., Blum, R.M. and Knauf, P.A., 1980, On the mechanism of Ca-dependent K transport in human red blood cells, in: "Membrane Transport in Erythrocytes," U.V. Lassen, H.H. Ussing, J.O. Wieth, eds., pp. 178-192, Munksgaard, Copenhagen.

King, Jr., L.E. and Morrison, M., 1977, Calcium effects on human erythrocyte membrane proteins, Biochem. Biophys. Acta, 471:162-171.

Kirkpatrick, F.H., Hillman, D.G. and La Celle, P.L., 1975, A23187 and red cells: Changes in deformability, K^+, Mg^{2+}, Ca^{2+} and ATP, Experientia, 31:653-654.

La Celle, P.L., Kirkpatrick, F.H., Udkow, M.P. and Arkin, B., 1973, Membrane fragmentation and Ca^{++}-membrane interaction: potential mechanisms of shape change in the senescent red cell, in: "Red Cell Shape," M. Bessis, R.I. Weed, P.F. Leblond, eds., pp. 69-78, Springer-Verlag, New York.

Lepke, S. and Passow, H., 1976, Effects of incorporated trypsin on anion exchange and membrane proteins in human red blood cell ghosts, Biochem. Biophys. Acta, 455:353-370.

Lew, V.L., Bookchin, R.M., Brown, A.M. and Ferreira, H.G., 1980, Ca-sensitivity modulation, in: "Membrane Transport in Erythrocytes," U.V. Lassen, H.H. Ussing, J.O. Wieth, eds., pp. 134-138, Munksgaard, Copenhagen.

Lew, V.L., Muallem, S. and Seymour, C.A., 1982a, Properties of the Ca^{2+} activated K^+ channel in one-step inside-out vesicles from human red cell membranes, Nature, 296:742-746.

Lew, V.L., Tsien, R.Y. and Miner, C., 1982b, The physiological $[Ca^{2+}]$ level and pump-leak turnover in intact red cells measured with the use of an incorporated Ca^{++} chelator, Nature, 298:478-481.

Lichtman, M.A. and Weed, R.I., 1973, Divalent cation content of normal and ATP-depleted erythrocytes and erythrocyte membranes, in: "Red Cell Shape," M. Bessis, R.I. Weed, P.F. Leblond, eds., pp. 79-93, Springer-Verlag, New York.

Lorand, L., Bjerrum, O.J., Hawkins, M., Lowe-Krentz, L. and Siefring, Jr., G.E., 1983, Degradation of transmembrane proteins in Ca^{2+}-enriched human erythrocytes, J. Biol. Chem., 258:5300-5305.

Lorand, L., Siefring, Jr., G.E. and Lowe-Krentz, L., 1978, Formation of γ-glutamyl-ε lysine bridges between membrane proteins by a Ca^{2+}-regulated enzyme in intact erythrocytes, J. Supramol. Struc., 9:427-440.

Palek, J., 1977, Red cell calcium content and transmembrane calcium movements in sickle cell anaemia, J. Lab. Clin. Med., 89:1365-1374.

Peters, R., 1983, Fluorescence microphotolysis, Naturwissenschaften, 70:294-302.

Ponnappa, B.C., Greenquist, A.C. and Shohet, S.B., 1980, Calcium-induced changes in polyphosphoinositides and phosphatidate in normal erythrocytes, sickle cells and hereditary pyropoikilocytes, Biochem. Biophys. Acta, 598:494-501.

Porzig, H., 1975, Comparative study of the effects of propranolol and tetracaine on cation movements in resealed human red cell ghosts, J. Physiol., 249:27-49.

Porzig, H., 1977, Studies on the cation permeability of human red cell ghosts, J. Memb. Biol., 31:317-349.

Porzig, H. and Stoffel, D., 1978, Equilibrium binding of calcium fragmented human red cell membranes and its relation to calcium-mediated effects on cation permeability, J. Memb. Biol., 40:117-142.

Schatzmann, H.J., 1973, Dependence on calcium concentration and stoichiometry of calcium pump in human red cells, J. Physiol., 235:551-569.

Schatzmann, H.J., 1983, The red cell calcium pump, Ann. Rev. Physiol., 45:303-312.

Schwarz, W. and Passow, H., 1983, Ca^{2+}-activated K^+ channels in erythrocytes and excitable cells, Ann. Rev. Physiol., 45:359-374.

Shields, M., La Celle, P.L., Peters, R. and Passow, H., 1985, Modification of the red cell membrane by internal Ca^{++} and trypsin: effects on mechanical properties and K^+ channels, in preparation.

Simons, T.J.B., 1976, The preparation of human red cell ghosts containing calcium buffers, J. Physiol., 256:209-225.

Simons, T.J.B., 1982, A method for estimating free Ca within human red blood cells, with an application to the study of their Ca-dependent K permeability, J. Memb. Biol., 66:235-247.

Steck, T.L., 1974, Organization of proteins in the human red blood cell membrane, J. Cell Biol., 62:1-19.

Szasz, I., Sarkadi, B., Schubert, A. and Gardos, G., 1978, Effects of lanthanum on calcium-dependent phenomena in human red cells, Biochem. Biophys. Acta, 512:331-340.

Weed, R.I., La Celle, P.L. and Merrill, E.W., 1969, Metabolic dependence of red cell deformability, J. Clin. Invest., 48:795-809.

Wiley, J.S. and Gill, F.M., 1976, Red cell calcium leak in congenital hemolytic anemia with extreme microcytosis, Blood, 47:197-210.

Yingst, D.R. and Hoffman, J.F., 1984, Ca-induced K transport in human red blood cell ghosts containing arsenazo III: Transmembrane interactions of Na, K and Ca and the relationship to the functioning Na-K pump, J. Gen. Physiol., 83:19-45.

Yingst, D.R. and Marcovitz, M.J., 1983, Effect of haemolysate on calcium inhibition of the (Na^+-K^+)-ATPase of human red blood cells, Biochem. Biophys. Res. Comm., III:970-979.

SPECTRIN AND THE MECHANOCHEMICAL PROPERTIES OF THE ERYTHROCYTE MEMBRANE

Arnljot Elgsaeter, Arne Mikkelsen and Bjørn T. Stokke

Division of Biophysics
University of Trondheim
Trondheim, Norway

INTRODUCTION

Spectrin was first isolated by Marchesi and Steers (1968) and by Mazia and Ruby (1968), from human and bovine erythrocytes, respectively. In human erythrocytes, it is now commonly believed that spectrin together with actin, protein 4.1, and possibly protein 4.9 constitute the major part of the membrane skeleton, and that this extrinsic membrane skeleton is crucial in maintaining the mechanochemical properties of these cells (for review see Evans and Skalak, 1979; Branton et al., 1981; Gratzer, 1981; Bennett, 1982; Cohen, 1983; Sheetz, 1983). For many years it was believed that spectrin was a protein present only in the membrane skeleton of the highly specialized erythrocyte membrane. However, a few years ago, Goodman et al. (1981) reported the presence of material immunologically cross-reactive to erythrocyte spectrin in nonerythrocyte tissue. Since then, studies in many laboratories have demonstrated the presence of a family of proteins related to erythrocyte spectrin in essentially every nonerythroid tissue examined (Bennett et al., 1982; Burridge et al., 1982; Glenney et al., 1982, 1983; Kakiuchi et al., 1982; Nelson and Lazarides, 1983; Repasky et al., 1982; Aster et al., 1984). We will refer to these proteins as nonerythroid spectrins.

Erythroid as well as nonerythroid spectrin consists of two different polypeptide chains; each has a molecular weight of 200-260 kdalton. Both erythroid and nonerythroid spectrin can exist in solution as $\alpha 2\beta 2$ heterotetramers. Drying of spectrin tetramers from glycerol solutions followed by platinum rotary replication yield electron micrographs indicating that spectrin tetramers consist of two flexible elongated subunits, each about 200 nm long, interconnected at both ends (Shotton et al., 1978; Glenney et al., 1982; Kakiuchi et al., 1982). Two spectrin dimers, each about 100 nm long and connected head to head, form the heterotetramer. In the human erythrocyte membrane skeleton, the elongated spectrin molecules appear to be interconnected into a continuous molecular network on the cytoplasmic surface of the membrane lipid bilayer. The prevailing view has been that the membrane skeleton of nonerythroid cells may be similar to that of the human erythrocyte, although this has recently been questioned (Mangeat and Burridge, 1984).

Our main interest has been related to the structure, physical properties and functional mechanisms of human erythrocyte spectrin and how the physical properties of the isolated spectrin molecules may account for important aspects of the erythrocyte membrane mechanochemical properties.

SPECTRIN FLEXIBILITY

One extreme is that the energy required to bend an elongated biological macromolecule is so high that the probability that this takes place under physiological conditions is vanishingly small. Even though such a macromolecule may bend reversibly when a sufficient amount of energy is supplied and therefore strictly speaking is flexible, such a macromolecule behaves, from a functional point of view, like a stiff rod. The other extreme is that the macromolecule consists of freely jointed segments in which a change of the angle between two adjacent segments is not associated with any change in the internal energy of the macromolecule. This theoretically important, but obviously unrealistic macromolecule model, is referred to as the freely jointed chain or random coil macromolecule model. The detailed conformation of the individual molecules in a solution of such chemically-identical molecules will be changing continuously because of thermal intramolecular motion. The probability that two such molecules have the same conformation at a given time is vanishingly small. Macromolecules with flexibility between stiff rods and random coils are commonly referred to as wormlike macromolecules. The mechanochemical properties of a membrane skeleton consisting of interconnected elongated macromolecules can be expected to depend critically on the physical properties of the macromolecules constituting such a submembrane macromolecular network. To understand the molecular basis for the mechanochemical properties of the erythrocyte membrane it is therefore important to establish where on the scale between the two extremes described above spectrin belongs.

Electron microscopy can be a beautifully direct method for studying the size, shape and flexibility of macromolecules. When it is known how to prepare the specimens free of artifacts, the results are very illustrative and can be interpreted with only a minimum of theoretical background (Fig. 1). However, electron microscopy can also be highly treacherous, particularly for flexible extended macromolecules. The misleading electron micrographs of isolated spectrin published prior to the work of Shotton et al. (1978) exemplify this point very well. We have therefore mainly studied spectrin flexibility and other physical properties using physical methods that allow studies of isolated spectrin while the macromolecules are kept in aqueous solution.

When light passes through a macromolecular solution, some of the light is scattered and the time average of the intensity of the scattered light normally depends on the scattering angle. This angular dependence, among other molecular parameters, contains information about the radius of gyration of the macromolecules in the solution (Zimm, 1948). Our results have shown that the radius of gyration of human erythrocyte spectrin was about 20 nm in 100 mM NaCl solution and increased with decreasing ionic strength. This finding strongly suggested that the highly elongated spectrin molecule in solution under physiological salt conditions was a wormlike molecule (Elgsaeter, 1978). This light scattering study indicated that the number of identical segments in the equivalent freely jointed chain molecule of spectrin dimers was roughly 5-7. We observed that the spectrin radius of gyration increases strongly as the ionic strength of the solution decreased. This indicated that the change in internal energy associated with bending became comparable to and begin to exceed the thermal energy when the ionic strength became less than about 5 mM. This change in internal energy appears to be mostly due to electrostatic intramolecular interactions. Such behavior is characteristic of flexible polyelectrolytic molecules.

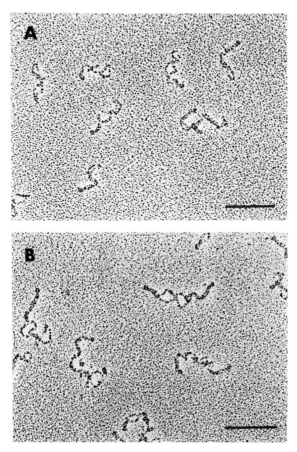

Fig. 1. Electron micrograph of isolated human erythrocyte spectrin dimers (A) and tetramers (B) dried from glycerol solution and rotary shadowed with platinum according to the procedure of Tyler and Branton (1980). Bar = 100 nm.

When an electric field is applied to a macromolecular solution, the molecules will normally tend to orient themselves relative to the direction of the electric field. This usually makes the solution birefringent. When the applied electric field is turned off abruptly, the birefringence will not disappear instantaneously but will in the simplest case follow an exponential decay determined by the rotational diffusion coefficient of the macromolecules. For stiff elongated macromolecules, the birefringence relaxation time is very sensitive to the length of the macromolecule: a doubling of the macromolecule length results in a six-fold increase in the birefringence relaxation time. If spectrin dimers were stiff molecules, spectrin tetramers would, therefore, exhibit a birefringence relaxation time six times that of the dimer provided the joint between the two dimers constituting the tetramers also was stiff. On the other hand, if the two dimers making up a tetramers were stiff, but freely jointed, the birefringence relaxation time of the tetramer would be about three times that of the dimers (Yu and Stockmayer, 1967). Measurements in our laboratory yielded a birefringence relaxation time for the spectrin dimer that is only one tenth of that expected theoretically for a 100 nm long stiff molecule. We further found no significant difference in the birefringence relaxation time of human erythrocyte spectrin dimers and tetramers (Mikkelsen and Elgsaeter, 1978, 1981). Using a lower exciting field strength, Roux and

Cassoly (1982) reported that the birefringence relaxation time of human spectrin tetramers was about 50 percent higher than that for the dimer. Both these observations indicated very strongly that spectrin is a flexible wormlike macromolecule. For macromolecules with the same contour length the birefringence relaxation time decreased as the persistence length of the molecule is decreased (Hagerman and Zimm, 1981). The measured birefringence relaxation times of human erythrocyte spectrin (Mikkelsen and Elgsaeter, 1978,1981; Roux and Cassoly, 1982) suggested a spectrin persistence length of approximately 20 nm when the ionic strength is 1 mM.

Intrinsic viscosity measurements of macromolecule solutions can also provide important information about the physical properties of elongated macromolecules. How strongly the intrinsic viscosity increases with increased macromolecular length yields information about the flexibility of the macromolecules. Because spectrin tetramers consist of two dimers connected head to head measurement of the intrinsic viscosity of dimers and tetramers can therefore provide important information about spectrin flexibility. Such measurements also clearly indicated that erythrocyte spectrin was a wormlike macromolecule (Ralston, 1976; Dunbar and Ralston, 1981; Stokke and Elgsaeter, 1981). We further found that the intrinsic viscosity increased strongly with decreasing ionic strength which is characteristic of flexible wormlike polyelectrolytes (Stokke and Elgsaeter, 1981). We have recently been studying the temperature dependence of human erythrocyte spectrin dimer intrinsic viscosity (Fig. 2). We found approximately a 30 percent reduction in spectrin dimer intrinsic viscosity when the temperature was increased from 4°C to 37°C which suggested that there was a substantial decrease in spectrin effective hydrodynamic volume when the temperature is increased. This indicated that the change in spectrin internal energy associated with bending of the molecule was comparable with the thermal energy and that the flexing of the elongated spectrin molecule due to internal thermal motion increased significantly when the temperature was increased from 4°C to 37°C.

All these physical studies of spectrin in solution support the view that human erythrocyte spectrin under physiological conditions is a flexible wormlike molecule undergoing substantial thermal intramolecular motion. Lemaigre-Dubreuil et al. (1980) and Lemaigre-Dubreuil and Cassoly (1983) came to the same conclusion using the saturation transfer electron paramagnetic resonance technique. In addition, these workers obtained data

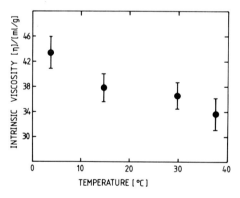

Fig. 2. Intrinsic viscosity of human erythrocyte spectrin dimers (in 10 mM NaCl and 1 mM Tris, pH 7.5) versus temperature. The measurements were carried out employing a low shear Cartesian diver viscometer (Mikkelsen, 1983) using isolated spectrin dimer off a gel filtration column and spectrin concentrations ranging from 70 to 650 µg/ml.

suggesting that human erythrocyte spectrin also preserves this flexible nature when the spin-labeled spectrin molecules are reassociated with the erythrocyte membrane.

THE ERYTHROCYTE MEMBRANE SKELETON

A swollen network of polyelectrolyte chain or wormlike macromolecules constitutes an ionic gel. The theory predicting the shear modulus and maximum extention ratio of such gels has been known for several decades (Flory, 1953; Treloar, 1975; Flory, 1976). However, a few years ago Tanaka et al. (1980) discovered experimentally that such ionic gels exhibit phase transitions and critical phenomena. They found that these phenomena are critically dependent on the environmental conditions such as pH, ionic conditions and temperature. These workers also showed that these very interesting and previously known phenomena in ionic gels can be accounted for by using an extension of the standard statistical thermodynamic theory of elastomers (Tanaka et al., 1980). From what is known about the topology of the erythrocyte membrane skeleton (Fig. 3) and the flexibility and the ionic nature (Elgsaeter et al., 1976) of the spectrin molecules constituting the major part of this skeleton, it appears reasonable to expect that the erythrocyte membrane skeleton constitutes an ionic gel. The startling new findings of Tanaka et al. (1980) may, therefore, open a completely new and so far unexplored perspective to our understanding of the molecular basis for the mechanochemical properties of the membrane skeleton of nonnucleated erythrocytes.

Standard gel theory predicts that for a two-dimensional gel in which the flexible molecules constituting a gel follow Gaussian chain statistics the gel elastic shear modulus equals

$$G = N k T \langle r^2 \rangle / \langle r^2 \rangle_0 \qquad (1)$$

where N is the number per unit area of flexible molecules with the two ends connected at two different junctions in the network, k is Boltzmann's constant, T is the absolute temperature, $\langle r^2 \rangle$ and $\langle r^2 \rangle_0$ are the mean square of the distance between the ends of the flexible molecules when the molecules are part of the gel and when they are unperturbed (free), respectively (Treloar, 1975; Flory, 1976). This expression is valid for gels with regular topology as well as randomly cross-linked two-dimensional gels. It is important to note that Eq. 1 does not rest on the assumption that the free energy of the flexible wormlike macromolecule is associated only with conformational entropy (Flory, 1976). However, in the extreme

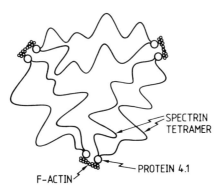

Fig. 3. A schematic illustration of the prevailing view of the topology of human erythrocyte membrane skeleton.

case where the gel is purely energetic, G is proportional to approximately the fifth power of N and not the first power, as given in Eq. 1 (Doi and Kuzun, 1980). Assuming that each of the spectrin tetramer subchains undergoes independent thermal movements and using published data on the erythrocyte total surface area (Canham, 1970) and the number of spectrin tetramers per cell (Steck, 1974) yield $N = 1560$ μm^{-2}. Because of the slow association equilibrium between the components of the erythrocyte membrane skeleton (Ungewickell and Gratzer, 1978; Liu and Palek, 1980) the ratio $<r^2>/<r^2>_0$ will be close to one under equilibrium conditions. The gel theory thus predicts that at room temperature the shear modulus of the erythrocyte membrane skeleton equals approximately 6×10^{-3} dyne/cm. Since the erythrocyte membrane is in the fluid state, this also equals the predicted value of the shear modulus for the whole erythrocyte membrane.

Several authors have measured the pressure required to aspirate a certain fraction of an erythrocyte into a microcapillary, and from these data have calculated the membrane elastic shear modulus. Waugh and Evans (1979) reported $G = (6.6 \pm 1.2) \times 10^{-3}$ dyne/cm at 25°C and Chien et al. (1978), $G = 4.2 \times 10^{-3}$ dyne/cm. The agreement between these experimental results and the value predicted by gel theory is remarkable. The former workers also measured G for the erythrocyte membrane as function of temperature and their data yielded approximately $(T/G) \partial G/\partial T = -(3 \pm 2)$ for the temperature range 5-35°C. The integrity of the membrane skeleton network and thus N in Eq. 1 depends on the following known noncovalent binding between the membrane skeleton components (Fig. 3): the spectrin dimer-dimer association, the spectrin-protein 4.1 association, the protein 4.1-F-actin association and the intramolecular associations of F-actin. The temperature dependence of the spectrin- spectrin association is partly known (Ungewickell and Gratzer, 1978) and indicates that this association contributes to a reduction in N and thus G when the temperature is increased. If the membrane skeleton stress relaxation is allowed to go to near completion before the measurements of G are done, the factor $<r^2>/<r^2>_0$ of Eq. 1 is expected to be temperature independent and close to one for all temperatures. However, if the measurement of G is done before the stress relaxation is completed the observed G will depend on the temperature dependence of $<r^2>/<r^2>_0$. For highly flexible and elongated macromolecules the intrinsic viscosity $[\eta]$ is proportional to $(<r^2>_0)^{3/2}$ (Bohdanecky and Kovar, 1981). Our data presented in Fig. 3 therefore indicate that the function $<r^2>/<r^2>_0$ will contribute to an increase in G when the temperature is elevated. However, because of all the lacking data on the various parameters involved in the integrity of the membrane skeleton, the reported temperature dependence of G for the erythrocyte membrane skeleton (Waugh and Evans, 1979) cannot yet be taken as evidence either for or against the ionic gel concept for the erythrocyte membrane skeleton.

When the flexible molecules constituting a gel have a finite contour length, the gel can only be extended to a certain maximum value. The average distance between nearest neighbor junctions in the erythrocyte membrane skeleton can be estimated from the data on cell surface area and number of tetramers per cell. Depending on the network junction functionality assumed, these data together with simple gel theory yield a maximum elastic extention ratio of 3-5, which is in excellent agreement with the experimental values for the maximum extention ratio of 3-4 (Evans and La Celle, 1975).

The theory of ionic gels predicts that a two-dimensional gel will exert an osmotic tension, π, and that this tension will depend strongly on the gel surface area, A, and the environmental conditions (Tanaka et al., 1980). The osmotic tension can be both positive and negative, and the gel theory further predicts that the modulus of gel area compressibility $(\partial \pi/\partial A)$ can vary several orders of magnitude when the environmental conditions are varied (Fig. 4). It is well known that human erythrocytes exhibit several different characteristic shapes and that transformations between these shapes can take place in response to changes in the environmental conditions. It can be shown using standard continuum mechanical methods that the favored erythrocyte shape depends both on π and $\partial\pi/\partial A$ and that change in the environmental conditions leading to change in these parameters therefore can lead to cell shape transformations (Stokke, Mikkelsen and Elgsaeter, in preparation). Molecular parameters such as net electrostatic charge and the association energy for contact between segments within the same macromolecule also enter the mathematical expression for π and $\partial\pi/\partial A$ given by the gel theory. When it has been demonstrated experimentally how these molecular parameters depend on the environmental conditions, the ionic gel theory can thus provide an up until now unexplored theoretical basis for relating changes in environmental conditions to cell shape transformations.

The available data on spectrin and the erythrocyte membrane skeleton indicate strongly that the erythrocyte membrane skeleton constitutes an ionic gel, but this remains to be verified experimentally. We have, therefore, undertaken to make in vitro three-dimensional networks of spectrin- protein 4.1-colloidal gold (Figs. 5 and 6) and have just begun a study of the various mechanochemical characteristics of such macromolecular networks intended to establish whether they exhibit mechanical properties similar to those reported by Tanaka et al. (1980) for ionic gels. Our preliminary data show that such in vitro spectrin networks exhibit a dynamic storage modulus (elastic shear modulus) at low frequency which is close to what is expected from gel theory and that the dynamic storage modulus in the same frequency region is much larger than the dynamic loss modulus. With our present instrumentation, we can measure the dynamic moduli for frequencies up to 2-3 Hz. In addition to such viscoelastic characterization of the network, we plan to carry out experiments intended to establish whether such spectrin networks undergo phase transitions of the kind reported for ionic gels by Tanaka et al. (1980).

CONCLUSION

The presently available data strongly suggest that spectrin is a flexible wormlike macromolecule. Together with the currently available information on the erythrocyte membrane skeleton topology and the ionic nature of spectrin, we feel that this strongly indicates that the erythrocyte membrane skeleton constitutes an ionic gel or "Tanaka gel." Because of the extensive theory describing such ionic gels, this so far unexplored view of the membrane skeleton opens a new and promising

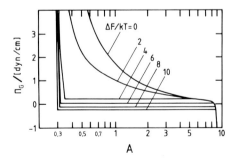

Fig. 4. Examples of osmotic surface tension π of a two-dimensional ionic gel with molecular parameters corresponding to those of the erythrocyte membrane skeleton versus gel surface area A for various values of association energy ΔF between the gel chain segments.

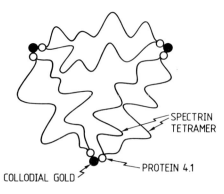

Fig. 5. Schematic illustration of the topology of the spectrin-protein 4.1-colloidal gold network. The spectrin-protein 4.1 binding is specific whereas the gold-protein 4.1 binding is unspecific.

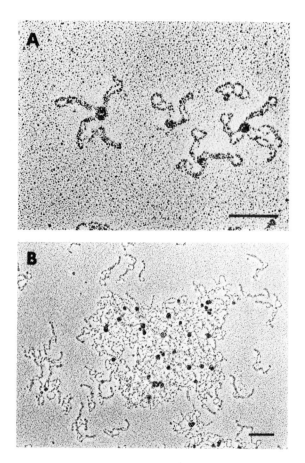

Fig. 6. Electron micrograph of protein 4.1-labeled colloidal gold bound to human spectrin dimers. Note that the protein 4.1-labeled colloidal gold binds only to one of the ends of the dimer (A). At increased concentration of spectrin tetramer and protein 4.1-labeled colloidal gold, formation of microgels takes place (B). The specimens were prepared for electron microscopy according to Tyler and Branton (1980). Bar = 100 nm.

perspective for a deeper understanding of the molecular basis for the mechanochemical properties of the erythrocyte membrane skeleton. The membrane skeleton of nonerythroid cells may be much more complicated than the case is for erythrocytes, but a better understanding of the molecular mechanisms of the erythrocyte membrane skeleton will most likely also contribute to an improved understanding of the functional mechanisms of membrane skeletons in general.

ACKNOWLEDGMENTS

We gratefully thank Dr. D. Branton for permitting one of us (B.T.S.) to stay in his laboratory to learn and practice the techniques used to prepare the components necessary to make the in vitro spectrin-protein 4.1-colloidal gold networks.

REFERENCES

Aster, J.C., Welsh, M.J., Brewer, G.J. and Maisel, H., 1984, Identification of spectrin and protein 4.1-like proteins in mammalian lens, Biochem. Biophys. Res. Comm., 199:726-734.
Bennett, V., 1982, The molecular basis for membrane-cytoskeleton association in human erythrocytes, J. Cell. Biochem., 18:49-65.
Bennett, V., Davis, J. and Fowler, W.E., 1982, Brain spectrin, a membrane associated protein related in structure and function to erythrocyte spectrin, Nature (London), 299:126-131.
Bohdanecky, M. and Kovar J., 1982, in: "Viscosity of Polymer Solutions," A.D. Jenkins, ed., pp. 44-60, Elsevier North-Holland, New York.
Branton, D., Cohen, C.M. and Tyler, J., 1981, Interaction of cytoskeletal proteins on the human erythrocyte membrane, Cell, 24:24-32.
Burridge, K., Kelley, T. and Mangeat, P., 1982, Nonerythrocyte spectrins: Actin-membrane attachment proteins occurring in many cell types, J. Cell Biol., 95:478-486.
Canham, P.B., 1970, The minimum energy of bending as a possible explanation of the biconcave shape of the human red blood cell, J. Theor. Biol., 26:61-81.
Chien, S., Sung, K.-L.P., Skalak, R., Usami, S. and Tozeren, A., 1978, Theoretical and experimental studies of viscoelastic properties of erythrocyte membrane, Biophys. J., 24:463-487.
Cohen, C.M., 1983, The molecular organization of the red cell membrane skeleton, Sem. Hematol., 20:141-158.
Doi, M. and Kuzun, N.Y., 1980, Nonlinear elasticity of rod-like macromolecules in condensed state, J. Polym. Sci., Polym. Phys. Ed., 18:408-419.
Dunbar, J.C. and Ralston, G.B., 1981, Hydrodynamic characterization of the heterodimer of spectrin, Biochim. Biophys. Acta, 667:177-184.
Elgsaeter, A., 1978, Human spectrin. I. A classical light scattering study, Biochim. Biophys. Acta, 536:235-244.
Elgsaeter, A., Shotton, D.M. and Branton, D., 1976, Intramembrane particle aggregation in erythrocyte ghosts. II. The influence of spectrin aggregation, Biochim. Biophys. Acta, 426:101-122.
Evans, E.A. and La Celle, P.L., 1975, Intrinsic material properties of the erythrocyte membrane indicated by mechanical analysis of deformation, Blood, 45:29-43.
Evans, E.A. and Skalak, R., 1979, Mechanics and thermodynamics of biomembranes, Parts 1 and 2, Crit. Rev. Bioeng., 3:181-418 (C.R.C. Press, Boca Raton).
Flory, P.J., 1953, in: "Principles of Polymer Chemistry," pp. 432-494, Cornell University Press, Ithaca, NY.
Flory, P.J., 1976, Statistical thermodynamics of random networks, Proc. R. Soc. London., Ser. A, 351:351-380.

Glenney, J.R., Glenney, P. and Weber, K., 1982, F-actin-binding and cross-linking properties of porcine brain fodrin, a spectrin-related molecule, J. Biol. Chem., 257:9781-9787.

Glenney, J.R., Glenney, P. and Weber, K., 1983, Mapping the fodrin molecule with monoclonal antibodies. A general approach for rod-like multidomain proteins, J. Mol. Biol., 167:275-293.

Goodman, S.R., Zagon, I.S. and Kulikowski, R.R., 1981, Identification of a spectrin-like protein in nonerythroid cells, Proc. Natl. Acad. Sci. USA, 78:7570-7574.

Gratzer, W.B., 1981, The red cell membrane and its cytoskeleton, Biochem. J., 198:1-8.

Hagerman, P.J. and Zimm, B.H., 1981, Monte Carlo approach to the analysis of the rotational diffusion of worm-like chains. Biopolymers, 20:1481-1502.

Kakiuchi, S., Sobue, K., Kanda, K., Morimoto, K., Tsukita, S., Tsukita, S., Ishikawa, H. and Kurokawa, M., 1982, Correlative biochemical and morphological studies of brain calspectrin: a spectrin-like calmodulin-binding protein, Biomed. Res., 3:400-410.

Lemaigre-Dubreuil, Y. and Cassoly, R., 1983, A dynamic study of the interactions between the cytoskeleton components in the human erythrocyte as detected by saturation transfer electron paramagnetic resonance of spin-labeled spectrin, ankyrin, and protein 4.1, Arch. Biochem. Biophys., 223:495-502.

Lemaigre-Dubreuil, Y., Henry, Y. and Cassoly, R., 1980, Rotation dynamics of spectrin in solution and ankyrin bound in human erythrocyte membrane, FEBS Letters, 113:231-234.

Liu, S.-C. and Palek, J., 1980, Spectrin tetramer-dimer equilibrium and the stability of erythrocyte membrane skeletons, Nature, 285:586-588.

Mangeat, P.H. and Burridge, K., 1984, Immunoprecipitation of nonerythrocyte spectrin within live cells following microinjection of specific antibodies: Relation to cytoskeletal structures, J. Cell Biol., 98:1363-1377.

Marchesi, V.T. and E. Steers, 1968, Selective solubilization of a protein component of the red cell membrane, Science, 159:203-204.

Mazia, D. and Ruby, A., 1968, Dissolution of erythrocyte membranes in water and comparison of the membrane proteins with other structural proteins, Proc. Natl. Acad. Sci. USA, 61:1005-1012.

Mikkelsen, A., 1984, Physical properties and functional mechanism of human erythrocyte spectrin and human bronchial mucin studied using electro-optic methods, viscometry and electron microscopy, dr.ing.thesis, University of Trondheim, pp. 2.1-2.13.

Mikkelsen, A. and Elgsaeter, A., 1978, Human spectrin. II. An electro-optic study, Biochim. Biophys. Acta, 536:245-251.

Mikkelsen, A. and Elgsaeter, A., 1981, Human spectrin. IV. A comparative electro-optic study of heterotetramers and heterodimers, Biochim. Biophys. Acta, 668:74-80.

Nelson, W.J. and Lazarides, E., 1983, Expression of the β subunit of spectrin in nonerythroid cells, Proc. Natl. Acad. Sci. USA, 80:363-367.

Ralston, G.B., 1976, Physio-chemical characterization of the spectrin tetramer from bovine erythrocyte membranes, Biochim. Biophys. Acta, 455:163-172.

Repasky, E.G., Granger, B.L., and Lazarides, E., 1982, Widespread occurrence of avian spectrin in non-erythroid cells, Cell, 29:821-833.

Roux, B. and Cassoly, R., 1982, Differences in the electric birefringence of spectrin dimers and tetramers as shown by the fast reversing electric pulse method, Biophys. Chem., 16:193-198.

Sheetz, M.P., 1983, Membrane skeletal dynamics: Role in modulation of red cell deformability, mobility of transmembrane proteins, and shape, Sem. Hematol., 20:175-188.

Shotton, D.M., Burke, B.E. and Branton, D., 1978, The shape of spectrin molecules from human erythrocyte membranes, Biochim. Biophys. Acta, 536:313-317.

Steck, T.L., 1974, The organization of proteins in the human red blood cell membrane, J. Cell Biol., 62:1-19.
Stokke, B.T. and Elgsaeter, A., 1981, Human spectrin. VI. A viscometric study, Biochim. Biophys. Acta, 640:640-645.
Tanaka, T., Fillmore, D., Sun, S.-T., Nishio, I., Swislow, G. and Shah, A., 1980, Phase transitions in ionic gels, Phys. Rev. Letters, 45:1636-1639.
Treloar, L.R.G., 1975, in: "The Physics of Rubber Elasticity," 3rd Edition, pp. 59-79, 101-127, Claredon Press, Oxford.
Tyler, J. and Branton, D., 1980, Rotary shadowing of extended molecules dried from glycerol, J. Ultrastr. Res., 71:95-102.
Ungewickell, E. and Gratzer, W., 1978, Self-association of human spectrin. A thermodynamic and kinetic study, Eur. J. Biochem., 88:379-385.
Waugh, R. and Evans, E.A., 1979, Thermoelasticity of red cell membrane, Biophys. J., 26:115-132.
Yu, H. and Stockmayer, W.H., 1967, Intrinsic viscosity of a once-broken rod, J. Chem. Phys., 47:1369-1373.
Zimm, B.H., 1948, The scattering of light and the radial distribution function of high polymer solutions, J. Chem. Phys., 16:1093-1099.

EFFECTS OF LOCAL ANESTHETICS AND pH CHANGE ON THE PLATELET "CYTOSKELETON"
AND ON PLATELET ACTIVATION

Vivianne T. Nachmias[*] and Robert M. Leven[+]

[*]Department of Anatomy
School of Medicine, University of Pennsylvania
Philadelphia, Pennsylvania, USA

[+]Specialized Center for Thrombosis Research
Temple University
Philadelphia, Pennsylvania, USA

INTRODUCTION

While platelets may at first appear to be very specialized fragments of cytoplasm, their cytoskeletons share many properties with that of more typical cells. We shall use the term cytoskeleton to mean the Triton-X 100 insoluble residue, obtained under defined conditions, which, when used on platelets attached to a surface, retains the shape of the cell. Most cells are in various states of activity. Since platelets circulate in a resting state, from which they are readily activated, they offer advantages for the study of factors that alter the state of the cytoskeleton. Also, since platelets lack intermediate filaments, properties of the cytoskeletal system that are observed in platelets cannot depend on the presence of the 10 nm filament system.

Morphologically, the resting state mentioned above is represented by a remarkable biconvex disc. The otherwise smooth surface membrane leads into tortuous channels that penetrate into the cytoplasm and are in close contact with a second membrane system, somewhat similar to the T tubule-sarcoplasmic reticulum arrangement of muscle (White, 1972). These channels may allow stimuli received at the membrane to be rapidly transmitted into the underlying cytoplasm. Structural changes occur in platelets within seconds after the addition of an agonist such as ADP or thrombin. Two changes can be distinguished. Filopodia, which can be as long as a micron or so, are formed. Second, the discoid cell body alters into an irregular spherical shape. The term "shape change" which is embedded in the literature is usually used to describe the second change in platelet morphology, since it can be detected by light scattering and even by eye as a change in the "swirling" character of the platelet suspension when it is stirred.

EFFECTS OF LOCAL ANESTHETICS

In our early efforts to examine the structural changes that occurred during filopodial extension and shape change, we concluded that it was first essential to obtain platelets in the fully resting state. Since the platelet cytoplasm is very dense, routine transmission electron microscopy provided little detail and needed to be supplemented by other techniques. It is extremely difficult to obtain platelets outside the body and separated from the blood plasma without causing some degree of activation: the first sign is the extension of filopodia. For example, if one gel filters the platelets at 37°C in suitable media and then subsequently leaves the platelets undisturbed for 10-15 minutes, only about 50 percent of the platelets are perfect discs, i.e., lack filopodia. This however is enough for the study of individual platelets, since the filopodia can be detected and one can select among the population for those which lack filopodia. Gently placing a drop of this suspension, which contains some 100-200 thousand platelets per cubic millimeter on a carbon filmed grid will allow some of the platelets to stick to the grid. They can then be treated with nonionic detergents, such as Triton-X 100, which solubilize the membrane lipids and integral membrane proteins, rinsed with various solutions which will solubilize more components, and then treated with a negative contrast agent such as uranyl acetate or other electron dense stains. Under favorable conditions, the density will surround the remaining structures and allow even relatively small proteins, such as actin filaments, to be detected as light elements surrounded by dense stain. Fig. 1 shows the appearance of such a platelet which lacks filopodia altogether. It can be seen that the cytoplasm on the grid, that lies around the remains of the microtubule, appears amorphous or with short fragments of microfilaments, but lacks long microfilaments, bundles of microfilaments or thick filaments. Nevertheless, the shape of the demembranated platelet is retained.

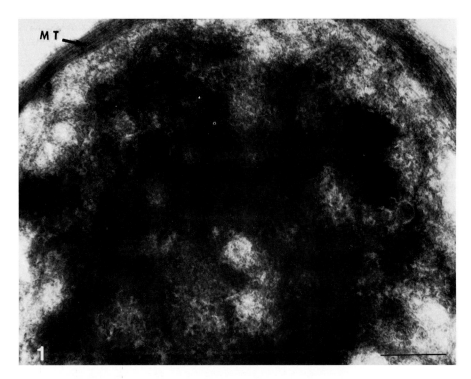

Fig. 1. Negatively stained, Triton-X 100 insoluble residue of the cytoskeleton of a perfectly discoid platelet resting on a carbon coated electron microscope grid. Note the lack of long microfilaments in the center and the subfilaments of the microtubule coil (MT) which is coursing around the periphery of the disc.

While this careful selective technique is useful for the examination of single platelets, it is not adequate if one desires to obtain the "cytoskeleton" in bulk in order to study its components. We discovered that with use of very low doses of local anaesthetics, such as 0.5-2 mM tetracaine (or higher doses of lidocaine) for short periods, up to 30 minutes, one could obtain a reversible retraction of the filopodia (Nachmias et al., 1977). The reversibility of the phenomenon should be emphasized, since with slightly higher concentrations (3-5 mM) and/or longer times (60-90 minutes) the effect becomes irreversible (Nachmias et al., 1979). An example will be given below. The structure of platelets treated briefly with low doses of tetracaine could not be distinguished from that of control platelets either by negative staining or by examining briefly-lysed whole mounts (Nachmias, 1980).

In order to obtain an entire population of platelets in the resting state, we added low doses of tetracaine directly to plasma, and then gel filtered the platelets in the presence of the anesthetic. We obtained a population of platelets that were no longer quite flat discs, but could be described best as ellipsoids, as shown in the scanning electron micrograph of Fig. 2. The retraction of filopodia, which is a striking feature of the population, was quite reversible. That is, when such platelets were gel filtered again in the absence of tetracaine, they reverted to the original state, discs, with filopodia. We lysed the treated platelets after the first gel filtration, i.e. in the filopodial suppressed state, to obtain a Triton-X 100 insoluble residue (Gonnella and Nachmias, 1981). Under our conditions (0.1 M KCl, 10 mM imidazole, 10 mM EGTA, pH 6.8, 5°C) a precipitate formed within 5 minutes. This had been observed first with untreated platelets by Lucas et al. (1978). This precipitate was examined by negative staining, and is shown in Fig. 3a. The character of this precipitate is granular, with very short fragments of microfilaments but no long filaments or bundles. We concluded that the use of low concentrations of tetracaine had allowed us to isolate in bulk the "resting cytoskeleton," and that it did not consist of long microfilaments, but of both amorphous material and of short pieces of microfilaments. When we prepared the precipitated material for SDS gel electrophoresis, we saw a very simple pattern in the high molecular weight region (Fig. 4). There were only two major bands with lower mobility than actin; one at 250,000 daltons has been shown to be an actin binding protein (Lucas et al., 1976) and the other, at 100,000 daltons, is an alpha-actinin (Langer et al., 1984).

When we incubated with tetracaine for longer periods, we observed two new effects. First, the SDS gels of the precipitate prepared in exactly the same way as before, showed that there was loss of the actin binding protein, with the appearance of a new band at about 130,000 daltons. In whole platelet lysates, a comparison of the time course of disappearance of the 250K dalton band, and also of a 235K dalton band, with the reversibility of filopodial extension is shown in Fig. 5. The actin binding protein was intact at 30 min and the platelets were still reversibly affected as indicated by (R). After 120 minutes or more, that band was strongly diminished and the alteration in platelet shape was irreversible (I). At this time the platelets became spherical, and a few began to show internal Brownian movement indicative of breakdown of structure. The 235K band parallels the 250K band.

In parallel with these changes, the cytoskeleton of single platelets lost the spherical shape of the platelet, and spread over the grid surface as shown in Fig. 6.

Since it has been shown that the actin binding protein can bundle actin, it is possible that its proteolysis prevents the formation of the microfilament bundles that make up the filopodia. What, then, causes the

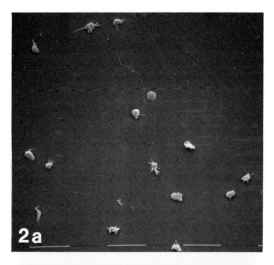
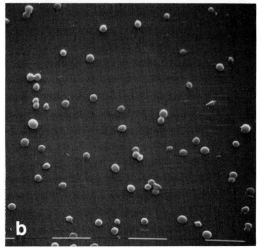

Fig. 2. Scanning electron micrographs of platelets treated for 10 minutes at 37°C in plasma with a) 0.5 percent dimethyl sulfoxide (DMSO) or b) 0.5 mM tetracaine in DMSO. The platelets were then gel filtered in Tyrode's buffer containing 0.5 percent DMSO or 0.5 mM tetracaine in DMSO, fixed and collected on a glass coverslip and prepared for scanning microscopy. Note the complete retraction of preparation-induced filopodia in the tetracaine treated sample. Bar = 10 μm.

proteolysis? It has been suggested for many years that anesthetics can cause the loss of or release of calcium from membranes. An identical proteolysis of the actin binding protein was shown by Phillips and Jakabova (1972), to be due to a calcium-activated protease. In intact platelets, this protease could be activated by an ionophore such as A23187. Interestingly, increases in calcium in purified extracts also activated a 90K protein which could then "cap," or limit the extent of actin filaments or, at high enough calcium levels, sever them (Wang and Bryan, 1981). This effect may explain the short filaments seen after A23187 (Fig. 3b) and the inability of platelets to reextend filopodia after longer tetracaine treatments. However this was not the case at early time points because the isolated cytoskeleton was capable of responding to form microfilament bundles as will now be described.

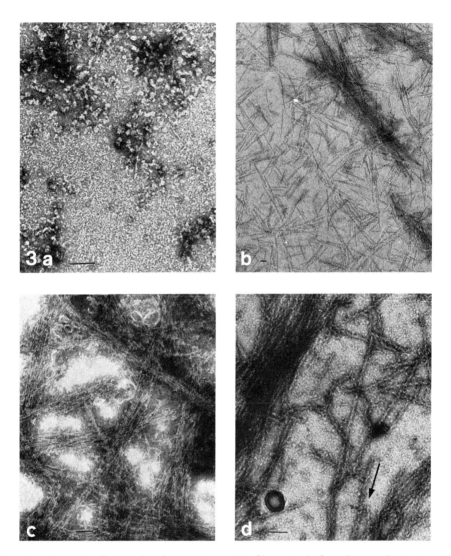

Fig. 3. Negatively stained samples of bulk cytoskeletal precipitates from a) platelets gel filtered in 1.5 mM tetracaine; b) platelets gel filtered and treated with 2 μM calcium ionophore A23187; c) residue as in a) further treated for 1 hr in the original buffer altered to contain 0.5 mM calcium chloride and to be pH 7.6; d) identical residue treated with heavy meromyosin subfragment 1 at 1 mg/ml -- note arrowheads. Bar = 0.1 μm. (Reprinted, in part, from Gonnella and Nachmias, 1981, with permission.)

Since we had obtained what we believed to be an analogue of the "resting cytoskeleton" in bulk, by the use of low, reversible concentrations of tetracaine, we thought it might be possible to alter it into a state resembling that in activated platelets. Most investigators believe that activation is due to a rise in free calcium levels within the platelet. The most straightforward experiment was to incubate the isolated resting cytoskeleton in isotonic buffer at pH 7.0, containing calcium at about 100 μM. These attempts met with very limited success. We were impressed by reports from studies of egg and sperm activation which showed that cytoplasmic alkalinization was a constant feature and that in fact

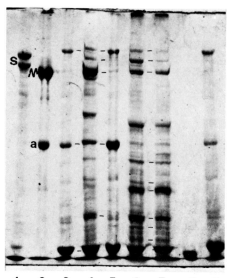

Fig. 4. Cytoskeleton as prepared in Fig. 3a, b and c and electrophoresed according to the Laemmli method on a 7 percent SDS gel. Lane 1, spectrin standard; lane 2, myosin and alpha-actinin; lane 3, cytoskeleton from platelets treated briefly with low doses of tetracaine as in Fig. 3a; lane 4, cytoskeleton of A23187-treated platelets as in Fig. 3b; lane 5, coelectrophoresis of lane 3 with authentic alpha-actinin - note enhancement of the a band; lanes 6 and 7, supernatants from preparations of lanes 3 and 4; lane 8, actin; lane 9, cytoskeleton after the transformation shown in Fig. 3c. Note the reduction in the alpha-actinin band. (Reprinted, from Gonnella and Nachmias, 1981, with permission.)

alkalinization alone could cause some, though not all, of the effects of egg and sperm activation (Tilney et al., 1978). Furthermore, Begg and Rebhun (1979) had shown that the cortex of a fertilized egg displayed microvilli and increased numbers of microfilaments compared to the unfertilized egg. If they isolated the cortex from an unfertilized egg in buffer at pH 7.6 in the presence of calcium, it partly resembled that of a fertilized egg, especially in the presence of increased numbers of microfilaments. We tried something similar, and were excited to find that the "resting cytoskeletons" were similarly altered by such treatments. When examined after 1 hr in the high pH buffer, such cytoskeletons now consisted of meshworks and bundles of microfilaments, and sometimes, remarkable paracrystalline structures which seemed to be made of very tightly compacted filaments since intermediate forms could be seen. Results of this type of experiment are shown in Fig. 3b, c and d. Since the microfilaments could be treated with heavy meromyosin subfragment one to form arrowhead structures, they were clearly composed, at least in part, of actin.

This transformation showed very clearly that tetracaine did not inhibit the actin and associated proteins from forming bundles. It served to distinguish the reversible effect from the subsequent irreversible effect. It has been hypothesized that an effect of local anesthetics is to dissociate the attachment of the cytoskeleton from the membrane (Woda et al., 1980). We have obtained some evidence related to this idea. We looked at the effect of 1-2 mM tetracaine on shape change as studied by spectrophotometry. After 30 <u>seconds</u> in the drug, there is no apparent inhibition of the rate or extent of

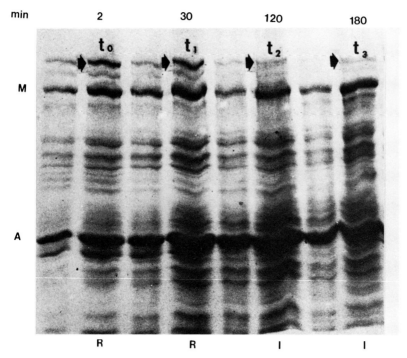

Fig. 5. SDS 5-20 percent gradient slab gel of whole platelet lysates showing the time course of proteolysis of the actin binding protein and the 235K band located just below it. Platelets were gel filtered, pelleted, and taken up into sample buffer after the period of incubation noted at the top. Ten and 20 µl were loaded for each time point. At 2 and 30 min, the tetracaine concentration was 1.5 mM; at 120 and 180 min, the level was 3 mM. R: platelets reverted to discs with filopodia when gel filtered into Tyrode's buffer without tetracaine; I: platelets remained spherical, with no filopodia.

shape change by either ADP (10 µM) or A23187 (0.13 µg/ml), each of which caused rapid increases in optical density, loss of oscillations, and morphological change as assessed by phase microscopy. However, after 90 seconds, by which time tetracaine itself has caused some loss of oscillations, ADP lost its ability to cause shape change, while the ionophore still had an effect, even when added after the ADP. This difference may indicate that the platelets could no longer respond to a signal that had to be transmitted through the plasma membrane, but could still respond, though more weakly than in the controls, to a signal which arrived directly by release of calcium from inner stores by the action of the calcium ionophore.

To summarize, our results suggest that tetracaine, lidocaine, and presumably other local anesthetics may have at least two sites of action. One site is probably at the plasma membrane itself, and causes loss of ability of the platelet to respond to signals, even when the internal structural aspects of the cytoplasm, i.e. the cytoskeleton, are still capable of responding to stimuli such as increased free calcium in the presence of increased pH. This early effect also is connected to retraction of filopodia. Whether or not it is related to an increase in membrane area, perhaps displacing receptors from ion channels, or related to an uncoupling from the cytoskeleton is not known. The second and later effect seems to be

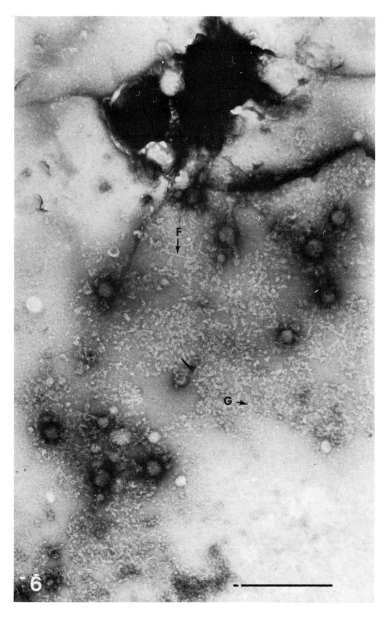

Fig. 6. Cytoskeleton of a platelet 120 min after tetracaine. Cf. Fig. 1; note loss of platelet shape, and both short microfilaments (F) and granular matrix (G). Bar = 0.5 μm. (Reprinted from Nachmias et al., 1977, with permission.)

related to release of calcium, presumably from internal stores, since our experiments were done in 10 mM EGTA. This effect rapidly lead to irreversible inactivation of the platelet by action of a calcium-activated protease, as deduced from the effects on the high molecular weight components of the whole platelet lysate or of the cytoskeleton itself. When A23187 was used at high enough doses to cause loss of proteolysis, it also caused platelet sphering and loss of structure. Thus, at high doses, the anesthetics and the ionophores share a similar mode of action.

pH CHANGES

Since pH increase appeared important for the change in the platelet cytoskeleton in vitro, we next tried to see whether increasing internal pH could cause shape change in intact platelets. At first, we were unsuccessful. Then, one of us (R.M.L.) found that this was because we had been looking for fast responses. When we suspended platelets in the presence of monensin or nigericin, monovalent cation ionophores that will exchange sodium for potassium or protons, shape change did occur, but at a very slow rate. What was especially interesting was that the platelets changed shape only if the external pH was relatively alkaline. The spectrophotometric results were confirmed by phase and scanning electron microscopy. Indeed, if the pH was lowered to 6.8 again during the shape change response, it stopped (Fig. 7). We determined that the shape change was not due to release of serotonin or ADP by the platelets due to the effect of monensin, since any serotonin release was below the threshold for activation, and since ATP, a competitive inhibitor of ADP induced shape change, did not affect the rate (Leven et al., 1983a). It also could not be due simply to depolarization of the membrane, since that would occur at neutral pH. Similar effects could also be shown for megakaryocytes (Leven et al., 1983b). These platelet precursor cells were isolated and cultured. When stimulated with the platelet agonists ADP or thrombin, they responded by forming a ruffled membrane, extending filopodia and spreading widely over the substratum in about 20 to 30 minutes. On the other hand, they could be stimulated to ruffle and spread in response to the calcium ionophore but only if an alkalinizing agent, such as methyl amine or monensin, at pH above 7, was also used.

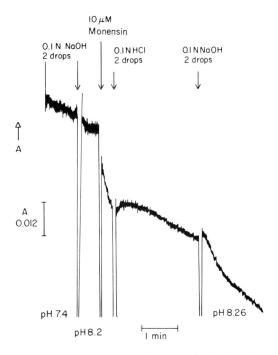

Fig. 7. Spectrophotometric tracing of monensin-induced shape change. The oscillations of discoid platelets at pH 7.4 are not altered after pH increase (first arrow) but are rapidly altered after monensin (second arrow). Platelets changed from discs to spheres as shown by scanning EM. (From Leven et al., 1983, with permission.)

The work of Simons et al. (1982) indicated that the internal pH of platelets rose upon activation with thrombin. However, platelet metabolism and proton extrusion were also stimulated, making it hard to determine whether such pH effects might be 'bystander' effects, i.e. unrelated to causal changes. Recently we have tested further the potential importance of pH by suspending platelets in weak acids, which, when the external pH is lowered to the range of 6.2-6.7, should enter the cells and cause decreased internal pH (Roos, 1975; De Hemptinne et al., 1983).

The results are striking. Shape changes at neutral pH were hardly altered from the normal response to 10 micromolar ADP when weak acids are used at pH 7.35-7.40. A typical response, using propionic acid, is shown in Fig. 8. However, when the pH was rapidly dropped, in this case to 6.25, the baseline showed a marked downward drift (see tracing at upper right). This correlated with an increase in cell volume, as shown by a double label technique and also by electron microscopy. If ADP was added now, the response showed an increased latency, a slowing in the rate of increase of optical density (after suitable correction for the downward drift) and a slowed diminution of the oscillations. The same phenomenon was observed with a wide variety of weak acids, both those which are metabolized, like

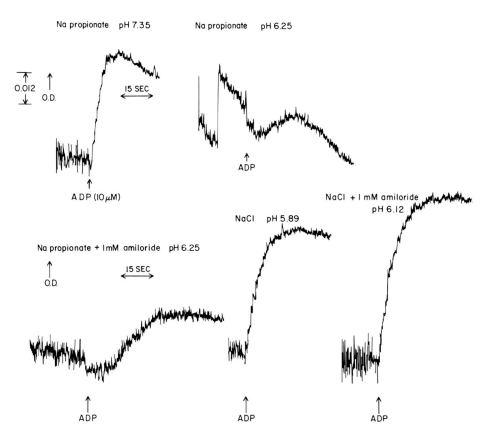

Fig. 8. Platelet rich plasma was diluted 1:1 in 135 mM NaCl, 5 mM KCl, 10 mM EGTA, 10 mM imidazole, pH 7.4, and incubated at 37°C. It was diluted further to about 10^8 platelets per ml with an equal volume of buffer in which the NaCl was replaced with Na propionate. Shape change was measured in a Gilford spectrophotometer at 609 nm, with constant stirring at a volume of 1.5 ml. ADP was added from a 1 mM stock solution. The pH was altered with a 1 M NaAc.

lactate, acetate and pyruvate, and those which are not, like propionic acid or dimethyl oxazolidinedione. Effects were seen at 22 mM acid or above.

What is most interesting about the weak acid effect is that the change in baseline seen here with propionic acid could be prevented completely by the addition of amiloride, which blocks sodium-proton exchange (Grinstein et al., 1984 and personal communication). Despite this stabilization of the baseline as shown in the left lower tracing of Fig. 8, the inhibition of shape change rate was not lost; in fact, it might be slightly more prominent. Therefore, the effect of the weak acids in inhibiting the rate of shape change can be separated from an effect due to cell swelling.

If the pH was lowered when the platelets were suspended in weakly buffered sodium chloride, there was a very transient decrease in the rate of platelet shape change, which for most donors was rapidly restored to essentially normal levels, as also shown in the central lower tracing (Fig. 8). In this diluted plasma, amiloride did not inhibit shape change (right lower tracing; Fig. 8). At these levels, the inhibition due to the weak acids was also observed when the calcium ionophore A23187 was used as an agonist (data not shown). This strongly suggests, but does not prove conclusively, that pH affected the cytoskeleton directly. Finally, if the suspending medium is first made acid and then rapidly realkalinized, the rate of shape change shows an overshoot. This is shown in the data of Table 1. This effect was not seen with hypotonic media in which platelets also swell, showing by a second criterion that the effects of platelet swelling were different from those of weak acids.

These data show that weak acids at low pH could inhibit both the rate of initiation of shape change and the rate of shape change. At pH lower than shown here (ca. pH 6.0), they could inhibit shape change completely. The existence of the rebound effect showed that the effect is reversible, and also suggested that the effect might depend not only on the absolute

Table 1. Relative Rates and Extents of Shape Change in Sodium Chloride (Cl), Sodium Acetate (Ac) or Hypotonic (H) Medium. Realkalinization is indicated by (R). ADP 10 μM; Chart, 1 inch = 5 sec

		pH	Rate	%	Height	%
1	Cl	6.20	52	100	55	100
2	Ac	6.20	26	50	35	64
3	H	6.17	23	44	31	56
4	Ac (R)	6.88	59	113	57	103
5	H (R)	7.4	35	67	27	49
6	H (R)	6.8	32	62	27	49
7	Ac (R)	7.05	51	98	46	84
8	Cl (R)	7.17	31	60	25	45

1-3 were tested 1 min after diluting from PRP into acidified media.

5-8 were treated as 1-3 for 80 sec (acidified) then re-alkalinized with 1 M NaOH. The rate was measured 40 sec later except for 7, which was 115 sec later.

Note that platelets in Ac (45 mM) overshoot the control rate (No. 2) when realkalinized; those in hypotonic medium (230 mosmoles) show a much smaller effect.

level of pH, but might be affected by the rate of which protons were lost from the cell. To distinguish between this and a rapid pH change which returns to normal with an overshoot it will be necessary to use probes which can record rapid pH changes. Although there are significant differences between the effects of weak acids and the effects of anesthetics, a recent most interesting study of lymphocytes (Corps et al., 1982) showed that dibucaine, at levels of 250-500 µM, caused a very large increase in the rate of lactate production. It may be one effect of the anaesthetics to cause the production of an internal weak acid, and hence to lead, at low doses, to the retraction of filopodia mentioned above, an effect which was also seen with weak acids. Finally, our data clearly show that there is now strong evidence that, while calcium release may be the primary event in causing platelet shape change, the role of pH in affecting or controlling the response of the cytoskeleton cannot be ignored.

ACKNOWLEDGMENTS

These investigations have been supported by Grant No. HL-15835 (Pennsylvania Muscle Institute).

REFERENCES

Begg, D.A. and Rebhun, L.I., 1979, pH regulates the polymerization of actin in the sea urchin egg cortex, J. Cell Biol., 83:241-248.
Corps, A.N., Hesketh, T.R. and Metcalfe, J.C., 1982, Limitations on the use of phenothiazines and local anaesthetics as indicators of calmodulin function in intact cells, FEBS Lett., 138:280-284.
De Hemptinne, A., Marrannes, R. and Vanheel, B., 1983, Influence of organic acids on intracellular pH, Am. J. Physiol., 245 (Cell Physiol. 14):C178-183.
Gonnella, P.A. and Nachmias, V.T., 1981, Platelet activation and microfilament bundling, J. Cell Biol., 89:146-151.
Grinstein, S., Cohen, S. and Rothstein, A., 1984, Cytoplasmic pH regulation in thymic lymphocytes by an amiloride-sensitive Na^+H^+ antiport, J. Gen. Physiol., 83:341-369.
Langer, B.G., Gonnella, P.A. and Nachmias, V.T., 1984, Alpha-actinin and vinculin in normal and thrombasthenic platelets, Blood, 63:606-614.
Leven, R.M., Gonnella, P.A., Reeber, M.J. and Nachmias, V.T., 1983, Platelet shape change and cytoskeletal assembly: effects of pH and monovalent cation ionophores, Thromb. Haemostas., 49:230-234.
Leven, R.M., Mullikin, W. and Nachmias, V.T., 1983a, Role of sodium in ADP-and thrombin-induced megakaryocyte spreading, J. Cell Biol., 96:1234-1240.
Lucas, R.C., Gallagher, M. and Stracher, A., 1976, Actin and actin-binding protein in platelets, in: Proceedings of the International Symposium on "Contractile Systems in Non-Muscle Tissues," S.V. Perry, A. Margreth and R.S. Adelstein, eds., pp. 133-139, Elsevier North-Holland Biomedical Press, Amsterdam.
Lucas, R.C., Rosenberg, S. and Shafig, S., 1978, The isolation and characterization of a cytoskeleton and a contractile apparatus from human platelets, Protides Biol. Fluids, 26:465-470.
Nachmias, V.T., 1980, Cytoskeleton of human platelets at rest and after spreading, J. Cell Biol., 86:795-802.
Nachmias, V.T., Sullender, J. and Asch, A., 1977, Shape and cytoplasmic filaments in control and lidocaine-treated human platelets, Blood, 50:39-53.
Nachmias, V.T., Sullender, J. and Fallon, J.R., 1979, Effects of local anesthetics on human platelets: filopodial suppression and endogenous proteolysis, Blood, 53:63-72.

Phillips, D.R. and Jakabova, M., 1972, Calcium dependent protease in human platelets, J. Biol. Chem. 252:5602-5605.

Roos, A., 1975, Intracellular pH and distribution of weak acids across cell membranes. A study of D- and L-lactate and of DMO in rat diaphragm, J. Physiol. (London), 249:1-25.

Simons, E.R., Schwartz, D.B. and Norman, N.E., 1982, Stimulus response coupling in human platelets:thrombin-induced changes in pH, in: "Intracellular pH. Its Measurement, Regulation and Utilization in Cellular Functions," R. Nuccitelli and D.W. Deamer, eds., pp. 463-482, Alan R. Liss, Inc., New York (Kroc Foundation Series).

Tilney, L.G., Kiehart, D.P, Sardet, C. and Tilney, M., 1978, Polymerization of actin IV. Role of Ca^{++} and H^+ in the assembly of actin and in membrane fusion in the acrosomal reaction of echinoderm sperm, J. Cell Biol., 77:536-550.

Wang, L-L. and Bryan, J., 1981, Isolation of calcium-dependent platelet proteins that interact with actin, Cell, 15:637-649.

White, J.C., 1972, Interaction of membrane systems in blood platelets, Am. J. Pathol., 66:295-312.

Woda, B.A., Yguerabide, J. and Feldman, J.D., 1980, The effect of local anesthetics on the lateral mobility of lymphocyte membrane proteins, Exp. Cell Res., 126:327-331.

INHIBITION OF MOTILITY BY INACTIVATED MYOSIN HEADS

Robin Jones and Michael Sheetz

Department of Physiology
University of Connecticut Health Center
Farmington, Connecticut, USA

INTRODUCTION

Muscle contraction and many other forms of cell motility, such as cell locomotion, cytoplasmic streaming, and cytokinesis, are believed to involve myosin dependent force generation. (For reviews, see Korn, 1978; Weeds, 1982). Our understanding of myosin structure and function has come largely from studies of movement of myosin relative to actin in muscle cells. The currently accepted model suggests that a myosin molecule moves on actin as it undergoes a conformational change of the head region while bound to actin (Huxley, 1969; Adelstein and Eisenberg, 1980). Myosin filaments translocate along actin filaments through a series of individual myosin movements (steps) along the actin filament. After each step, the heads release from the actin so that other mysoin heads can move the myosin filament forward. In most cells, it is currently impossible to quantify myosin movement on actin, and there has been no assay for myosin-based motility in vitro. Recently, however, an in vitro assay was developed (Sheetz and Spudich, 1983a,b) for measuring the movement of nonmuscle as well as muscle myosins on actin. With this assay, it is now possible to correlate myosin motility in vitro with muscle contraction and myosin ATPase activity. In the past year, the in vitro assay has been useful in defining a number of important aspects of myosin motility.

In this chapter we present studies on the inhibition of myosin motility by myosin and a major fragment of myosin, heavy meromyosin (HMM), with modified sulfhydryls. Sulfhydryl modification was chosen because oxidants are a major cause of myosin inactivation in vitro and probably in vivo as well. In addition, sulfhydryl-modified myosins have been utilized to test for myosin involvement in cell function (Meusen and Cande, 1979). Thus, an understanding of how sulfhydryl modified myosins inhibit myosin motility can aid in understanding the mechanisms of oxidant induced alteration of cell motility. Three distinct mechanisms of altering motility are evident from these and previous studies which are defined in the Discussion.

The in vitro motility assay utilizes myosin-coated polymer beads to follow the myosin position as it moves on actin cables of the giant algal cell, Nitella (Sheetz et al., 1984). Myosin is assembled into filaments at low ionic strength and covalently attached to the bead surface. In the intact Nitella, actin cables are aligned in a parallel array on the cytoplasmic surface of chloroplasts which coat the plasma membrane (Kersey and Wessells, 1976; Kersey et al., 1976); these cables are believed to support the rapid cytoplasmic streaming observed in Nitella (Kamiya, 1981). Although the Nitella actin may differ from other actins in some aspects, we have found that it will bind a number of actin binding proteins including the spectrin-like protein from chicken gut (Vale, Mooseker and Sheetz, unpublished results) and rabbit skeletal troponin-tropomyosin complex (Vale, Szent-Gyorgyi and Sheetz, unpublished results) in a normal manner. Thus it is possible to utilize this assay to study actin-dependent as well as myosin-dependent regulation of motility.

The effects of sulfhydryl group modification on myosin activity have been studied extensively, particularly in muscle cells. Modification of four myosin sulfhydryls (two on each myosin head) with an alkylating agent such as N-ethylmaleimide (NEM) is required to block myosin's actin-activated ATPase activity (Bremel et al., 1972). In that state, myosin will still bind to actin but it no longer releases in the presence of ATP. NEM-inactivated myosin has been used to modify muscle contractility (Bremel et al., 1972) but the tendency of intact myosin to form filaments under cellular ionic conditions has limited its usefulness in other cells. The heavy meromyosin (HMM) fragment of myosin remains soluble at low ionic strength and contains two myosin heads and actin binding sites. NEM inactivated HMM has been shown to inhibit cell division and has been used generally to test for myosin involvement in cell function (Meusen and Cande, 1979). We have compared the inhibition of bead motility produced by NEM-inactivated myosin on the surface of the bead with that produced by NEM-HMM added to Nitella actin.

MATERIALS AND METHODS

Materials

Nitella axillaris was cultured from original stocks provided by Dr. L. Taiz (University of California, Santa Cruz).

Skeletal myosin was prepared from rabbit muscle by the method of Kielly and Harrington (1959) and was stored at 4°C as a stock solution in 0.6 M KCl, 50 mM potassium phosphate, pH 6.5, 0.5 mM DTT, and 0.5 mM EDTA. HMM was prepared according to the procedure of Lowey et al. (1969).

Myosin-Bead Preparation

For the normal assay, Covaspheres MX particles (0.97 µm in diameter, 2.5×10^{10} beads/ml; Covalent Technology Corp., Ann Arbor, MI) were mixed with a myosin stock solution and water in the ratio of 1:1:8. The final buffer conditions were 0.5 mM potassium phosphate, pH 6.5, 50 mM KCl, 0.2 mM DTT, and 0.03 mM EDTA. Most often the beads aggregate into groups about 5-10 µm in diameter, but the rates of movement have been shown to be independent of bead diameter between 0.6 and 120 µm (Sheetz and Spudich, 1983b).

Nitella Dissection

Nitella internodal cells (2-4 cm in length and about 0.7 mm in diameter) were trimmed free of branch cells. Cells were rinsed briefly with distilled H_2O and placed in dissection buffer (10 mM imidazole, pH 7.0, 10 mM KCl, 4 mM $MgCl_2$, 2 mM EGTA, 50 mM sucrose, and 1 mM ATP), 3 ml for HMM studies and 10 ml for myosin studies. The dissection was performed at 22°C in 50 mm plastic petri dishes on a layer (2-3 mm thick) of Sylgard (Dow Corning). First a cell was secured at both ends with pins of tungsten wire (1-3 mm in length, 0.003 inch in diameter) sharpened at one end by electrolysis in 2 M NaOH. Next, the cell was opened with a transverse cut using microscissors (Moria, MC1036 from Fine Science Tools Ltd., North Vancouver, B.C., Canada), and then opened along its whole length taking care not to disturb chloroplasts in the process. Then, transverse cuts were made at each end of the lengthwise cut, and the central portion was opened and pinned flat. The cytoplasm was washed away largely in the process of cutting and pinning the cell to the Sylgard. Occasionally, vesicular materials from the Nitella continued to move along the chloroplast rows after the cell was cut open, but this material ran off the ends of the pinned substratum within a few minutes.

Motility Assay

Myosin-coated beads were mixed in a ratio of 1:1 with dissection buffer containing 0.3 M sucrose. Samples were then drawn into a microcapillary tube and 0.1-1 µl was applied to a limited region of the dissected Nitella. The assay was performed at 23°C. In the figures, each data point represents the average of 5-15 beads or distinct bead aggregates.

Bead movement was recorded with a video camera mounted on an Olympus BH-4 microscope and connected to a Gyyr video cassette recorder (Model NV-8050). A 50x water-immersion objective lens (Leitz) was used; the magnification on the monitor screen was 2000x. The beads were viewed using bright field optics. Analysis of movement was performed on replay from the video cassette recorder where it was possible to freeze the motion. The positions of actively moving beads were traced on a piece of plastic placed on the monitor screen; the tape was then advanced and the new positions of the beads were recorded. Since the time is also recorded on the tape, the velocities of the beads were easily determined.

RESULTS

Addition of NEM-HMM to the dissection buffer inhibited bead movement. The degree of inhibition was strongly dependent upon concentration and the time of reaction of NEM with HMM. We determined the time course of reaction of NEM with HMM by measuring the degree of inihibition of myosin-coated bead mobility by NEM-HMM added to the Nitella dissection buffer. HMM (4.93 mg/ml) was incubated with 1 mM excess NEM for periods ranging from one hour to 48 hours and as shown in Fig. 1, inhibition of motility was strongly dependent on the time of reaction. Maximal inhibitory activity of NEM-HMM was seen in the interval between 15 and 24 hours of reaction and further reaction resulted in a reduction in inhibition. In subsequent studies, therefore, the 24 hour reacted NEM-HMM was used.

The concentration dependence of NEM-HMM inhibition of myosin bead motility was explored as shown in Fig. 2. A concentration of NEM-HMM required for total inhibition of bead motility was 50 µg/ml. The extent of inhibition was nearly linear with concentration of NEM-HMM. The ionic strength of the medium is known to have a dramatic effect on the affinity of HMM for actin (Greene, 1981). We therefore explored the effect of higher KCl concentrations on the NEM-HMM dependent inhibition of motility. As seen in Fig. 2, the increase in ionic strength from 10 to 50 mM KCl causes a dramatic increase in the concentration of NEM-HMM required to inhibit bead motile activity. The inhibition by NEM-HMM was linear with concentration in 50 mM KCl but 100 µg/ml was required for complete inhibition.

In addition, we examined whether the inhibition was caused by potassium or chloride by substituting potassium acetate for potassium chloride and found that 50 µg/ml of NEM-HMM completely inhibited movement in 50 mM potassium acetate. To control for an effect of NEM-HMM on total ATP levels, perhaps through an effect of contaminating active HMM, fully active HMM was added to the bead assay. If NEM-HMM were inhibiting movement by decreasing the ATP concentration, fully active HMM would lower the ATP level further.

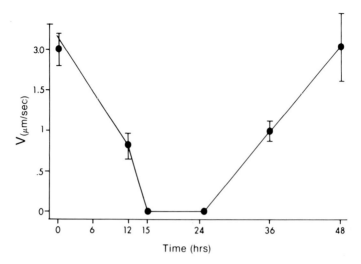

Fig. 1. Time dependence of NEM treatment of HMM on bead velocity. One ml of HMM (4.9 mg/ml) was incubated at 0°C with 1 mM excess of freshly prepared NEM. At indicated times, 200 μl aliquots were removed and the reaction stopped with 3 mM DTT. Motility assays were performed in dissection buffer containing 10 mM imidazole, 0.5 mM EGTA, 10 mM KCl, 4 mM MgCl$_2$, pH 7.0 and 0.05 mg/ml NEM-HMM.

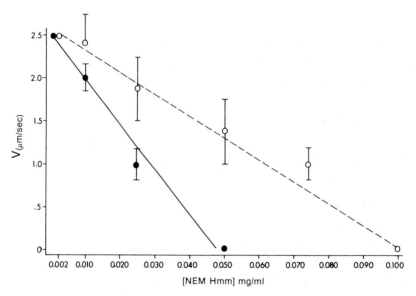

Fig. 2. The effect of NEM-HMM concentration on bead velocity. Conditions as described in Materials and Methods. NEM-HMM, at concentrations indicated, was placed in 3 ml dissection buffer containing 10 mM imidazole, 0.5 mM EGTA, 10 mM KCl, 4 mM MgCl$_2$, pH 7.0 (●) or 3 ml dissection buffer containing 10 mM imidazole, 0.5 mM EGTA, 50 mM KCl, and 4 mM MgCl$_2$, pH 7.0 (○).

Fully active HMM, however, had no effect on bead motile activity at 50 μg/ml. We also added NEM-HMM to the beads at an inhibitory concentration of 100 μg/ml and saw much less inhibition of bead motile activity, suggesting that NEM-HMM was indeed acting by binding to the Nitella actin. These latter experiments further control for the possibility that NEM-HMM is binding to myosin and acting through a myosin dependent interaction as well.

NEM-myosin was mixed with fully active myosin on the bead surface. As seen in Fig. 3, a 20 percent addition of NEM-myosin to active myosin caused a 50 percent inhibition of bead velocity and almost total inhibition occurred with 70 percent NEM-myosin. Inhibition by NEM-myosin on the bead surface, however, was not strongly dependent upon ionic strength as shown in Fig. 3.

DISCUSSION

As noted above, there are three major mechanisms whereby inactivated myosin can alter motile function. In earlier studies of muscle, where calcium dependence is conferred by the troponin-tropomyosin complex, the binding of low amounts of NEM-inactivated heads relieves the inhibition of troponin-tropomyosin even in the absence of calcium by preventing the troponin-tropomyosin complex from binding appropriately to the actin. Because the troponin-tropomyosin complex is not thought to be a primary controlling mechanism for nonmuscle motility (Hitchcock, 1977), we believe the above mechanism is not likely to be important for nonmuscle systems. There are two other mechanisms of altering motile function which are illustrated by NEM inactivated myosin heads in the present studies. In one case, the NEM-inactivated myosin heads coat the cellular actin and block the binding of active heads (Fig. 4a); in the other case inactive heads on the structure which is moving (for example myosin filaments or myosin-coated beads) bind to actin and act as a drag on movement (Fig. 4b.).

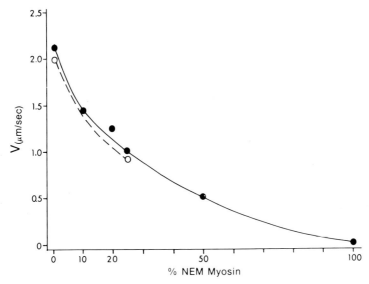

Fig. 3. The effect of NEM-myosin concentration on bead velocity. Assay conditions as described in Materials and Methods. Rabbit skeletal myosin (1 mg/ml) was incubated with 2 mM NEM for 16 hrs and the reaction stopped by 3 mM DTT. Mixtures of active rabbit skeletal myosin and 16-hr-treated myosin were used in the assay. All experiments ●----● done with 0.2 mM ATP in 10 ml of dissection buffer containing 10 mM imidazole, 0.5 mM EGTA, 10 mM KCl, 4 mM $MgCl_2$, pH 7.0. All experiments o----o done with 1 mM ATP in 10 mM imidazole, 0.5 mM EGTA, 50 mM KCl, 4 mM $MgCl_2$, pH 7.0.

Inhibition by NEM-HMM

NEM-HMM inhibits movement of myosin-coated beads on Nitella actin apparently by coating the Nitella actin cables. Earlier studies have shown that HMM will decorate the Nitella actin (Kersey et al., 1976a) and motility of the myosin-coated beads on Nitella actin has been shown to require myosin-actin interaction (Sheetz and Spudich, 1983a). Further, our evidence confirms that NEM-HMM does not act by binding myosin or the myosin-coated beads nor is it reducing ATP levels within the medium required for motile activity.

The affinity of NEM-HMM for actin is the major factor in determining its inhibitory potential in these experiments since there is always an excess of NEM-HMM for the Nitella actin. By altering the ionic conditions, therefore, to decrease the affinity of NEM-HMM for actin, we may decrease the inhibitory potential of NEM-HMM. THe HMM fragment has two heads, whereas the moving structure, the myosin-coated bead, has many heads which are interacting with the Nitella actin. Although raising the salt concentration is expected to lower the affinity of all myosin heads for actin, the structure with many heads (i.e., the bead) will retain a high affinity overall for actin. Because the bead can remain bound to the actin, it can move forward as actin binding sites are exposed when NEM-HMM molecules release and return to the medium. In fact, it was observed that at higher KCl concentrations, higher concentrations of NEM-HMM were required to inhibit motile activity. From the K^+ acetate experiments, it appeared that chloride and not potassium caused the weakening of myosin head and actin interaction. Within nonmuscle cells the ratio of myosin to actin is very low and high concentrations of NEM-HMM have to be added to block motility either by microinjection of cells or by perfusion in permeabilized cell systems (Meusen and Cande, 1979).

NEM-Myosin Inhibition of Motility

As discussed below, the inhibition of motility by NEM-myosin might be relevant to the alteration of motility in vivo by virtue of damage to myosin by oxidants. It was previously observed (Sheetz et al., 1984), that the velocity of myosin-coated beads is decreased by 50 percent by the presence of only 20 percent NEM-myosin on the bead surface. When NEM-myosin is present on the bead, it will not release from actin with ATP; bead movement forward then necessitates the breaking of the NEM-myosin-actin interaction. For the same amount of NEM-inactivated heads in the medium, therefore, the presence of the NEM heads on the bead surface is a much more effective inhibitor than their presence on the actin filaments themselves. This concept is suggested by Fig. 4, i.e., with multiple myosin heads on a bead, a free region of actin may be found when the actin is partially coated whereas for the beads that contain NEM-myosin to move forward all the myosin heads must release from actin and an attachment will significantly slow the bead movement.

Inhibition of Motility by Oxidation of Motile Proteins

It is perhaps worthwhile to speculate about the types of alterations in cell motile function that oxidation can produce. Although we have not discussed them previously, there are many actin-binding proteins in cells which could interfere with myosin movement on actin (Weeds, 1982). Free radicals and other reactive species could produce covalent linkages between these proteins and actin thereby coating portions of the actin filaments. As we have seen here, the coating of portions of the actin filaments only inhibits motile activity in proportion to the fraction of actin coated. More significant inhibition could be produced by the crosslinking of actin-binding proteins to moving structures. In that situation the actin-binding

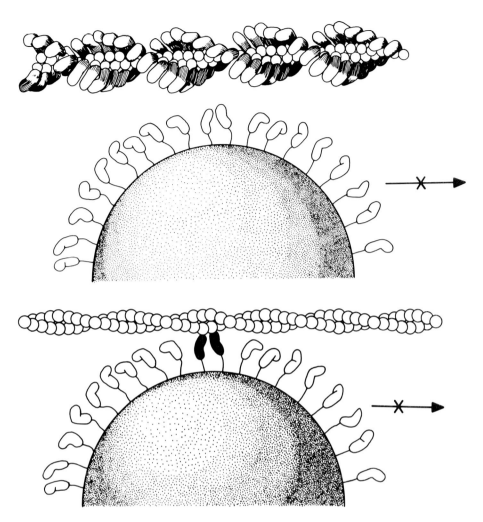

Fig. 4. Two different mechanisms of inhibition of motile activity by NEM-inactivated myosin heads are shown. In panel A, the actin filaments are coated with NEM-HMM and active myosin on the bead is unable to interact with the actin. In panel B, NEM-myosin heads are bound to actin but do not release with ATP; therefore, force generated by active myosins is expended in breaking the NEM-myosin bond with actin.

proteins would act as drags like the NEM-myosin because they would release from actin only slowly as the myosin driving force exceeded the strength of the bond interaction between actin and the actin-binding protein. Finally, as discussed above, myosin itself could be oxidized and inhibit motility. With both actin-binding protein attachment to the moving structure and myosin oxidation, the percentage inhibition of motility could far exceed the percentage of total protein undergoing oxidation.

ACKNOWLEDGMENTS

This work was supported by NIH grant GM-33351. M.P.S. is an established investigator of the American Heart Association.

REFERENCES

Bremel, R., Murray, J. and Weber, A., 1972, Manifestations of cooperative behavior in the regulated actin filament during actin-activated ATP hydrolysis in the presence of calcium. Cold Spring Harbor Symp. Quant. Biol., 37:267-275.

Greene, L.E., 1981, Comparison of the binding of heavy meromyosin and myosin subfragment 1 to F-actin, Biochemistry, 20:2120-2126.

Hitchcock, S.E., 1977, Regulation of motility in nonmuscle cells, J. Cell Biol., 74:1.

Huxley, H.E., 1969, The mechanism of muscular contraction, Sci. 164:1356-1366.

Kamiya, N., 1981, Physical and chemical basis of cytoplasmic streaming, Ann. Rev. Plant Physiol., 32:205-236.

Kersey, Y.M. and Wessells, N.K., 1976, Localization of actin filaments in internodal cells of Characean algae, J. Cell Biol., 68:264-275.

Kersey, Y.M., Hepler, P.K., Palevitz, B.A. and Wessells, N.K., 1976a, Polarity of actin filaments in Characean algae, Proc. Natl. Acad. Sci., 73:165-167.

Kielly, W.W. and Harrington, W.F., 1959, A model for the myosin molecule, Biochim. Biophys. Acta, 41:401-411.

Korn, E.D., 1978, Biochemistry of actomyosin-dependent cell motility, Proc. Natl. Acad. Sci., 75:588-599.

Lowey, S., Slayter, H.S., Weeds, A.G. and Baker, H., 1969, Substructure of the myosin molecule. I. Subfragments of myosin by enzymatic degradation, J. Mol. Biol., 42:1-29.

Meusen, R.L. and Cande, W.Z., 1979, N-ethylmaleimide-modified heavy meromyosin. A probe for actomyosin interactions, J. Cell Biol., 82:57-65.

Sheetz, M.P. and Spudich, J.A., 1983a, Movement of myosin-coated fluorescent beads on actin cables in vitro, Nature, 303:31-35.

Sheetz, M.P. and Spudich, J.A., 1983b, Movement of myosin-coated structures on actin cables, Cell Motility, 3:485-488.

Sheetz, M.P., Chasen, R. and Spudich, J.A., 1984, ATP-dependent movement of myosin in vitro: Characterization of a quantitative assay, J. Cell Biol. (in press).

Weeds, A., 1982, Actin binding proteins. Regulators of cell architecture and motility, Nature, 297:811-815.

INTRACELLULAR TRANSLOCATION OF INORGANIC PARTICLES

Arnold R. Brody, Lila H. Hill, Thomas W. Hesterberg,
J. Carl Barrett and Kenneth B. Adler[*]

Laboratory of Pulmonary Pathobiology
National Institute of Environmental Health Sciences
Research Triangle Park, North Carolina, USA

[*]Department of Pathology
University of Vermont
Burlington, Vermont, USA

INTRODUCTION

A variety of inorganic particles may be inhaled during environmental or occupational exposures. A number of studies with animals have shown that inhaled toxic particles such as silica (Brody et al., 1982) and asbestos (Brody et al., 1981; Brody and Hill, 1982) accumulate in the lung interstitium soon after exposure. The presence of interstitial asbestos is known to induce the progression of fibrotic scarring (Selikoff and Lee, 1978) by undefined mechanisms.

As a part of ongoing studies on the cellular mechanisms through which inhaled particles cause interstitial lung disease, we have attempted to determine the pathways of particle translocation from air spaces to interstitium. We propose that actin-containing microfilaments of the alveolar epithelium play a role in moving inhaled particles to the lung interstitium where the particles accumulate and induce fibrotic disease. Furthermore, it is also likely that cytoskeletal components are integral in the translocation of phagocytized intracellular particles to a perinuclear location where abnormal chromosomal segregation and incomplete cytokinesis have been documented. A combination of in vivo and in vitro studies have been carried out to investigate the interactions of cytoskeletal elements with inorganic particles.

METHODS AND MATERIALS

In Vivo Studies

White rats were exposed to aerosolized chrysotile asbestos for 1 or 3 hrs and sacrificed at 0, 24 or 48 hrs post-exposure. The lungs were fixed by perfusion through the vasculature with a glutaraldehyde paraformaldehyde solution and prepared for transmission and scanning electron microscopy (SEM) as previously described (Brody et al., 1981).

The bifurcations of alveolar ducts, where the asbestos particles originally deposited (Brody and Roe, 1983), were dissected from the lung parenchyma according to established techniques (Brody and Roe, 1983; Brody, 1984; Warheit et al., 1984). To stabilize microfilaments in the alveolar epithelium, lung tissue was dissected into 3 mm^3 blocks and immersed in 25 percent glycerol:75 percent KCl for 3 hrs. Then, the tissue was placed in 0.5 percent KCl containing dialyzed heavy-meromyosin for 18 hrs. Blocks were fixed and embedded for conventional electron microscopy (Adler et al., 1981).

In Vitro Studies

Epithelial cells from a tracheobronchial cell line (Mossman et al., 1980) and fibroblasts from Syrian hamster embryos (SHE) (Barrett et al., 1977) were plated on plastic cover slips and maintained in culture. Chrysotile asbestos fibers or silica crystals were added to cell cultures at a concentration of ~0.1 mg/ml of culture medium. After 2, 24, and 48 hrs of dust exposure, the cultures were fixed in glutaraldehyde and prepared for SEM by critical point drying. The numbers of particles on the cell surface, in the cytoplasm and surrounding the nucleus were determined by secondary and backscattered electron imaging (Warheit et al., 1983).

SHE cells were studied after microfilament stabilization with heavy meromyosin (HMM) and surface membrane extraction with Triton X-100. Cultured SHE cells containing asbestos were treated with HMM at a concentration of 1.5 mg/ml in PBS. Cell membranes were removed by treatment with a 0.1 percent solution of Triton in 10 mM Hepes buffer. The cells were fixed in glutaraldehyde and critical point dried for SEM. High resolution SEM was carried out on a JEOL 100CX TEMSCAN in order to visualize the 50-70 Å filaments of the cytoskeleton.

RESULTS AND DISCUSSION

In Vivo Studies

Asbestos inhaled by the rats was deposited initially on the surface of alveolar duct bifurcations (Fig. 1). With time, numerous fibers were taken up by the underlying Type I epithelial cells (Fig. 2). Subsequently, asbestos fibers were translocated to the underlying basement membranes and interstitial cells and connective tissue (Brody et al., 1981; Brody and Hill, 1982). Microfilaments measuring 40-80 Å were recognized in clear association with the intracellular asbestos fibers (Fig. 3). The HMM-treated lung tissue revealed networks of microfilaments in apposition to asbestos fibers (Fig. 3). Interestingly, some asbestos fibers in the interstitium exhibited complexes of microfilaments (Brody and Hill, 1982).

The preliminary data support the hypothesis that microfilaments of the alveolar epithelium could play a role in translocating asbestos fibers through the cytoplasm (Brody et al., 1983). This was suggested by the finding of ultrastructural associations between the fibers and HMM-stabilized filaments which measure 40-80 Å. Actin-containing microfilaments reportedly are stabilized by treatment with HMM prior to aldehyde fixation (Trotter, 1981). Actin-containing microfilaments are known to participate in the movement of intracellular organelles such as lysosomes (Hartwig et al., 1977) and mucin granules (Adler et al., 1981). In addition, it is particularly significant that Tsilibary and Williams (1983) have clearly demonstrated the presence of actin-containing microfilaments in alveolar epithelial cells of the rat lung. Further studies will be necessary to prove that such microfilaments play a role in translocating inhaled toxic particles to the lung interstitium where the fibrotic response is manifested.

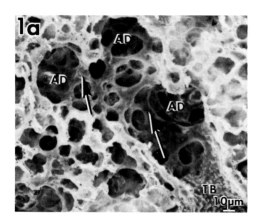 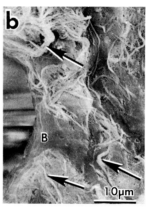

Fig. 1. a) Scanning electron micrograph of lung tissue from a rat. The terminal bronchiole (TB) leads to alveolar ducts (AD) which divide at bifurcations (arrows); b) Higher magnification SEM of first alveolar duct bifurcation (B) from a rat exposed to chrysotile asbestos for 1 hr. The duct surface is littered with large numbers of chrysotile fibers (arrows) which will be cleared by macrophages (Warheit et al., 1984) or by the alveolar epithelium (see Fig. 2) (Brody et al., 1981).

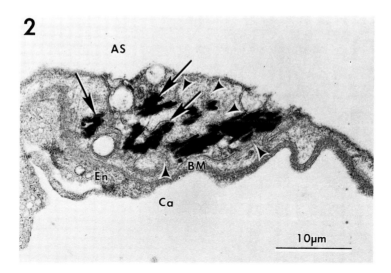

Fig. 2. Electron micrograph of Type I alveolar epithelium containing asbestos fibers (arrows) phagocytized during the first hour of exposure. Networks of microfilaments (arrowheads) surround many of the fibers. The capillary (Ca) space is lined by endothelium (En), and a thin basement membrane (BM) separates the epithelium and endothelium. Epithelium, basement membrane, and endothelium comprise the "alveolar-capillary membrane" which lines alveolar air spaces (AS).

In Vitro Studies

Both tracheal epithelial cells and fibroblasts phagocytized asbestos fibers in culture (Fig. 4). Previous studies have shown that, in epithelial cells, this uptake of fibers caused an increased ratio of filamentous

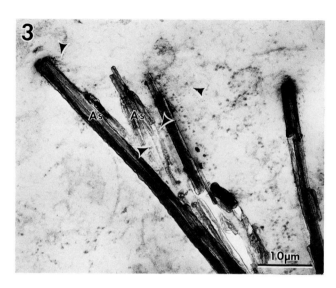

Fig. 3. Electron micrograph of asbestos (As) in alveolar epithelial cells from lung tissue pretreated with heavy meromyosin (HMM). Thin filaments (arrowheads) stabilized with HMM are associated with asbestos fibers.

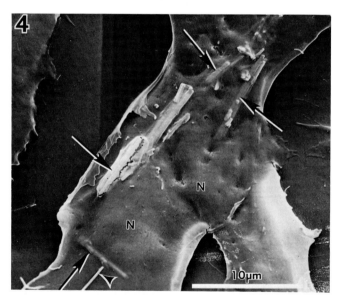

Fig. 4. Scanning electron micrograph of a Syrian hamster fibroblast which has phagocytized numerous asbestos fibers (arrows). A single thin fiber (arrowhead) remaining on the cell surface provides comparison for fibers beneath the cell membrane (arrows). Most of the fibers have accumulated around the nuclear region (N).

(polymerized) to nonpolymerized actin (Brody et al., 1983). Further studies are ongoing to corroborate these findings and to establish whether or not actin polymerization is a generalized response to particle uptake.

SHE fibroblasts phagocytized numerous asbestos fibers. Twenty-four hrs after treatment with 1 $\mu g/cm^2$ of crocidolite asbestos, the cells exhibited an average of 20 fibers per cell. Seventy percent of these were totally embedded and 16 percent of the fibers were associated with the nucleus, arranged in a perinuclear fashion (Fig. 4). The percentage of perinuclear fibers increased over time from 20 percent at 2 hrs after exposure to over 65 percent by 48 hrs post-treatment. Previous studies have clearly demonstrated that the SHE cells undergo neoplastic transformation as a result of asbestos uptake (Hesterberg et al., 1982; Hesterberg and Barrett, 1984). The fibers apparently induce cytogenetic effects (Oshimura et al., 1984) which are exemplified by binucleate aneuploid cells with lagging chromosomes and micronuclei. These nuclear anomalies were likely to be caused by the perinuclear distribution of the internalized asbestos (Figs. 4 and 5). Thus, when the cells underwent division, fibers that had been translocated to the nucleus were in a position to interfere with normal chromosomal segregation and cytokinesis, possibly through interactions with cytoskeletal components. In preliminary studies to address this issue, we have shown that HMM-stabilized, Triton-extracted SHE cells exhibited a familiar organization of the cytoskeleton (Fig. 5). After asbestos treatment, it appeared that perinuclear fibers were intimately bound between the cytoskeleton and the nucleus (Fig. 5). Individual actin filaments appeared to be attached to the fiber surfaces (Fig. 6). Further studies are ongoing to determine whether cytoskeletal elements actually bind to particle surfaces and if this reaction causes alterations in normal cell functions.

SUMMARY

Actin-containing microfilaments appear to interact with inorganic particles in a variety of cell types both in vivo and in vitro. The filaments may play an integral role in translocating inhaled particles through alveolar epithelial cells to the interstitium where fibrotic lung disease is manifested. In addition, cytoskeletal elements are likely to

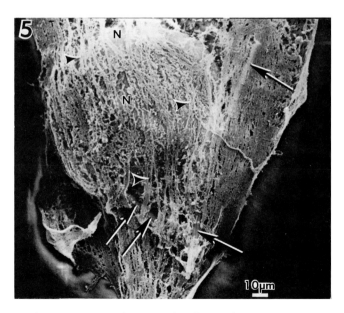

Fig. 5. Scanning electron micrograph of a triton-extracted HMM- treated, triton-extracted SHE cell. Removal of the cell membrane reveals the linear pattern of the cytoskeleton (arrowheads) as bundles of filaments pass over and around the nucleus (N). Two asbestos fibers (arrows) are observed within the cytoskeletal matrix adjacent to the nucleus.

Fig. 6. High magnification scanning electron micrograph of asbestos fibers (arrows) within the cytoskeletal matrix of a triton extracted, HMM-treated SHE cell. Individual actin filaments (arrowheads) appear to be attached to the asbestos fibers.

play a major role in transporting intracellular asbestos particles to a perinuclear location; this may in turn result in chromosomal aberrations and abnormal cytokinesis. The preliminary findings presented here provide a basis for ongoing studies on the mechanisms of intracellular particle translocation.

REFERENCES

Adler, K.B., Brody, A.R. and Craighead, J.E., 1981, Studies on the mechanism of mucin secretion by cells of the porcine tracheal epithelium, Proc. Soc. Exp. Biol. Med., 166:96-106.

Adler, K.B., Craighead, J.E., Vallyathan, N.V. and Evans, J.N., 1981, Actin-containing cells in human pulmonary fibrosis, Amer. J. Pathol., 102:427-437.

Barrett, J.C., Crawford, B.D., Grady, D.L., Hester, L.D., Jones, P.A., Benedict, W.F. and Ts'o, P.O.P., Temporal acquisition of enhanced fibrinolytic activity by Syrian hamster embryo cells following treatment with benzo(a)pyrene, Cancer Res., 37:3815-3823.

Brody, A.R., 1984, The early pathogenesis of asbestos-induced lung disease, Scan Elect. Mic., 1:167-171.

Brody, A.R. and Hill, 1982, Interstitial accumulation of inhaled chrysotile asbestos fibers and consequent formation of microcalcifications, Amer. J. Pathol., 109:107-114.

Brody, A.R. and Roe, M.W., 1983, Deposition pattern of inorganic particles at the alveolar level in the lungs of rats and mice, Amer. Rev. Resp. Dis., 128:724-729.

Brody, A.R., Hill, L.H., Adkins, B. and O'Connor, R.W., 1981, Chrysotile asbestos inhalation in rats: Deposition pattern and reaction of alveolar epithelium and pulmonary macrophages, Amer. Rev. Resp. Dis., 123:670-679.

Brody, A.R., Roe, M.W., Evans, J.N. and Davis, G.S., 1982, Deposition and translocation of inhaled silica in rats: quantification of macrophage participation and particle distribution in alveolar ducts. Lab. Invest., 47:533-542.

Brody, A.R., Hill, L.H. and Adler, K.B., 1983, Actin-containing microfilaments of pulmonary epithelial cells provide a mechanism for translocating asbestos to the interstitium, Chest, 83:11-12.

Craighead, J.E. and Mossman, B.T., 1982, The pathogenesis of asbestos-associated diseases, N. Eng. J. Med., 306:1446-1455.

Hartwig, J.H., Davies, W.A. and Stossel, T.P., 1977, Evidence for contractile protein translocation in macrophage spreading, phagocytosis and phagolysosome formation, J. Cell Biol., 75:956-967.

Hesterberg, T.W. and Barrett, J.C., 1984, Dependence of asbestos and mineral dust-induced transformation of mammalian cells in culture on fiber dimension, Cancer Res., 44:2170-2180.

Hesterberg, T.W., Cummings, T., Brody, A.R. and Barrett, J.C., 1982, Asbestos induces morphological transformation of Syrian hamster embryo cells in culture, J. Cell Biol., 95:449.

Mossman, B.T., Ezerman, E.B., Adler, K.B. and Craighead, J.E., 1980, Isolation and spontaneous transformation of cloned lines of hamster tracheal epithelial cells, Cancer Res., 40:4403-4409.

Oshimura, M., Hesterberg, T.W., Tsutsui, T. and Barrett, J.C., 1985, Correlation of asbestos-induced cytogenetic effects with cell transformation of Syrian hamster embryo cells in culture. Cancer Res. (in press).

Selikoff, I.J. and Lee, D.H.K., 1982, "Asbestos and Disease," Academic Press, New York.

Trotter, J.A., 1981, The organization of actin in spreading macrophages, Exp. Cell Res., 132:235-248.

Tsilibary, E.C. and Williams, M.C., 1983, Actin in peripheral rat lung: S_1 labeling and structural changes induced by cytochalasin, J. Histochem. Cytochem., 11:1289-1297.

Warheit, D.B., Hill, L.H. and Brody, A.R., 1983, Pulmonary macrophage phagocytosis: Quantification by secondary and backscattered electron imaging, Scan. Elect. Mic., 4:431-437.

Warheit, D.B., Chang, L.Y., Hill, L.H., Hook, G.E.R., Crapo, J.D., and Brody, A.R., 1984, Pulmonary macrophage accumulation and asbestos-induced lesions at sites of fiber deposition, Amer. Rev. Resp. Dis., 129:301-310.

RECEPTOR CYCLING IN PITUITARY CELLS: BIOLOGICAL CONSEQUENCES OF INHIBITION OF RECEPTOR-MEDIATED ENDOCYTOSIS

Patricia M. Hinkle, Jane Halpern, Patricia A. Kinsella and Risa A. Freedman

Department of Pharmacology and the Cancer Center
University of Rochester, School of Medicine and Dentistry
Rochester, New York, USA

INTRODUCTION

The anterior pituitary gland of mammalian species secretes six major trophic hormones: adrenocorticotropic hormone, two gonadotropins, thyrotrophin, prolactin and growth hormone. Synthesis and secretion of these pituitary hormones is regulated in part by hypothalamic factors which reach the pituitary gland by a portal system and which act upon specific target cells to stimulate or inhibit hormone production. In addition, pituitary function is regulated by circulating hormones including steroids and peptides.

Pituitary cell lines have proven extremely valuable in studying the biochemical mechanisms of control by peptide and steroid hormones. The most widely studied are pituitary cell lines that secrete two hormones, prolactin and growth hormone. There are a number of clones in use, and since there are no important differences in the mechanisms of peptide hormone action in these lines, they are referred to collectively here as GH cells. GH cells offer the advantage of a homogeneous population; the cells are available in theoretically unlimited quantities and there is a large literature describing their properties. Regulation of hormone production by GH cells has been the subject of a number of recent reviews (Tashjian, 1979; Tixier-Vidal and Gourdji, 1981; Gourdji et al., 1981; Hinkle, 1984).

GH cells synthesize prolactin and growth hormone, package the hormones in secretory granules, and secrete them at a relatively constant rate in the absence of added regulators. It takes 15 min, from the time a molecule of prolactin or growth hormone is synthesized, for the first newly-synthesized hormone to appear in the culture medium, representing the time needed to package the hormone into a granule, move the granule to the plasma membrane and secrete its contents. The quantities of prolactin and growth hormone stored intracellularly are equivalent to the quantities secreted in two hours. It is not known whether prolactin and growth hormone are contained in the same granules in GH cells.

The focus of the work summarized in this chapter has been the receptors for peptide hormones on pituitary cell lines. Peptide hormones initiate their actions on target cells by binding to specific sites on the plasma membrane. Their receptors are of high affinity, restricted to target cells, and exhibit a high degree of selectivity for the ligand. Peptide hormones controlling GH cells include thyrotropin-releasing hormone (TRH), epidermal growth factor (EGF), insulin, somatostatin, bombesin, and vasoactive intestinal peptide. These hormones can affect the rate of prolactin and growth hormone release, the rate of de novo hormone synthesis, or both. They bind to independent sites on the membrane and their receptors have been characterized; these studies have been reviewed recently (Tixier-Vidal and Gourdji, 1981; Hinkle, 1984). Table 1 summarizes the biological activities of peptide hormones regulating GH cell function and key features of their receptors.

In recent years, it has become apparent that many peptide hormone receptors do not remain fixed in one position on the plasma membrane, but instead, undergo aggregation and receptor-mediated endocytosis after hormone binding. Ligand binding by itself may trigger the degradation of receptors, perhaps by an "auto-feedback" mechanism regulating response. It has been suggested that molecules that interact with surface receptors can be divided into two groups (Kaplan, 1981). One group contains receptors for peptide hormones; these hormones bind to receptors diffusely distributed over the cell surface, and the receptors then aggregate in coated pit regions and are taken into the cell by receptor-mediated endocytosis. There is uncertainty about whether the receptors enter the cell in coated vesicles, or whether the vesicles shed their clathrin coat while still attached to the membrane. In any case, within a few minutes, the receptors are found in smooth membraned vesicles which Willingham and Pastan (1980) have termed receptosomes. The receptosomes fuse with Golgi elements and lysosomes after 15-30 min, resulting in degradation of ligand and receptor. Peptide hormone receptors cycle to the surface a number of times without degradation in some cell types. In the fibroblast model, fluorescence microscopy studies with rhodamine- and fluorescein-labeled hormones has shown that the receptors for insulin, α_2-macroglobulin, L-triiodothyronine and EGF are co-internalized and are found in the cell in the same receptosome (Pastan and Willingham, 1981). The second group of cell surface receptors may be viewed as serving a transport function; they bind ligands such as low density lipoprotein or immunoglobins, and carry them into or through the cell. These receptors

Table 1. Peptide Hormone Responses and Receptors on GH Cells*

Peptide	Prolactin		Growth Hormone		Receptors	
	Release	Synthesis	Release	Synthesis	#/cell	Kd(nM)
Bombesin	Increase		Increase		3,600	1.2
EGF		Increase		Decrease	34,000	0.5
Insulin		Increase			10,000	2.0
Somatostatin	Decrease	Decrease	Decrease	Decrease	13,000	0.6
TRH	Increase	Increase	Increase	Decrease	130,000	10.0
VIP	Increase	Increase	Increase			

*The biological activities and properties of receptors for peptide hormones are summarized. EGF = epidermal growth factor; TRH = thyrotropin-releasing hormone; VIP = vasoactive intestinal peptide. References for the individual hormones can be found in reviews by Tashjian, 1979; Tixier-Vidal and Gourdji, 1981; Gourdji et al., 1981; and Hinkle, 1984.

are localized in coated pit regions whether occupied or not, and they cycle continually through the cell with relatively little degradation (Goldstein et al., 1979).

These generalized models of the cycling of peptide hormone receptors have provided a useful framework for studies in the field, but it has become apparent that there are significant differences in the pathways of receptor-mediated endocytosis in different cell types (Pastan and Willingham, 1981; Goldstein et al., 1979; King and Cuatrecasas, 1981). We chose to study receptor cycling in pituitary cells because a number of receptors had been characterized, and because the biological consequences of peptide binding were well established and could be correlated with receptor events. The objectives of the experiments described below were twofold: first, to characterize the cycling of several receptors on GH cells; and second, to find drugs capable of blocking receptor-mediated endocytosis and study hormone responsiveness when receptor cycling could not take place.

METHODS

GH_4C_1 cells were grown in monolayer culture in a humidified atmosphere of 5 percent CO_2, 95 percent air in Ham's F10 medium containing 2.5 percent fetal calf serum and 15 percent horse serum as previously described (Tashjian et al., 1968). For use in experiments, equal aliquots from a single donor culture were inoculated into 35 mm culture dishes and the cells were grown for 3-10 days until they were in late log phase growth.

The specific binding of [^{3}H]TRH and [^{125}I]EGF was measured as described previously (Hinkle and Tashjian, 1973; Halpern and Hinkle, 1983). Wheat germ agglutinin and transferrin were iodinated by the procedure of Hunter and Greenwood (1962), using 2.5 µg of chloramine T, and their binding was measured by the same procedures as those described for TRH and EGF. In brief, serum-free medium containing radioactive hormone was added to the dishes and incubation carried out at the temperatures noted in the text. The dishes were washed 3 or 4 times with 0.15 M NaCl to remove free hormone. Nonspecific binding controls were included in all experiments; nonspecific binding, which rarely exceeded 10 percent of total, was subtracted from all data. In experiments measuring the fraction of internalized hormone, 1 ml of an acid/salt buffer (0.2 M acetic acid, 0.5 M NaCl) at 0°C was added for 5 min. Radioactivity extracted by the acid/salt buffer corresponded to surface bound hormone. Radioactivity in the cells corresponded to internalized hormone (Haigler et al., 1980). An aliquot of the cell suspension was reserved for the determination of cell protein. Details of the methods can be found in Hinkle and Kinsella (1982) and Halpern and Hinkle (1983). Data were corrected for differences in cell protein among dishes when these were significant. In order to measure biological responses, cells were rinsed twice and then incubated in fresh medium with or without TRH. The culture medium was removed and the concentration of prolactin measured by specific radioimmunoassay using reagents and standards supplied by the Hormone Distribution Program of NIH.

RESULTS

Fluorescence microscopy with a fluorescein-labeled bioactive TRH derivative revealed that this tripeptide binds initially to diffusely localized receptors on the cell surface and that the receptors aggregate rapidly after binding at 37°C (Halpern and Hinkle, 1981). However, since GH cells are rounded, it was not possible to determine whether label subsequently internalized, although it clearly was not densely packed in vesicles like those seen in fibroblasts. The TRH tripeptide lacks functional groups that would allow it to be fixed for autoradiography and the TRH receptor does not survive fixation and sectioning. For these reasons, it has been necessary to approach the question of internalization using a biochemical method such as that described by Haigler et al., (1980), in which cells are rinsed with a low pH buffer after binding to extract cell surface, but not internal, hormone.

In this way we found that TRH, EGF and transferrin were all internalized in the first 10 minutes after binding (Fig. 1). The internalization reaction was temperature dependent, with little receptor-mediated endocytosis at $10°C$ or below, and maximal internalization at $20°C$ (Fig. 2). These data suggest that these peptides might have been internalized together. However, the intracellular fate of the hormones and the receptors differed. The TRH tripeptide and the TRH receptor were not degraded in the first hour after binding (Hinkle and Tashjian, 1975a, 1975b). In contrast, the EGF peptide and EGF receptor were extensively broken down after internalization (Halpern and Hinkle, 1983). It seemed unlikely that these two receptors were in a vesicle that fused with a lysosome, since it would be difficult to explain how the TRH receptor escaped degradation. It is more reasonable to postulate that the two peptides were either internalized in different vesicles, or that sorting of the two receptors occurred after endocytosis and before fusion with lysosomes.

The goal of these experiments was to find a way to block receptor-mediated endocytosis under conditions that permit biological responses to be measured. Light fixation of the cells blocked internalization, which was certainly expected to require a mobile membrane (Fig. 3). However, glutaraldehyde also prevented secretion and was not useful for studying the importance of internalization in the TRH response. Internalization of the TRH receptor complex was not inhibited by high concentrations of microtubule and microfilament poisons, by the calmodulin inhibitor trifluoperazine, or by inhibitors of glycolysis and oxidative phosphorylation (Table 2). Similar data have been obtained for EGF.

One proposal that has been put forth to explain internalization is that receptors are crosslinked first in a reaction involving the enzyme transglutaminase (Pastan and Willingham, 1980). Transglutaminase is inhibited by a number of drugs including monodansylcadaverine, bacitracin and chloroquine; these drugs reportedly inhibit the internalization of some hormone-receptor complexes in fibroblasts. In order to test this hypothesis in pituitary cells, we studied the effects of drugs that inhibit trans-glutaminase on the internalization of TRH and EGF. As shown in Table 3, there was no effect on either the total amount of [^{3}H]TRH binding to GH cells or on the fraction of specifically bound hormone that was internalized. More detailed analysis has indicated that monodansylcadaverine and chloroquine did not alter the rate of internalization either.

The results for [^{125}I]EGF were somewhat different. Drugs with an amine group increased the total amount of EGF specifically associated with the pituitary cells, though they either had no effect on or increased the fraction internalized. The amount of cell-associated radioactivity increased with the time of incubation. The drugs listed in Table 3 did inhibit transglutaminase but they are also lysomotropic, i.e., they accumulate in acidic vesicles such as lysosomes and raise their pH, thereby inhibiting proteolytic enzymes (Seglen, 1983). We have found that the amine reagents caused increased accumulation of [^{125}I]EGF inside the cells by preventing degradation (Halpern and Hinkle, 1983); [^{125}I]EGF is normally degraded quite rapidly in GH cells and the radioactive product of this degradation, [^{125}I]monoiodotyrosine, is extruded from the cell. Inhibition of the degradation process therefore caused an increase in cell associated radioactivity.

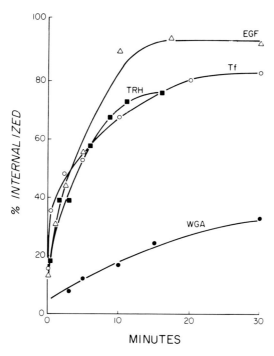

Fig. 1. Rates of ligand internalization. GH_4C_1 cells were incubated at 37°C with concentrations of $[^3H]TRH$, $[^{125}I]EGF$, and $[^{125}I]$ transferrin (Tf) sufficient to saturate approximately 50 percent of receptors. At intervals, the fraction of specifically bound ligand internalized was determined. The rate of internalization of $[^{125}I]$wheat germ agglutinin (WGA), which binds nonspecifically to surface carbohydrates, is shown for comparison. Data for TRH and wheat germ agglutinin are from Hinkle and Kinsella (1982), and those for EGF from Halpern and Hinkle (1983).

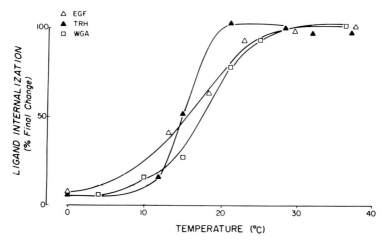

Fig. 2. Temperature dependence of ligand internalization. GH_4C_1 cells were incubated with radiolabeled hormones for 30 min in L15 medium in room air at the temperatures shown. Abbreviations are as in Fig. 1.

Table 2. Effect of Drugs on TRH Binding and Internalization*

Drug	[^{3}H]TRH Bound % Control	Internalization %
Control	100	75 - 85
100 µM Colchicine	140	66
80 µM Vinblastine	161	77
50 µg/ml Cytochalasin B	119	65
13 µM Trifluoperazine	113	62
10 mM NaCN +100 mM 2-Deoxyglucose	86	81

*GH$_4$C$_1$ cells were treated for 30 min with drugs, and 20 nM [^{3}H]TRH was added then for 30 min incubation in the continued presence of drug. Data are from Hinkle and Kinsella (1982).

Table 3. Effect of Drugs on TRH and EGF Binding and Internalization*

Drug	[^{3}H]TRH Binding		[^{125}I]EGF Binding		
	Total 30 min	Internal	Total 30 min	3 h	Internal 30 min
	% Control		% Control		%
None	100	71	100		77
100 µM Monodansylcadaverine	123	80	112	200	78
100 µM Ethylamine	85	80	123	76	
1 mM Bacitracin	100	72	113	158	74
100 µM Chloroquine	102	74	64	259	73
10 µM Quinacrine				329	

*To determine the effect of drugs on [^{3}H]TRH binding, GH$_4$C$_1$ cells were preincubated with drugs for 30 min. The cells were incubated then in the continued presence of the drugs with [^{3}H]TRH (20 nM) for 30 min. Data are from Hinkle and Kinsella (1982).

To determine the effects of drugs on EGF binding, cells were incubated with drugs and [^{125}I]EGF (50,000-100,000 cpm/dish) for 30 min or 3 hr. The fraction of hormone internalized was measured after 30 min; at 3 hr essentially all cell associated [^{125}I]EGF is internal. Some of the values are from Halpern and Hinkle (1983).

The values shown are the mean of duplicate or triplicate determinations; SE's averaged less than 5 percent.

Phenylarsene oxide (arsenosobenzene) is a poison reported to block internalization of immunoglobulins. Phenylarsene oxide inhibited potently the binding of TRH and EGF to their respective inhibitors on pituitary tumor cells, and also prevented internalization of the hormone receptor complexes

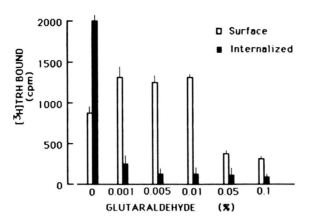

Fig. 3. Effect of fixation on TRH internalization. GH_4C_1 cells were treated for 10 min at 37°C with glutaraldehyde, and 5 nM [^{3}H]TRH was then added for 60 min at 37°C. The open bars show surface-bound [^{3}H]TRH, the solid bars, internalized [^{3}H]TRH. Values are the mean and range of duplicate determinations.

(Fig. 4). Phenylarsene oxide reacts preferentially with viscinal dithiol groups in proteins (Webb, 1966), as shown below.

$$\phi\text{-As=O} + \begin{matrix} \text{HS-}| \\ \text{HS-}| \end{matrix} \quad \phi\text{-As}\begin{matrix} \text{S-}| \\ \text{S-}| \end{matrix} + H_2O$$

Other drugs reacting with viscinal dithiols are cadmium and diamide (azodicarboxylic acid bis[dimethylamide]). Cadmium sulfate and diamide did not exhibit the potent inhibitory actions of phenylarsene oxide on TRH and EGF internalization (Figs. 5 and 6).

If the toxic effects of phenylarsene oxide were due to its activity as a sulfhydryl reagent, then addition of thiol reagents should prevent its effects. This was the case, as shown in Fig. 7. BAL (British anti-Lewisite, or 2,3-dimercapto-1-propanol), blocked the inhibition of internalization in approximately stoichiometric amounts with phenylaresene oxide. Other thiol reagents, such as Cleland's reagent (dithiolthreitol) and β-mercaptoethanol, prevented phenylarsene oxide toxicity at ten times higher concentrations. If cells were first incubated with phenylarsene oxide and then treated with BAL, TRH binding and internalization could be partially but not fully restored. The toxicity of phenylarsene oxide was markedly temperature dependent; whereas 0.1 μM blocked TRH binding at 37°C, 10 μM had little effect at 0°C (Fig. 8).

It would be of great interest to block internalization of hormone-receptor complexes and measure the capacity of the cells to respond. The effects of phenylarsene oxide on both total TRH binding and TRH internalization were not overcome at high TRH concentrations (Fig. 9). GH cells appear to have "spare receptors" for TRH since the concentration of the peptide yielding a half-maximal biological response, 2 nM, was below the concentration half-saturating receptors, 10 nM (Hinkle and Tashjian, 1973). Because the cells have excess receptors, it seemed possible that cells might still respond to TRH in the presence of phenylarsene oxide, even though total receptor binding was reduced. To test this hypothesis, we treated cells with different concentrations of phenylarsene oxide and measured biological response to TRH-stimulation of prolactin secretion. Because

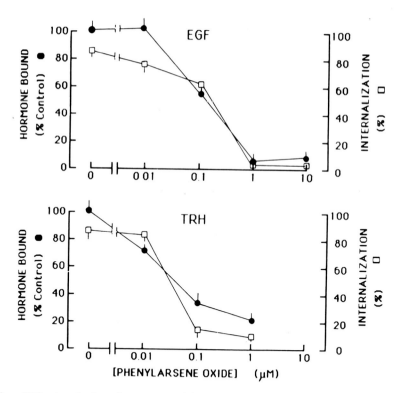

Fig. 4. Effect of phenylarsene oxide on hormone internalization. GH_4C_1 cells were incubated at 37°C for 1 hr with phenylarsene oxide and either [^{125}I]EGF (300,000 cpm/ml) or [^{3}H]TRH (10 nM). (□) Total specific [^{125}I]EGF or [^{3}H]TRH binding, as percentage of control; (●) Percentage of specifically bound hormone internalized. Control dishes bound 4602 ± 330 cpm [^{125}I]EGF and 4991 ± 221 cpm [^{3}H]TRH. Values are the mean ± range of duplicate dishes.

phenylarsene oxide was severely toxic and cells did not remain viable after exposure, as judged by trypan blue exclusion and plating efficiency, only the acute response to TRH could be measured.

In control cultures, TRH stimulated a twofold rise in prolactin secretion by GH cells (Fig. 10). TRH also increased prolactin synthesis slightly in the 30 min of this experiment, as shown by the fact that total prolactin in the cells plus medium was higher in TRH-treated cultures. Phenylarsene oxide reduced prolactin secretion by the cells in a dose-dependent fashion. Nonetheless, TRH still caused a twofold increase in prolactin secretion in cultures exposed to 1 μM phenylarsene oxide. At this concentration of toxin, the internalization reaction was inhibited by 90 percent (see Fig. 4). The implication of this result is that receptor-mediated endocytosis need not occur for TRH to cause an acute rise in the rate of prolactin secretion from pituitary cells. This conclusion was corroborated by the finding that TRH stimulated a burst of hormone release within 15 sec, before internalization had occurred (Aizawa and Hinkle, 1984).

SUMMARY AND CONCLUSIONS

The results described above are summarized schematically in Fig. 11. Specific peptide hormone receptors on pituitary tumor cells are processed in

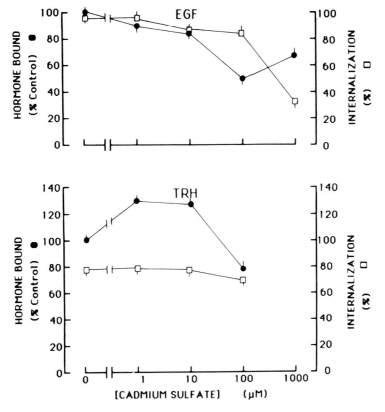

Fig. 5. Effect of cadmium sulfate on hormone internalization. The protocol was the same as that described in Fig. 4. Cells treated with 1 mM cadmium sulfate bound too little [^{3}H]TRH to measure. Control dishes bound 8952 ± 703 cpm of [^{125}I]EGF and 6300 ± 900 cpm [^{3}H]TRH. Values are the mean and range of duplicate determinations.

unique ways after their respective ligands bind to them. Receptors for TRH, EGF and bombesin were internalized by receptor-mediated endocytosis in a reaction taking about 10 minutes (Westendorf and Schonbrunn, 1983; Hinkle, 1984). In contrast, receptors for somatostatin were not internalized by the cells (Presky and Schonbrunn, 1983). After internalization, the TRH receptor cycled back to the surface without degradation of either the TRH tripeptide or the receptor (Hinkle, 1984). EGF and bombesin were degraded, as were their receptors (Westendorf and Schonbrunn, 1983; Hinkle, 1984). Since the degradation could be prevented by lysomotrophic drugs, it is likely that it occured via a lysosomal pathway. Further work is needed to determine the mechanisms of receptor cycling in pituitary cells.

Receptor-mediated endocytosis can be blocked by low temperature, glutaraldehyde and phenylarsene oxide. TRH stimulated the release of prolactin from GH cells when internalization was blocked by phenylarsene oxide, suggesting that this important biological activity of the peptide resulted from events triggered by the hormone-receptor complex at the plasma membrane of the cell.

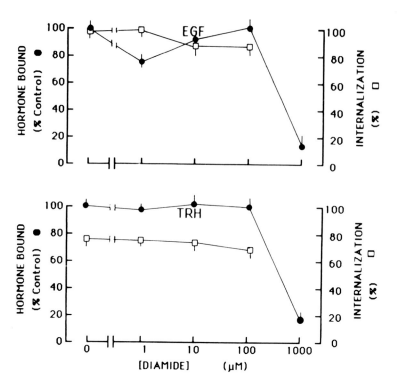

Fig. 6. Effect of diamide on hormone internalization. The protocol was the same as that described in Fig. 4. Total hormone binding at 1 mM diamide was so low that the fraction internalized could not be determined accurately. Control dishes bound 8952 ± 703 cpm [^{125}I]EGF and 1040 ± 50 cpm [^{3}H]TRH. Values are the mean and range of duplicate determinations.

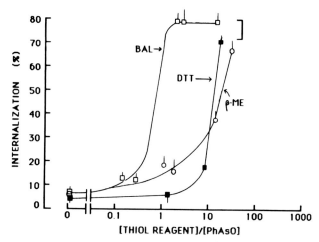

Fig. 7. Effect of thiol reagents on phenylarsene oxide toxicity. Phenylarsene oxide and thiol reagent were mixed prior to addition to GH_4C_1 cells. [^{3}H]TRH binding was measured as in Fig. 4. The fraction of specifically bound hormone internalized is shown. Thiols tested were BAL (British anti-Lewisite, 2,3-dimercapto-1-propanol), DTT (Cleland' reagent, dithiolthreitol) and β-mercaptoethanol. Phenyl-arsene oxide was tested at 5 μM with DTT and 50 μM with BAL and ME. Values are the mean and range of duplicate determinations.

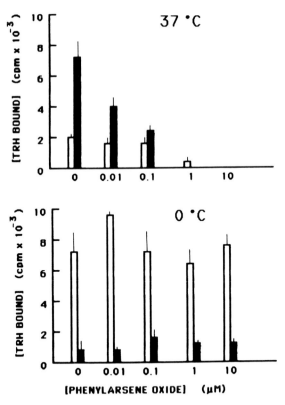

Fig. 8. Temperature dependence of phenylarsene oxide inhibition. GH_4C_1 cells were incubated with phenylarsene oxide and 10 nM [^{3}H]TRH in L15 medium in room air at either 37°C (upper panel) or 0°C (lower panel) for 1 h as described in Fig. 4. The open bars show cell surace [^{3}H]TRH and the solid bars internalized hormone. Values are the mean and range of duplicate determinations.

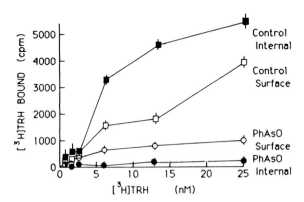

Fig. 9. Effect of [^{3}H]TRH concentration on phenylarsene oxide toxicity. GH_4C_1 cells were incubated with or without 50 μM phenylarsene oxide and [^{3}H]TRH at the concentrations shown for 60 min. Values are the mean and range of duplicate determinations for [^{3}H]TRH specifically bound: (□,○) cell surface, (■,●) internalized.

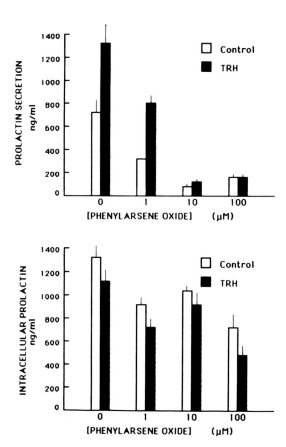

Fig. 10. Effect of phenylarsene oxide on TRH response. GH_4C_1 cells were rinsed twice in serum-free medium and then incubated for 1 h with phenylarsene oxide at the concentrations shown and either no additions (open bars) or 100nM TRH (solid bars). Values shown are the mean and range of triplicate determinations.

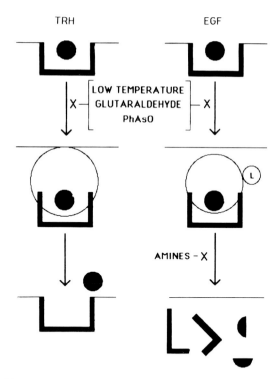

Fig. 11. Summary of drug effects on receptor cycling.

ACKNOWLEDGMENTS

This research was supported in part by NIH Grant AM-19974 and Cancer Center Core Research Grant CA-11198. J.H. was supported in part by Training Grant GM-07141, R.A.F. by Training Grant ES-07026, and P.M.H. by Research Career Development Award AM-00827.

REFERENCES

Aizawa, T. and Hinkle, P.M., 1985, TRH rapidly stimulates a biphasic secretion of PRL and GH in GH_4C_1 rat pituitary tumor cells, Endocrinology, 116:73-82.

Goldstein, J.L., Anderson, R.G.W. and Brown, M.S., 1979, Coated pits, coated vesicles, and receptor-mediated endocytosis, Nature, 279:679-685.

Gourdji, D., Tougard, C. and Tixier-Vidal, A., 1981, Clonal prolactin stains as a tool in neuroendocrinology, in: "Frontiers in Neuroendocrinology," vol. 7, W.F. Ganong and L. Martini, eds., pp. 317-357, Raven Press, New York.

Haigler, H.T., Maxfield, F.R., Willingham, M.C. and Pastan, I.H., 1980, Dansylcadaverine inhibits internalization of ^{125}I- epidermal growth factor in BALB 3T3 cells, J. Biol. Chem., 255:1239-1241.

Halpern, J. and Hinkle, P.M., 1981, Direct visualization of receptors for thyrotropin-releasing hormone with a fluorescein labeled analog, Proc. Natl. Acad. Sci., 78:587-591.

Halpern, J. and Hinkle, P.M., 1983, Binding and internalization of epidermal growth factor by rat pituitary tumor cells, Mol. Cell Endocrinol., 33:183-196.

Hinkle, P.M., 1984, Interaction of peptide hormones with rat pituitary tumor cells in culture, in: "Secretory Tumors of the Pituitary Gland," P. Mcl. Black, N.T. Zervos, E.C. Ridgway and J.B. Martin, eds., pp. 25-43, Raven Press, New York.

Hinkle, P.M. and Kinsella, P.A., 1982, Rapid temperature dependent transformation of the thyrotropin-releasing hormone receptor complex in rat pituitary tumor cells, J. Biol. Chem., 257:5462-5470.

Hinkle, P.M. and Tashjian, A.H., Jr., 1973, Receptors for thyrotropin releasing hormone in prolactin producing rat pituitary cells in culture, J. Biol. Chem., 148:6180-6186.

Hinkle, P.M. and Tashjian, A.H., Jr., 1975a, Thyrotropin-releasing hormone regulates the number of its own receptors in the GH3 strain of pituitary cells in culture, Biochem., 14:3845-3851.

Hinkle, P.M. and Tashjian, A.H., Jr., 1975b, Degradation of thyrotropin-releasing hormone by the GH3 strain of pituitary cells in culture, Endocrinology, 97:324-331.

Kaplan, J., 1981, Polypeptide-binding membrane receptors: analysis and classification, Science, 212:14-20.

King, A.C. and Cuatrecasas, P., 1981, Peptide hormone induced receptor mobility, aggregation and internalization, New Engl. J. Med., 305:77-88.

Pastan, I.H. and Willingham, M.C., 1981, Receptor-mediated endocytosis of hormones in cultured cells, Ann. Rev. Physiol., 43:239-250.

Presky, D.H. and Schonbrunn, A., 1983, Receptor-bound somatostatin (SRIF) and epidermal growth factor (EGF) are processed differently by GH_4C_1 pituitary cells. Program of the 65th Annual Meeting of the Endocrine Society, A 564, p. 221.

Seglen, P.O, 1983, Inhibitors of lysosomal function, Methods Enzymol., 96:737-764.

Tashjian, A.H., Jr., 1979, Clonal strains of hormone-producing pituitary cells. Methods Enzymol., 58:527-535.

Tashjian, A.H., Jr., Yasumura, Y., Sato, G.H, Parker, M.L., 1968, Establishment of clonal strains of rat pituitary tumor cells that secrete growth hormone, Endocrinology, 82:342-352.

Tixier-Vidal, A. and Gourdji, D., 1981, Mechanism of action of synthetic hypothalamic peptides on anterior pituitary cells, Physiol. Rev., 61:974-1011.

Webb, J.L., 1966, "Enzyme and Metabolic Inhibitors," Chapter 6, pp. 595-647, Academic Press, New York.

Westendorf, J.M. and Schonbrunn, A., 1983, Characterization of bombesin receptors in a rat pituitary cell line, J. Biol. Chem., 258:7527-7536.

Willingham, M.C. and Pastan, I., 1980, The receptosome: an intermediate organelle of receptor-mediated endocytosis in cultured fibroblasts, Cell, 21:67-77.

CONFERENCE SUMMARY

N. Karle Mottet

Department of Pathology
University of Washington School of Medicine
Seattle, Washington, USA

INTRODUCTION

Two decades ago many scientists regarded the cell cytoplasm as an ill-defined sol or gel containing a few organelles. Biochemists distinguished structural proteins and enzymes but without specification of the former. Beginning in the late 1960s more detailed features of the cytoskeleton began to be defined. This body of knowledge has grown rapidly so that now not only are the principal structural protein components of microfilaments, intermediate filaments and microtubules well defined but also many of the modulating proteins are known as reviewed in Chapter 1 of this volume. This conference has focused on the role of the cytoskeleton as a target for toxic agents and in this summary I have tried to condense and highlight the important and most interesting features of individual reports.

CYTOSKELETON AND CELL PROLIFERATION

At the outset of this conference structural and functional features of the cytoskeleton were reviewed by Dr. J.B. Olmsted of the University of Rochester. Toxicologists and other biologists have investigated the effects of a variety of agents on the cytoskeleton as a subcellular mechanism for altering cell function. One may classify these toxic agents as those that are developed as a useful probe in the study of the structure and function of the cytoskeleton and those that are significant occupational or environmental hazards. A system of classification of these toxic agents was proposed by Dr. L. Wilson. He proposed that the toxic agents that affect microtubules, which now number more than 20 distinct classes of drugs, be divided into substances that (1) react with specific binding sites and those (2) with little or no specificity, such as the sulfhydryl reactive agents. Examples of the drugs with high affinity for a specific site include vinca alkaloids and colchicine. Weak affinity compounds include benomyl, diphenylhydantoin and chlorpromazine. Colchicine is a drug which has a high affinity for a specific site in tubulin from mammalian cells but lower affinity for tubulins from lower species cells. This represents one line of evidence for the existence of some species differences in reaction to toxic agents on microtubules, which was further discussed by Dr. Gull. Dr. Wilson noted that there is one colchicine binding site per dimer, that the binding is noncovalent and there was no chemical alteration of colchicine in the process. Formation of the colcicine-tubulin complex is strongly temperature dependent, proceeds slowly, and the binding action is not affected by pH. The binding constants of colchicine by different tubulins from the same species of animal, differ as do those for tubulins from different species. Colchicine interacts with the ends of microtubules, and therefore a small amount may totally block assembly. Vinblastine has a very different effect from colchicine. At low concentration it completely inhibits polymerization and at intermediate concentrations it causes a rearrangement of tubulin molecules to form a crystalline structure. At higher concentrations (1 mM), one gets a peeling of the ends of the microtubules and the formation of spiral rings. Thus, the sensitivity of microtubules to drugs, such as colchicine and vinblastine, is related to the type of tubulin that is used in the experiment and the specific agent. There are also species differences. Implicit in the dose-effect studies was the fact that chemical agents may produce a subtle injury to the cell.

Taxol, as reported by Dr. S.B. Horwitz, is another interesting probe for the study of the structure and function of microtubules. First isolated from plants in 1971, it has a strong cytotoxic effect inhibiting the growth of cells. Taxol is an antimitotic agent that binds to microtubules and stabilizes them somewhat against depolymerization both in cells and in vitro. The drug increases both the rate and extent of microtubule assembly. By apparently strengthening the interaction between tubulin dimers they act as nucleating sites and shift the equilibrium to favor polymer formation. Taxol is effective in blocking the growth of some mouse tumors such as melanomas, mammary tumors, and leukemias. Remarkably, no neurotoxicity has been found for taxol as yet. Flow cytometry revealed that cells in vitro tended to accumulate in G2 and M phases of the cell cycle. There was no effect on DNA synthesis or thymidine transport in the S phase. Electron microscopy revealed bundles of microtubules in the cytoplasm of taxol treated cells in addition to the unusual mitotic spindles, but it was unclear as to how these bundles were formed. They appeared to be associated with the rough endoplasmic reticulum. In nerve cells, there were spiral arrays of microtubules in the cytoplasm and attached to nucleopores. The use of a series of taxol resistant cell lines should be of value in probing the unique mechanism of action of this drug.

The report by Dr. K. Gull was another using carbamate pesticides as a probe for the structure and function of microtubules. His report added further evidence that there are species differences in the response of microtubules to some toxic agents. The benzimidazole carbamate drugs are widely used as systemic fungicides, broad spectrum antihelmintics against cestode and nematode infestations, and as antitumor agents. Many of these compounds are readily available and their effectiveness for these uses is well established. Recent investigations show that the drugs have a singular action (i.e., they bind to tubulin selectively) and that some organisms are insensitive to the drug whereas others are not. These drugs have a wide range of biological action depending on the concentration of the drug used and the target organism. In fungi, the drugs act to produce mitotic spindle instability at sublethal concentrations and completely inhibit growth at higher concentrations by blocking the formation of mitotic spindles. The microtubule assembly necessary to form a proper spindle is disrupted resulting in improper movement of the chromosomes thus leading to growth inhibition. Treatment of helminths with these drugs can kill vulnerable larval stages or eggs, but have no effect on adults. It has been shown that mebendazole treatment impairs the secretion of digestive enzymes in the intestines of some worms and that this effect is due to a loss of cytoplasmic microtubules from the gut epithelium. This further restricts the parasite by decreasing adhesive capacity. In mammalian liver cells, the benzimidazole carbamates have no effect on the synthesis of albumin or lipids, but there is a decrease in cell secretion. Some of the analogues have no effect, whereas others completely inhibit liver cell secretion.

Thus, the different drugs of this family do not affect microtubules in fungae, worm, or mammalian cells in the same way. There are species differences in the tubulin structures. For example, methylbenzimidazol-2-yl carbamate (MBC) has less effect on mammalian cells tubulin but has a major effect on fungal tubulin. Parbenzadole has a major effect on mammalian cells and helminths, but little effect on fungi. In addition to the difference in tubulins, microtubules from different species, are structurally different having 9, 12, 13, 14 subunits.

Whereas the foregoing reports used toxic agents as a probe to study the biology of the cytoskeleton, the report by Dr. P.R. Sager used the system to study the effects of methylmercury, an important toxic sulfhydryl-group-binding, metal teratogen, on the development of the nervous system in mice. It has long been established that at high doses methylmercury is cytoclastic, whereas at lower doses it produces a reduction in size of the offspring. This reduction has been shown to effect major visceral organs and the brain. The subcellular mechanisms for this reduction have not been completely defined. Dr. Sager reviewed the evidence that several metal ions are capable of disruption of microtubules in vitro, but only methylmercury appeared to selectively affect the microtubules in cultured cells. In vitro assembly of microtubules is inhibited by methylmercury in a concentration dependent fashion. This inhibition of polymerization is due to an interaction of methylmercury with tubulin sulfhydryl groups. Equimolar concentrations of methylmercury are necessary to inhibit tubulin assembly into microtubules. She used immunofluorescence to visualize the microtubules in cultured human fibroblasts incubated in a methylmercury-containing medium. Methylmercury causes disassembly of microtubules in a time and concentration dependent fashion. This effect was reversed by the addition of sulfhydryl containing compounds such as DMSA and cysteine. In a second line of investigation, the effects of methylmercury on the developing mouse brain were tested by measuring mitotic activity. Tubulin is an abundant protein in the brain, comprising about 10 percent of the total brain protein around the time of birth. The external granular layer of the cerebellum was chosen for analysis. Two-day-old animals were administered methylmercury and then sacrificed 24 hours later. Cell number and mitotic index and the fraction of late mitotic figures were counted. The mitotic activity was suppressed, especially in males, and these changes in the mitotic activity were consistent with the failure of cells to complete mitosis into the anaphase and telophase. Decreases in cell numbers were still apparent, especially in males, up to 19 days after treatment.

CYTOSKELETON AND THE NERVOUS SYSTEM

Six participants at the conference discussed the effects of several toxic agents on the function of the neuronal cytoskeleton, especially intermediate filaments, microtubules and their associated proteins, in the axonal transport and resulting axonal lesions. First among these was the report of Dr. S. Papasozomenos who discussed the effects of β,β'-iminodipropionitrile (IDPN) on the reorganization of axonal cytoskeleton. IDPN is a synthetic compound closely related to beta-aminopropionitrile. Intoxication of various experimental animals with IDPN produces excitement, circling, and a 'waltzing syndrome', a permanent symptom complex indicative of irreparable damage to the nervous system. Amorphous bodies develop along the axons of the anterior horn and other large neurons. The axons form balloons connected to the cell body by a normal initial segment. Their structure suggests a stasis caused by a plug of particulate organelles in the axon, producing the swelling. It has been shown by Dr. J.W. Griffin and co-workers that IDPN severely impairs the transport of neurofilament protein, and to a lesser degree tubulin and actin, while the rate of antegrade and retrograde axonal transport of cellular constituents remains normal. The impairment of the transport of neurofilaments results in the accumulation in the proximal axon. In small neurons with fewer neurofilaments, the transport of tubulin and actin was not affected. Only the transport of neurofilament proteins was selectively impaired. Electron microscopy revealed a segregation of axonal cytoskeleton with displacement of neurofilaments towards the periphery and of microtubules together with mitochondria and smooth endoplasmic reticulum towards the center, according to the investigations of Dr. Papasozomenos. Side arms between the neurofilaments and "wispy" material among microtubules were easily recognized. No excessive cross-linking or other structural abnormality of the neurofilaments was evident. His immunohistochemical studies used antibodies against tubulin to determine the spatial and temporal evolution of the segregation of microtubules from neurofilaments in the lumbar segment of the spinal cord ventral roots and along the sciatic nerve. The segregation of neurofilaments from microtubules seen clearly by immunohistochemistry, occurred along the entire length of the axon. This was detectable four days after IDPN exposure and was more discrete by six weeks and disappeared after 6-16 weeks in a proximal to distal pattern. Further changes occurred in the ensuing four months. The axonal balloons in the most proximal segment, first noticed at two weeks after IDPN intoxication, were still present after one year. They were the only histological abnormality present and were filled with neurofilaments. In the posterior columns of the lumbar spinal cord almost all of the axons had a ring like appearance about two weeks after IDPN intoxication by immunofluorescent staining of neurofilaments. Electron microscopic morphometric analysis of cross sections of small, medium, and large sized axons revealed no significant increase in neurofilaments or microtubules in IDPN axons, and the ratio remained the same. To investigate the pathogenesis of the microtubule-neurofilament segregation, he used two monoclonal antibodies against the MAP2. Spinal and other motor axons contained detectable amounts of MAP2 and in the central portion of IDPN axons both antibodies localized with the microtubules, whereas in the peripheral portion one localized with the neurofilaments and the other with microtubules. Direct injection of IDPN into the endoneural space indicated that IDPN or its metabolite exerted a direct local effect on both the peripheral and central axons. The molecular basis of the microtubule-neurofilament segregation remains to be determined. The data presented by Dr. Papasozomenos so far suggest that the MAP2, at least partially, mediates the reaction. Distal axonal atrophy takes place both in acute and chronic IDPN intoxications. In the chronic experiment, a reduction in the delivery of neurofilaments has been suggested as a cause of the atrophy, however, no morphometric analyses have been done.

Dr. J.W. Griffin presented a regional classification of neurofibrillary pathology produced by axonal transport blocking agents. Agents such as aluminum, colchicine, and vinca alkaloids for example, produce changes in the vicinity of the perikaryon, whereas IDPN intoxication produces proximal axonal accumulations. A variety of agents such as acrylamide, carbon disulfide, and 2,5 hexanedione (2,5 HD) produce distal axon lesions. He also used acrylamide to probe other features of the axonal transport problem. He particularly pointed out the usefullness of dividing the toxic effects into those that are direct upon the transport system, and those that are secondary, altering neurofibril flow from the perikaryon. The direct effects are generally produced by single doses, are selective and can be measured by the velocity of movement of neurofibrils in IDPN or dimethyl-2,5-hexanedione (DMHD) treated animals. Other common neurofibrillary toxins, such as hexanedione, cause almost identical changes to those of IDPN. Neurofibrillary changes are the hallmark of a wide variety of neurotoxic and degenerative disorders. Much of the progress in the last five years in reconstructing the pathogenesis of neurofibrillary changes has resulted from correlations of ultrastructural changes with axonal transport studies.

Maldistribution of neurofilaments can cause changes in the caliber of the affected neurons. The neurofilaments in normal axons regulate the axonal caliber. The volume changes seen in these disorders are the product of this function of neurofilaments. The volume changes are easily quantified within axons because of the parallel cylindrical geometry of nerve fibers and the simplicity of the organization of the axonal filaments. The selectivity of the various transport abnormalities produced by different toxic agents varies widely. Administration of a single high dose of acrylamide results in a 20 percent reduction in transport of both neurofilament proteins and tubulin. The slow transport defect was nonselective with all slow component proteins retarded equally. There was a modest increase in the caliber of the large axons, as well as an increase in neurofilament density. Some fibers showed typical giant axonal swellings and this increase in neurofilament numbers was presumably a morphologic correlate of the impairment of slow transport. It reflects a mismatch between delivery of neurofilaments to the axon from the nerve cell body and the capacity of the axon to carry them centrifugally. Animals which received repeated smaller doses of acrylamide showed a different alteration in slow transport and in the morphology of the proximal motor roots. Slow component proteins including tubulin and neurofilament proteins were increased. Morphologically, there was a striking decrease in the caliber of axons in the proximal ventral roots. This suggests that this change reflects a secondary response of the neuron to the axonal injury and resembles closely the lesions produced by nerve section.

Dr. P. Gambetti studied the effects of 2,5 HD, the mono and dimethyl analogues of 2,5 HD, and carbon disulfide to further refine the information on axonal transport. These agents produce swelling lesions in the peripheral portion of the axon in contrast to the IDPN, which produces a proximal lesion. He observed numerous swellings after five days with no axonal degeneration, but the balloon distention was filled with neurofilaments. 2,5 HD accelerated the speed of migration of neurofilaments but tubulin and actin transport was unchanged. An analogue of 2,5 HD is 3,4-dimethyl-2,5-hexanedione and is much more toxic. The evidence suggests that the rate of neurofilament transport is increased between the cell body and axonal enlargement, whether the latter is proximal, intermediate or distal on the axon. The enlargement is related to neurofilaments and the location of the enlargement is related to the number of neurofilaments being transported.

Dr. V.-P. Lehto reported on the effects of some dithiocarbamates and their metabolic products on the cytoskeleton of nerve cells. With the carbamates studied, very slight changes involving swelling but no neurofilament or microtubule changes, were detectable by electron microscopy. Studies using monoclonal antibodies against tubulin revealed no alteration of neurofilaments or microtubules. There was no segregation of these components. Again, there was slight axonal swelling, with some degenerative changes in the myelin sheath, but no changes in the cytoskeleton. He also searched for similar effects on cultured cells treated with disulfiram. There was a significant decrease in the microtubule to neurofilament ratio in treated animals. The reason for this was apparently a decrease in the number of microtubules, because the number of neurofilaments seems to be about the same in treated and control animals. The molecular mechanism for the decrease in microtubules by disulfiram is not known. To some extent, the phenomenon resembles the effects of IDPN; however, there were no signs of microtubule segregation. A study of the effects of 2,5 HD on cultured mouse neuroblastoma cells revealed vimentin-type intermediate filament increase but not neurofilaments. This suggests that a reorganization of the intermediate filaments can take place without morphological alterations in the microtubules. Thus, the primary target of the drug may be the intermediate filaments themselves. Other investigators have shown that normal human fibroblasts seem to respond in a similar manner to 2,5 HD.

Dr. M. Shelanski reviewed the current state of our knowledge on the role of aluminum on neurofibrillary degeneration. Even though aluminum has not been considered a neurotoxin, it remains controversial whether dietary aluminum is involved in the cause or development of neurological disease. When aluminum is topically or intrathecally applied to the brain neurofibrillary changes similar to the neurofibrillary tangles of Alzheimer's disease are seen, whereas ultrastructurally they are different. The aluminum induced filaments are unpaired 10 nm diameter filaments whereas in Alzheimer's they are paired and have a node visible every 80 nm. Electron microscopy has revealed the importance of cross-linking between adjacent filaments and tubules and also of filaments and tubules with other organelles. Many of the links are composed of proteins, perhaps MAPs. Some of these proteins form lateral projections or arms on the microtubules and are capable of binding to neurofilaments as well as to microtubules and appear to mediate filament to microtubule interaction in vitro. MAP-mediated interactions between actin and microtubules and between MAPs and the alpha subunit of spectrin have been reported. The diversity of these links may have a role in the stability/lability of the cytoskeletal interconnections. Therefore, potential sites for the action of toxins on the neuronal cytoskeleton are numerous. Unlike the colchicine effect, with aluminum encephalopathy, the microtubules remain intact. Also, the colchicine effect is seen in all cells and is not limited to neurons as is the aluminum effect. The aluminum effect may be directly on the link molecule rather than the microtubule itself.

Because neurons synthesize proteins only in the cell body and dendritic cytoplasm, elaborate transport mechanisms convey these products down the axon to the nerve ending. Cytoskeletal elements are both transported and involved in the transport machinery. Neurofilament effects could be due to (1) disruption of the relationship between filaments and other organelles, (2) transport blockade, (3) filament overproduction, (4) decreased filament degradation, or (5) a combination of these factors. An increased amount of neurofilaments has been shown in aluminum treated cells and neurofilament transport inhibition has been shown.

Little is known about the association of metallic elements with the filaments of human neurofibrillary diseases. In Alzheimer's disease the paired helical filaments differ markedly from the normal neurofilaments. Although the analytical association of aluminum with the lesions of Alzheimer's is uncertain, electron microprobe studies have revealed elevated levels of aluminum within the neurofibrillary tangles. Whether the aluminum is a cause or merely an accompaniment of the neurofibrillary tangle remains moot.

Dr. D. Graham, the last speaker in this section, discussed the steps of pathogenesis that lead to axonal degeneration adding to and completing the picture presented by the previous discussants. He noted that 2,5 HD is the toxic metabolite of the n-hexane and the methyl n-butylketone molecules. When 2,5 HD reacts with lysyl residues of proteins, it forms an imine that cyclizes to a pyrrole. The pyrrole then auto-oxidizes to form an orange chromophore and a stable protein crosslinkage forms with consequent disruption of axonal transport. This is followed by the cessation of transport in the presence of protein synthesis leading to the accumulation of neurofibrillary elements forming the balloon enlargement. He noted that the positioning of the swelling was determined by the dose and the rate of administration of the 2,5 HD. He proposed the hypothesis that the development of the balloon swelling was related to the rate of cross-linking. As the cross-linking progresses, the mass enlarges and is stopped at constrictions in the axon at the node of Ranvier. With larger doses, the blockage may become more proximal in position.

The neurotoxicity of 3,4-dimethyl-2,5-hexanedione (DMHD) was investigated to determine whether pyrrole formation and auto-oxidation were necessary steps in the pathogenesis of diketone neuropathy. DMHD was found to be 20-30 times more potent than HD. The axonal swellings were proximal in the DMHD treated animals whereas HD produced distal swelling. Also the position varied with the rate of intoxication. Rapid intoxication by DMHD produced more proximal swelling. These results were similar to those found with IDPN. Both markedly retarded axonal transport. HD, carbon disulfide and acrylamide produced swellings in the distal axon. This led to the conclusion that the rate at which the neurofilaments were aggregated determined the location of the swelling. Protein cross-linking occurred in vivo with HD and DMHD. Comparing the relative insensitivity of mice to rats to gamma diketone neuropathy, results from mice suggest that these axons are too short to develop large aggregates of neurofilaments and too small in diameter to have significant constrictions at nodes of Ranvier. The axonal swelling itself does not cause weakness, but rather degeneration of the distal axon is necessary.

THE CYTOSKELETON AND MEMBRANE RELATED EVENTS

This section was composed of reports describing the use of widely different approaches to the study of the relationship of cytoskeletal structure to the action of plasma membranes. Dr. A. Elgsaeter described several different biophysical approaches used to study the molecular basis for the mechano-chemical properties of the erythrocyte membrane, particularly as they related to the membrane skeletal component spectrin. Since spectrin was first identified in 1968 the protein has been found, along with actin, to be the major component of the erythrocyte membrane skeleton. Spectrin has subsequently been found also in virtually all nonerythroid cells. It consists of two protomers making up heterodimers about 100 nm long. Actin appears to participate in the linking of spectrin into a molecular network. The physical methods described by Dr. Elgsaeter yield information about the molecular dynamics and native configuration of spectrin while the macromolecules are in solution. The method based on studying the relaxation of electrically induced birefringence in spectrin solutions yields valuable information about the flexibility of the highly elongated spectrin molecules. The results of such measurements strongly suggest that spectrin is a highly flexible molecule. Along with other currently available information on the erythrocyte membrane structure, this indicates that the erythrocyte membrane skeleton constitutes an ionic gel or "tanaka gel". Because the remarkable properties of such gels are well accounted for theoretically, this opens a new and promising perspective for a deeper understanding of the molecular basis of the mechano-chemical properties of the erythrocyte membrane skeleton and potentially also the membrane skeleton of other cells.

Dr. H. Passow discussed the effects of low levels of intracellular calcium on the structure and function of the human erythrocyte membrane. Under physiological conditions the intracellular activity of ionized Ca^{++} in human erythrocytes is below 0.4 μmoles/l whereas the extracellular concentration of free Ca^{++} is about 1200 μmoles/l. This enormous difference in concentration across the plasma membrane is maintained by a powerful ATP-driven Ca^{++} pump. Augmentation of the intracellular Ca^{++} activity above its normal physiological value leads to dramatic alterations of the plasma membrane structure and function. The effects range from increased Ca^{++} pump activity and increased K^{+} permeability to decreased erythrocyte deformability and shape changes. High Ca^{++} concentrations (0.1-1.0 μmoles/l) induce microvesiculation, protein cross-linking and proteolytic destruction of membrane proteins. It is therefore very important for normal erythrocyte structure and function that intracellular Ca^{++}

activity is maintained at the normal physiological level. Dr. Passow presented experimental data from a study of the effects of low levels of intracellular Ca^{++} on K^+ efflux, and on the mechanical properties of the erythrocyte membrane. In this study Dr. Passow and co-workers used the patch clamp technique to measure K^+ efflux and found that most of the erythrocytes had one to five Ca^{++} activated K^+ channels. The mechanical properties of the erythrocyte membrane was studied using micropipette aspiration of unswollen cells to determine the elastic shear modulus of plasma membrane and using photobleaching to study the lateral diffusion of the band 3 protein. The results suggest that there is no correlation between the effect of Ca^{++} on the cytoskeleton and on the K^+ selective channels. The experiments further show that the changes in the mechanical properties of the cell membrane are a direct consequence of Ca^{++} binding to the membrane at μmolar concentrations without the participation of other cytosol components because of the elevated intracellular Ca^{++} levels. Increasing concentrations of lead have essentially the same effects on opening the membrane channels. However, lead at very high levels began to inhibit its own effect. Further, lead inhibited the calcium induced channels at high concentrations.

Using a markedly different approach, Dr. V. Nachmias studied the use of anesthetics to control platelet activation which involves cyoskeletal changes. After first studying the platelet cytoskeleton with the electron microscope, she studied the effects of anesthetic agents on the activation of the cytoskeleton. She later studied the effects of calcium and pH as activating agents. Platelets are very specialized fragments of megakaryocyte cytoplasma but their cytoskeleton shares many properties with those of most typical cells. Platelets circulate in the resting state as biconave discs from which they can be activated. This provides an advantage for the study of factors that may alter the state of the cytoskeleton. Also, platelets lack intermediate filaments; thus properties of the cytoskeletal system can be observed in platelets that do not depend on the presence of 10 nm intermediate filaments. The smooth surface membrane leads into torturous channels that penetrate deep into the cytoplasm and are in close contact with a second membrane somewhat similar to the T-tubule reticulum in muscle. These channels allow stimuli to be received at the membrane and rapidly transmitted into the underlying cytoplasm. Structural changes occur in platelets within seconds after the addition of an agonist, such as ADP or thrombin. Two changes can be distinguished: filopodia form and the discoid cell body changes into an irregular sphere. In resting platelets, the cytoplasm lies around the remains of microtubules which appears amorphous or with short fragments of microfilaments. It lacks long microfilaments, bundles of microfilaments, or thick filaments. The addition of a low dose of a local anesthetic, such as tetracaine or lidocaine, for short periods (up to 30 minutes) reversed the activation of the platelets resulting in a retraction of the filopodia. When the anesthetic agent was removed the platelets reverted to their previous state--discs with filopodia. The local anesthetics apparently have two effects on platelets. The first is on the plasma membrane, causing a loss of the ability of the platelet to respond to external signals even when the internal structural aspects of the cytoplasm (cytoskeleton) is still capable of responding. The second action results in the retraction of filopodia. This second and later effect seems to be related to the release of calcium from internal stores, leading to irreversible inactivation of the platelet by calcium activated proteases. She further found that weak acids can inhibit both the rate of initiation of the shape change and the rate of the shape change itself.

Dr. M. Sheetz discussed the movement of myosin-coated beads on actin cables. Our understanding of myosin structure and function has come largely from studies of myosin relative to actin in muscle cells. In these cells the myosin filaments translocate along actin filaments through a series of individual myosin movements. In most cells, however, it is currently impossible to quantify myosin movement on actin. It therefore represented a major advance when Drs. Sheetz and Spudich recently succeeded in developing an in vitro assay for measuring the movement of nonmuscle as well as muscle myosin on actin filaments. The method consists of linking myosin to beads large enough to be seen in the light microscope and adding these myosin coated beads to the system to be studied. The initial study was done on the giant alga cell Nitella where all the actin filaments in different local regions along the approximately 10 cm long cell are known to have the same polarity. Dr. Sheetz and co-workers found that as soon as the myosin coated beads bound to the actin filaments they started to migrate. The speed of migration was 4-5 μm/second independent of the bead size. The direction of migration was dependent on the polarity of the actin fibers. This also represents an elegant method for studying factors affecting the rate of myosin movement on actin filaments. Dr. Sheetz presented studies on the inhibition of myosin motility by antibodies specific to certain locations on myosin, and by myosin and heavy meromyosin with modified sulfhydryl groups. Such studies can provide important information about the functional mechanisms of myosin dependent force generation.

Dr. A. Brody presented some interesting findings on the pathogenesis of inhaled asbestos particle lesions in the lung and particle transport through the cell membranes into the interior of the cell. He showed scanning electron micrographs to indicate that the particles of asbestos deposit at the crotch of the bifurcation of the bronchioles. This is the site to which macrophages migrate and pick up the asbestos particles. The migration of the macrophages is the product of the activation of complement by the asbestos to produce chemotaxis. The ingested fibers are picked up by type I alveolar pneumocytes. The fibers are subsequently phagocytosed by macrophages. The type I pneumocytes react to the asbestos, becoming thicker and creating folds which surround the asbestos fibers. The asbestos becomes located in the cytoplasm of the type I cells. They are also transported into connective tissue fibroblasts, which may ultimately result in fibrosis. The asbestos subsequently accumulates in the vicinity of the nucleus. There appear to be alterations in the pattern of the cytoskeleton as the fibers are transported within the cells. Cytoplasmic asbestos fibers are associated with defective cytokinesis.

Dr. P. Hinkle discussed surface receptor cycling as seen in pituitary cells. These cells synthesize prolactin and growth hormone, package the hormones in secretory granules, and secrete at a relatively constant rate in the absence of regulators. The rate of release and the rate of de novo synthesis of these are controlled by the peptide hormones thyrotropin-releasing hormone (TRH), epidermal growth factor (EGF), insulin, somatostatin, bombesin, and vasoactive intestinal peptides which all have independent receptors on the cell surface. Receptors for TRH, EGF, and bombesin are internalized by receptor mediated endocytosis in a reaction taking about ten minutes. In contrast, the receptor for somatostatin is not internalized. After the TRH receptor is internalized its cycles back to the cell surface without degradation of either the TRH tripeptide or the receptor. Because degradation can be prevented by lysosomatrophic drugs it is likely that it occurs via a lysosomal pathway. Receptor mediated endocytosis can be blocked by low temperature, or addition of glutaraldehyde or phenylarsene oxide. TRH can stimulate the release of prolactin from cells studied even when internalization is blocked by phenylarsene oxide, suggesting that this important biological activity of the peptide results from events triggered by the hormone receptor complex at the plasma membrane of the cell.

PERSPECTIVES

Tore L.M. Syversen[a], Polly R. Sager[b] and Thomas W. Clarkson[c]

[a]Department of Pharmacology and Toxicology
University of Trondheim, School of Medicine
Trondheim, Norway

[b]Department of Physiology
University of Connecticut Health Center
Farmington, Connecticut, USA

[c]Division of Toxicology
University of Rochester School of Medicine
Rochester, New York, USA

This conference has given an overview of the rapidly expanding knowledge of how certain chemicals interact with cytoskeletal components. These developments have given both a better understanding of cell biology as well as details on mechanisms of toxic actions. Areas where studies of the cytoskeleton might be of particular importance to the development of toxicology are the nervous system, reproduction and development.

A number of examples have been presented throughout this book that demonstrate the importance of the cytoskeleton in proper neuronal function. Certain toxic agents have been identified which act specifically on components of the neuronal cytoskeleton and these have been useful in defining the role of cytoskeleton in neuropathological processes. For example, decreased transport of neurofilaments and subsequent axonal swellings are observed after experimental administration of certain forms of hexacarbon solvents which are known also to cause neuropathies in exposed workers. Other chemicals or environmental and industrial interest may act by similar mechanisms. For example, many organic solvents produce acute reversible effects on the central nervous system and there is growing concern that long-term exposure may eventually lead to irreversible damage. Thus, other solvents may interact with the cytoskeleton, as has been demonstrated for γ-diketones; this possible target bears systematic investigation.

Reproductive toxicology is another area where considerable benefits might be gained from studying toxic effects on the cytoskeleton. This may provide a better understanding of some of the mechanisms of damage leading to reproductive loss and to teratogenicity. The cytoskeleton plays a critical role in a number of processes which are of particular importance in reproduction and during development, including meiotic progression, sperm motility, cell proliferation, cell migration and changes in cell shape which accompany differentiation. Recently, there has been an increased interest in mechanisms involving cytoskeleton elements. One example is anomalies in chromosome number which result from non-disjunction. This area is of interest both to the cell biologist and the developmental toxicologist. The questions of microtubule nucleation and chromosome attachment may provide interesting basic information on the function of kinetochores. To the toxicologist, elucidation of mechanisms underlying these anomalies could provide information useful in screening for developmental toxicants which act in this way and thus the early identification of possible human teratogens.

Another intriguing area of toxiciolgy involves the cellular specificity of toxic compounds. In many cases, specificity is a result of the distribution of the compound. However, there are other examples of toxic compounds causing damage in organs and cells where the concentration of the toxicant may be far less than in other parts of the organism. Components of the cytoskeleton appear to differ in certain cells, tissues, and species, not only in their distribution and function but also in molecular structure. For example, a number of different tubulins have been identified in the brain and other tissues by means of electrophoretic techniques. Thus, it might be possible to explain tissue and cell specificity based on differences in the molecular composition of the cytoskeleton.

Future research should provide both basic and practical contributions to toxicology. A better understanding of the molecular mechanisms involved in the toxic action of agents primarily affecting the cytoskeleton will be, in itself, an important contribution to the basic science of toxicology. It should lead to a better understanding of species differences, of selective action at the tissue and cellular levels and of the degree of reversibility of the toxic action. From the practical standpoint, it should be possible to develop both in vitro and animal tests that provide a more reliable basis for prediction of effects in man. New procedures may be developed that provide more sensitive means of detection of early effects in exposed humans and more selective and informative diagnostic tests. In short, both from the practical and more basic viewpoint, research into the cytoskeleton as a target for toxic agents should enhance our ability to protect public health by a more informed regulation of certain chemicals in the environment.

AUTHOR INDEX

Adler, K.B., 221-227
Anthony, D.C., 167-174
Autilio-Gambetti, L., 129-142
Barrett, J.C., 221-227
Brody, A.R., 221-227
Byard, E.H., 83-96
Cavanagh, J.B., 3-21, 25-34
Clarkson, T.W., 3-21, 25-34, 253, 254
Elgsaeter, A., 3-21, 25-34, 187-197
Freedman, R.A., 229-242
Gambetti, P., 129-142
Gold, B.G., 119-127
Gottfried, M.R., 167-174
Graham, D.G., 167-174
Grygorczyk, R., 177-186
Griffin, J.W., 119-127
Guldberg, H.C., 3-21, 25-34
Gull, K., 83-96
Halpern, J., 229-242
Hesterberg, T.W., 221-227
Hill, L.H., 221-227
Hinkle, P.M., 229-242
Hoffman, P.N., 119-127
Horwitz, S.B., 53-65
Jones, R., 213-220
Kinsella, P.A., 229-242
La Celle, P., 177-186
Lee, S.D., 3-21, 25-34
Lehto, V.-P., 143-158
Leven, R.M., 199-211

Lichtman, M.A., 3-21, 25-34
Manfredi, J.J., 53-65
Mellado, W., 53-65
Mikkelsen, A., 187-197
Milam, L., 167-174
Monaco, S., 129-142
Mottet, N.K., 3-21, 25-34, 243-251
Nachmias, V.T., 199-211
Olmsted, J.B., 3-21, 25-34
Papasozomenos, S. Ch., 67-82
Parness, J., 53-65
Passow, H., 177-186
Peters, R., 177-186
Price, D.L., 199-127
Roy, S.N., 53-65
Sager, P.R., 3-21, 25-34, 97-116, 253, 254
Savolainen, K., 143-158
Sayre, L.M., 129-142
Schiff, P.B., 53-65
Schwarz, W., 177-186
Sheetz, M., 213-220
Shelanski, M.L., 159-166
Shields, M., 177-186
Stokke, B.T., 187-197
Syversen, T.L.M., 3-21, 25-34, 97-116, 253, 254
Szakal-Quin, G., 167-174
Virtanen, I., 143-158
Wilson, L., 37-52

SUBJECT INDEX

Acrylamide
  neurofibrillary neuropathy, 31, 119-123, 136, 137, 168, 246, 247
Actin, 4, 8-10, 14-18, 46, 213, 221, 224
  axonal transport, 130, 131
  erythrocyte, 11-113, 17, 183, 192
  Nitella, 213-215, 217-219, 250, 251
  platelet, 201, 202, 204
Actin-associated proteins, 4, 5, 13, 16, 18
Actin-binding protein, 5, 13, 130, 201, 202, 205, 219
Actin-myosin interactions, 18, 213, 217, 218
α-Actinin, 5, 11, 13, 201, 204
Acumentin, 4, 14, 18
Aluminum, 37
  human neurological disease, 162, 163
  neurofibrillary degeneration, 159-162, 248
  neurofibrillary pathology, 30, 119, 246
Alzheimer's disease, 159, 168, 248
  and aluminum, 30, 159, 163
Ankyrin, 11, 13, 179, 183
Anthelmintics, benzimidazole compounds, 83-86, 90
Antimicrotubule drugs, 4, 6, 17, 18, 25, see also specific drugs
  and receptor internalization, 233, 234
Asbestos
  and cytoskeleton, 32
  intracellular translocation, 221-226, 251
Assembly, polymerization
  actin, 4, 5, 14, 18, 32
  microtubules in vitro, 6, 39-47, 57, 60, 61, 63, 98-102, 114
Axons, 6, 8, 10, 11, 17
  distal segment neurofibrillary pathology, 123, 134, 138, 139, 168
  IDPN, 68-82

Axons (continued)
  intermediate segment neurofibrillary pathology, 129, 130, 138, 139
  microtubule-neurofilament organization, 26, 30
  proximal segment neurofilament pathology, 30, 120, 123, 126, 129, 136, 168
Axonal caliber, diameter, 81, 126, 247
  intermediate filaments, 7, 10, 15
  neurofilaments, 119-121, 123
Axonal cytoskeleton, reorganization following IDPN intoxication, 67-82
Axonal length and diameter
  vulnerability to γ-diketone neuropathy, 170-172
Axonal swelling, enlargements, 28, 119, 136
  acrylamide, 121, 136, 137
  carbon disulfide, 134, 135
  dithiocarbamates, 147
  γ-diketone compounds, 31, 138, 139, 167, 168
  2,5-hexanedione, 123-125, 132-134, 155, 172
  IDPN, 31, 67, 80, 168
Axonal transport, 7, 8, 10, 15-17, 28, 31, 38, 130
  changes in neurofibrillary disorders, 119-126
  fast, 29, 31, 81, 123, 130, 137
  slow, 30, 76, 120-122, 126, 129-136, 247

Band III protein, 11
Band 3 protein, 179
  lateral diffusion, 183, 184
Band 4.1, see protein 4.1
Benomyl, 84, 86, 91, 243
Benzimidazole carbamates, 26, 27, 37, 38
Benzimidazole carbamate pesticides, interactions with microtubules, 83-93, 244
Benzimidazole drugs, compounds
  biological action, 84-87

257

Benzimidazole drugs (contd),
  effects in vitro, 87-91
  genetics of resistance, 91-93
  interactions with microtubules, 83-93

Calcium (Ca^{++})
  effects on red blood cell
    membrane, 177-184, 249, 250
    keratin filaments, 26
    membrane viscoelasticity, 179-181
    modification of K^+ efflux, 179-181
    platelets, 202-206, 250
    taxol-microtubules, 57
Calcium pump, red blood cell
  membrane, 177-179
Calmodulin, 16, 130, 177
Cambendazole, 84, 90
Carbon disulfide, 28, 154
  neurofibrillary pathology, 31, 129, 134-136, 168, 246, 247
Cell shape, 14, 15
Cellular specificity of toxic
  compounds, 254
Cerebellum, development and
  methylmercury, 32, 110-112
Cestodes, benzimidazole drugs, 84-86
Cilia, 6, 14
Clathrin, 16, 130, 131, 230
Colcemid, 6, 8, 26
Colchicine, 6, 8, 11, 15, 18, 25-27, 35, 38, 86-90, 161, 243, 248
  mechanism of action on
    microtubules, 37, 39-44
  neurofibrillary changes, 119, 246
  neuronal microtubules, 29
  receptor internalization, 233, 234
  taxol-microtubules, 55
  tubulin binding, 27, 39, 40, 243, 244
Colchicine binding site, tubulin, 37, 39, 57
Cross-link formation, neurofilaments, 137, 167, 168, 170, 172, 248, 249
Cultured cells
  cytoskeleton organization, 8-10
  effect of methylmercury, 102-110, 114
  effects of taxol, 53, 55-57
Cysteine, 245
  methylmercury effects on
    microtubules, 103-106, 108-110
Cytochalasins, 5, 15, 17, 18, 25, 26, 37
Cytokeratin, 9, see also keratin
Cytokinesis, 4, 10
  asbestos, 32, 221, 225, 226, 251

Cytoskeleton
  components, 3-8
  function, 14-18, 243
  organization, 8-14
  structure and function, 1-18
Cytoskeleton as a target for toxic
  agents, 23-33

Dendrites, 6, 10
Desmosomes, 10, 14
Dialysis dimentias, 162, 163
γ-Diketone, 7, 136, 137
  interspecies differences in
    vulnerability, 170-172
  neuropathy, 167, 168, 172
Dimercaptosuccinic acid (DMSA) and
  methylmercury effect on
  microtubules, 101-109, 245
3,4-Dimethyl-2,5-hexanedione
  (DMHD), neurofibrillary
  pathology, 119, 124, 136-138, 168, 170, 172, 246, 249
Diphenylhydantoin, 37, 38, 243
Distal neuropathy, 31, 246
Disulfiram (DSF), effects on
  cytoskeleton of neuronal
  cells, 143, 156
Dithiocarbamates, effect on
  cytoskeleton of neuronal
  cells, 143-156, 247
DMHD, see 3,4-dimethyl-2,5-
  hexanedione
DMSA, see dimercaptosuccinic acid
Dynein, 14, 16

Endocytosis, 6, 17, 18, 229-241
Epidermal growth factor (EGF), 230-234, 237, 251
Erythrocytes, 11-13, see also red
  blood cell
  membrane skeleton, 11-13, 17, 187-195, 249, 250
Ethylenethiourea (ETU), effect on
  cytoskeleton of neuronal
  cells, 144-149, 152
Exocytosis, 6, 18
External granular layer,
  cerebellum, 110-113

Fenbendazole, 84, 89, 90
Ferban (fer), 144, 145, 147-149, 151, 152, 154
Fibroblasts
  and asbestos, 223
  cell shape, 15
  microtubules and methylmercury, 103-110
Filamin, 5
Fimbrin, 5, 14
Flagella, 6, 39, 40, 61
Fodrin, 5, 11, 130

Fungicides
    benzimidazole compounds, 83-85
    dithiocarbamates, 143

Gelsolin, 5, 11, 13, 14, 18
Glial fibrillary acidic protein, 7
Glioma cells, microtubules and
    methylmercury, 103, 104,
    106-110
Glutathione and methylmercury
    effects on microtubules,
    103-110, 114
Growth hormone, 229, 230, 251

Heavy meromyosin (HMM), 203, 213,
    222, 224-226
Hexacarbon compounds neuro-
    fibrillary pathologies, 30, 31
n-Hexane, 31, 132, 143, 155, 161,
    167
2-5-Hexanedion, (2,5HD, HD), 26, 28
    effect on cytoskeleton of
        neuronal cells, 143, 156
    neurofibrillary pathology, 31,
        119, 120, 123-125, 129, 132-
        134, 137, 138, 167, 168, 170

IDPN, see β,β-iminodipropionitrile
β,β-Iminodipropionitrile (IDPN),
    11, 26, 28, 29
    axonal cytoskeleton
        reorganization, 67-82, 115,
        245, 246
    neurofibrillary pathology, 30,
        31, 119, 138, 168, 172, 246
Inorganic particles, intracellular
    transport, 112-226
Intermediate filaments, 3, 4, 6-8,
    15
    axonal transport, 15, 16
    cultured cells, 8-10, 149, 152,
        156, 159, 161
    membrane function, 17
    neurons, 159, 160, 162, 245,
        247
    specific probes, 26
Intracellular transport, 41, 15-17
    asbestos fibers, 32
    neurons, 10, 15-17

K^+ selective channels, Ca^{++}
    effects, 179-182, 184, 250
K^+ efflux, Ca^{++}-stimulated,
    179-180, 182, 184, 250

Leukocytes, 13, 14
Local anesthetics, platelet
    cytoskeleton and activation,
    199-205, 210, 250
Lymphocytes, 13

Maneb (ma), 144, 145, 147, 149
MAP2-neurofilament association, 8,
    11, 26, 76, 79, 80
MBC, see methyl benzimidazole-2-yl
    carbamate
Mebendazole, 84, 86, 89, 90, 244
Membrane-associated cytoskeleton,
    erythrocytes, 11-13, 17,
    191-195, 249, 250
Membrane properties and functions,
    17, 18
Membrane topology regulation, 4,
    6, 17
Membrane viscoelasticity, Ca^{++}
    modification, 179, 181-183
Methyl benzimidazole-2-yl
    carbamate (MBC), 84-88, 245
3-Methyl-2,5-hexanedione
    neurofibrillary pathology,
    129, 136-138
Methylmercury, 32, 37, 245
    disruption of mitosis in
        developing brain, 110-114
    and microtubules, 97-110, 114
Microfilaments, 3-6, 15-18, 25
    and asbestos fibers, 32, 222-
        225
    cultured cells, 8-10
    intracellular transport, 15-17,
        221
    platelets, 200-202, 204, 250
    selective probes, 26
Microtubules, 3, 4, 6-8, 13-18,
    25, 243-245
    axons, 147, 148, 150, 152, 154-
        156, 160
    and benzimidazole carbamate
        pesticides, 83-93
    cultured cells, 8-10
    fugal, 88, 89
    IDPN axons, 68-73, 76-82
    mechanisms of action of
        colchicine and vinblastine,
        37-52
    and methylmercury, 32, 97-114
    nematode, 86, 88
    neurons, 10, 11, 245-247
    platelets, 200
    selective probes, 26
    and taxol, 53-63
Microtubule assembly, see assembly
Microtubule-associated proteins,
    4, 6, 10, 16, 161, 162
    axonal transport, 130
    MAP1, 6, 10, 11
    MAP2, 6, 10, 11, 26, 76, 79,
        80, 246
    tau, 6, 11, 16
Microtubule-disrupting agents,
    toxicological effects, 38, 39

Microtubule organizing center
    (MTOC), 8, 9, 10, 55
Microtubule poisons, 28
Microtubule protofilaments, 86, 90
Microvilli, 5, 14, 18
Mitosis, 10, 15, 16, 18
    benzimidazole drugs, 84, 86
    disruption in developing brain,
        110-113, 245
    methylmercury, 97, 102
    taxol, 53, 55, 56
Mitotic spindle, 6, 9, 10, 16, 39,
        114, 244
    block by benzimidazole
        compounds, 85, 86
    poisons, 27, 32, 37
Motility, 4, 18
    inhibition by inactivated myosin
        heads, 123-219, 250, 251
Motive force generation, 4, 13,
        15, 16, 213
Mutagenic effects, microtubule-
    disrupting substances, 39
Myosin, 4, 5, 13-15, 18, 204, 213
    motility assays, 214, 250, 251

Nabam (na), 144, 145, 147, 149, 151
NEM, see N-ethyl-maleimide
Nematode
    and benzimidazole drugs, 84-86
    microtubules, 86, 88, 89
N-ethyl-maleimide (NEM)
    heavy meromyosin, 214-219
    microtubule assembly, 99
    myosin, 217-219
Neurite formation, growth, 15, 25,
        149
Neuroblastoma cells, 6, 15
    cytoskeleton, 144-149, 151-156
    microtubules and methylmercury,
        104, 106-110
Neurofibrillary disorders, changes
    in axonal transport, 119-126
Neurofilaments, 7, 10, 11, 15, 16,
        28, 29, 145, 146, 160, 171,
        172, 247, 248
    cross-links, 137, 167, 168,
        170, 172
    dithiocarbamates, 147, 148
    IDPN axons, 68-81
    neurofibrillary pathology, 29-
        31, 119
Neurofilament association with
        microtubules, 7, 8,
    association with MAP2, 11, 26
    axons, 26, 30, 31, 130, 131,
        138, 139, 156
    neurons, 10, 11, 160, 161, 248
    segregation by IDPN, 68-73, 76-
        81, 246

Neurofilament-containing axonal
    enlargements, pathogenesis,
    137-139
Neurofilament subunits, 68, 132, 136
Neurofilament transport, 119, 130,
        168, 245, 246
    acceleration by chemical
        neurotoxins, 129-139
    acrylamide, 120-123, 137
    aluminum, 162
    carbon disulfide, 134-136
    2,5-hexanedione, 123-125, 132-
        134
    IDPN, 67, 168
    pathogenesis, 137-139
    response to toxins, 120-125
Neurofilament triplet, 7, 11, 16,
        146, 161, 162
Neurofilamentous neuropathies
    pathogenesis, 167-172
Neuronal cytoskeleton, 10, 11, 38,
        245, 253,
    microtubules, 29
    neurofibrillary pathology, 29-
        31
    sites of toxic action, 161, 162
    toxic agents, 28-32
Neurotoxins, mechanisms of action
    of chemicals, 137-139
Neutrophil, 13, 17, 18
Nitella motility assay, 213-219
Nocodazole, 8, 15, 26, 27, 84-89

Oxfendazole, 84, 86, 89, 90

Parbendazole, 84, 87, 88, 90, 245
Peptide hormone receptors
    pituitary cells, 229-231, 236
Perikaryon, neurofilament
    pathology, 30
pH changes, platelet cytoskeleton
    and activation, 199, 205,
    207-210, 250
Phagocytosis, 13, 17, 18
Phalloidin, 5, 18, 26
Phenylarsine oxide and receptor
    internalization, 234-240, 251
Pituitary cell lines, receptor
    cycling, 229-241
Platelets, 12, 13, 199-210
Podophyllotoxin, 26, 37, 38
Pox (C,R), 114
Profilin, 5, 13, 18
Prolactin, 229, 230, 236, 251
Protein 4.1, 11, 12, 179, 182,
        183, 187, 192-194
Protein 4.2, 182
Protein 4.9, 5, 11, 12, 187
Protofilament peeling and
    vinblastine, 27, 46, 48
Proximal neuropathies, 30, 246

Pyrrole formation, γ-diketones, 137, 167, 168, 170, 248

Receptor cycling, pituitary cells, 229-241, 251
Receptor internalization, 17, 231-239
Receptor-mediated endocytosis and receptor cycling, 229-241, 251
Red blood cells, see also erythrocyte
  membrane effects of calcium, 177-184
Reproductive toxicology and cytoskeleton-related events, 253, 254

Shape changes, platelets, 199-210, 250
Somatostatin, 230, 237 251
Spectrin, 5, 11, 13, 16, 17, 161, 179, 183, 187, 249
  flexibility, 188-191
  nonerythroid, 145, 146, 148, 187
Stress fibers, 5, 10
Structural support, cells and tissues, 14, 15
Substoichiometric microtubule poisons, 26, 40
Sulfhydryl modification, heavy meromyosin and motility, 213, 214

Tau proteins, 6, 11, 16
Taxol, 61, 26-28, 37, 53-63, 244
  effects in cells, 53, 55-57, 59, 60, 62
  effects on microtubules in vitro, 57, 60-62
  microtubule binding, 27, 57
Taxol-resistant cells, 62, 63
Teratogenic agents, 15, 32
Teratogenic effects, 97, 245
  microtubule disrupting substances, 39
Tetracaine, 201-206
Thiabendazole, 84, 89, 90
Thiram, (thi), 144, 145, 147-149, 151, 152, 154
Thyrotropin-releasing hormone (TRH), 230-237, 239, 251
Toxic agents
  acting on neuronal cytoskeleton, 28-33
  acting on non-neuronal cytoskeleton, 32
  tools for studying cytoskeleton, 25-28
Toxic chemicals, deposition, primary reaction, effects, 28, 29

Transferrin, 231-233
Treadmilling of microtubules, 6, 57
Tropomyosin, 5, 11, 12
Tubulin, 4, 6, 9, 15, 16, 26, 32, 62, 97, 135, 146, 243, 244
  axonal transport, 120, 123
  binding sites, 27, 37-39, 46-48, 57
  and colchicine, 39-45
  fungal, 84, 85, 89-91
  genes and benzimidazole resistance, 91, 92
  mammalian and benzimidazole drugs, 89-91
  nematode, 86, 88-91
  neuroblastoma cells, 149, 151-153
  neurons, 146, 148, 246
  species differences, 90
  sulfhydryl groups, 100-102, 114
  and vinblastine, 44-49

Vanadate, 7, 26
Villin, 5, 14
Vimentin, 7, 8, 26, 145, 146, 148, 149, 151-153, 156
Vinca alkaloids, 6, 15, 26, 27, 29, 37, 44, 243, 246
Vinblastine, 26, 27, 37, 38, 161, 244
  mechanism of action on microtubules, 44, 46-49
  microtubule binding, 44, 47-49, 57
  receptor internalization, 233, 234
Vinblastine-tubulin crystals, 27, 46, 47
Vincristine, 29, 37, 38
Vinculin, 5
Vindesine, 37

Zineb (zin), 144, 145, 147-149
Ziram (zir), 144, 145, 147-149, 151, 154